Science Between Truth and Ethical Responsibility

Mario Alai · Marco Buzzoni · Gino Tarozzi
Editors

Science Between Truth and Ethical Responsibility

Evandro Agazzi in the Contemporary Scientific and Philosophical Debate

Springer

Editors
Mario Alai
Department of Basic Sciences
and Foundations
University of Urbino "Carlo Bo"
Urbino
Italy

Gino Tarozzi
Department of Basic Sciences
and Foundations
University of Urbino "Carlo Bo"
Urbino
Italy

Marco Buzzoni
Department of Humanistic Studies
University of Macerata
Macerata
Italy

ISBN 978-3-319-36467-4 ISBN 978-3-319-16369-7 (eBook)
DOI 10.1007/978-3-319-16369-7

Springer Cham Heidelberg New York Dordrecht London

Softcover reprint of the hardcover 1st edition 2015

Printed on acid-free paper

Springer International Publishing AG Switzerland is part of Springer Science+Business Media
(www.springer.com)

Preface

This volume is a tribute to Evandro Agazzi and his work by Italian and international scholars, former direct or indirect pupils, friends, colleagues or associates who esteem him and are grateful to him for long years of discussions, advice and fruitful philosophical exchanges. These essays were first presented at the international congress "Science Between Truth and Ethical Responsibility. Evandro Agazzi in the Contemporary Philosophical and Scientific Debate", held in Cesena and Urbino (Italy) from 22 to 24 April 2014. That congress, like this volume, intended to celebrate his 50 years of academic activity, by offering a systematic study and critical discussion of his many and often pioneering contributions to a wide spectrum of philosophical issues.

Agazzi constitutes an extraordinary example of rigorous and original thought, successful professional leadership and organizing capacities at the international level. His teaching spans widely on scientific knowledge, its nature, limits and requirements, as well as on the connected questions of ethical responsibility, on its anthropological presuppositions and metaphysical backgrounds. The exemplar clarity of his explanations and lectures helped us and many others to see their own way into these difficult problems and to progress along directions he has indicated or suggested.

The papers collected here express, each in its own way, admiration and gratitude for his work, in the conviction that paying homage to a great philosopher means analysing, interpreting and disseminating his ideas, but even more importantly critically discussing them and taking advantage of their fecundity as a starting point for further advancements in the field.

Evandro Agazzi graduated from the Catholic University of Sacred Heart in Milan, under the supervision of Gustavo Bontadini. He then studied physics at the State University of Milan, philosophy of science in Oxford and mathematical logic in Münster. After becoming "libero docente" in philosophy of science and in mathematical logic, he taught various disciplines, both scientific and philosophical, in a number of universities (sometimes even simultaneously): those where he lectured for longer periods are the Catholic University in Milan, the University of Genoa, the Superior Normal School in Pisa, the Università Vita Salute San

Raffaele in Milan, the University of Fribourg (Switzerland), the Universidad Autònoma Metropolitana in Mexico City and the Panamerican University in Mexico City. In these and other universities he gave courses in philosophy of science, theoretical philosophy, philosophical anthropology and philosophy of nature, as well as mathematical logic, advanced geometry and complementary mathematics.

The development of Agazzi's philosophy is clearly explained and related to its historical context by Fabio Minazzi's article "Evandro Agazzi Philosopher. An overview". But at the same time Agazzi was also very active as an editor, organizer and cultural leader: in Italy, in 1978, he founded the journal *Epistemologia. An Italian Journal for the Philosophy of Science*; soon thereafter, for the publisher Franco Angeli in Milan, he started the collection "Epistemologia"; he then chaired the Centre of Studies on Contemporary Philosophy of the Italian National Council of Research; he became president of the Italian Philosophical Society and of the Italian Society of Logic and Philosophy of the Sciences. Furthermore, he chaired the most important international philosophical societies and academies: the International Academy of Philosophy of Sciences (from 1978 to now); the International Federation of Philosophical Societies (as president from 1988 to 1993, then as honorary president); and the International Institute of Philosophy (as president from 1993 to 1998, then as honorary president).

Agazzi's earliest researches concerned the foundations of mathematics and logic, on which he wrote *Introduzione ai problemi dell'assiomatica* (Agazzi 1961) and *La logica simbolica* (Agazzi 1964). He rejected a purely formal viewpoint, holding that what human thought can discover or "see" in these areas exceeds what can be proved (Agazzi 1961, p. 199). An "eidetic meaning" is needed not just for interpreting a formal system, but also for laying down the composition and transformation rules, since we must understand what they prescribe. This already sets clear limits to the possibilities of artificial intelligence, as he argued in later works (Agazzi 1967, 1981a). These ideas have been further developed and systematically argued for in his recent book *Ragioni e limiti del formalismo. Saggi di filosofia della logica e della matematica* (Agazzi 2012).

Empirical sciences, however, represent the main subject of Agazzi's vast philosophical enquiries. Here he always held that science can aim at truth in the realist sense, avoiding both the Scylla of scepticism on the possibility of reaching the truth, and the Charybdis of the dogmatic illusion that truth has already been completely achieved. A key role is played in this respect by his distinctions between reality and objectivity, and between two senses of objectivity: as in-principle intersubjectivity and as reference to the object. On the one hand, he notices that the agreement among people about cognitive content does not hinge on their "private" data, but on the public actions they perform. On the other hand, operations constitute the specific "objects" of each particular discipline. Thus, the very conditions that define a science by structuring its proper objects also provide intersubjective knowledge of those objects (Agazzi 1969, Chap. 10, 2014, Chaps. 1 and 2).

More recently, at least since the volume *Il bene il male e la scienza* (Agazzi 1992 [2004]) and up to *Scientific Objectivity and Its Contexts* (Agazzi 2014),

Agazzi has studied scientific objectivity in its relations to the social reality, from a system theoretic viewpoint: the scientific-technological system is fully autonomous as to its cognitive value, but it is an "open adaptive social system", interacting with other social systems; as such, it cannot just aim at maximizing its own internal goals, but must respect the constraints provided by different systems, such as the economic, political or energetic system. Among them, it must also respect the system of moral norms and values.

These central issues in Agazzi's philosophy are discussed by many contributions to the present volume. The main features of his general philosophy of science are analysed by Craig Dilworth ("The Perspectivist Conception of Science"), Marco Buzzoni ("Science and Operationality"), Mario Alai ("The Issue of Scientific Realism"), Vincenzo Fano and Giovanni Macchia ("Scientific Progress and Verisimilitude").

On the basis of this general theoretical framework, over the years Agazzi has carried out special researches on a number of particular issues. Some (though not possibly all) of them are accounted for by other papers of this volume. To begin with, he has offered important contributions to the foundations of the special sciences: contemporary physics, in particular quantum mechanics (discussed by Gino Tarozzi's "Philosophy of Physics"); mathematics, (a subject explored by Marco Borga's "Foundations and Philosophy of Mathematics"); artificial intelligence (analysed by Mariano Bianca's "Artificial Intelligence"); sociology (the topic of Giuliano Di Bernardo's "From Physics to Sociology"); and education (accounted for in detail by Giuseppe Bertagna in "Between Education and Pedagogy").

According to Agazzi, the requirements of objectivity and rigour, characteristic of the natural sciences, can be satisfied also by the human sciences, since they are independent of quantitative methods. Besides, deductive rationality must be supplemented by argumentative and hermeneutic rationality (Agazzi 1979). Furthermore, when it comes to the psychological and pedagogical sciences, a key role is played by the principle of dignity of the human person. Agazzi devoted long reflections to this principle and to pedagogical theories, and he founded and directed for many years *Nuova Secondaria*, the main Italian journal for high-school teachers and administrators.

Of course, he developed his philosophy of science in connection with deep considerations on other closely related philosophical disciplines, which are the focus of a third group of contributions: Pierluigi Graziani's "Philosophy of Mathematics and Logic", Antonio Livi's "The Issue of Alethic Logic", Jure Zovko's "Interpretation and Hermeneutical Judgement" and Carlo Penco's "Philosophy of Language".

But philosophy of science cannot be detached from an even wider theoretical horizon, including the anthropological, historical and more widely cultural dimensions of science, and Agazzi's philosophical contributions span over all of them. Anthropology is dealt with in "Philosophical Anthropology", by Matteo Negro; the historical and complex dimensions of science are discussed by Giuseppe Gembillo in "Science, Historicity and Complexity"; the strict relations between philosophy and history of science is the subject of Flavia Marcacci's "Epistemology and History of Science"; and the cultural and intercultural dimension of philosophy is the topic of the essay ("Contributions to Latin American Philosophy") in which Lourdes Velàzquez examines the impact of Agazzi's thought in shaping the philosophical landscape of a whole continent.

A further decisive dimension of science is the ethical one, which is also presupposed by (and presupposes) anthropology, education and history. To ethics Agazzi has dedicated a great amount of work in recent years, variously dealing with the moral issues raised both by science and technology as human practices (ethics of science and technology), and by the ever more advanced forms of control and intervention on human and animal life that scientific and technological progresses make possible (bioethics). His contributions to the former area are examined by Boris Yudin in "Ethics of Science and Technology", and Alfredo Marcos in "The Autonomy of Science in a System Theoretic Approach".

There is a bidirectional relation between science and technology on the one side and ethics and anthropology on the other. On the one hand, in fact, as noticed above, science and technology are an "open adaptive social system", which must harmonize with other social systems, including the system of moral norms and values (Agazzi 1992 [2004], Chap. 14). On the other hand, however, both the natural and the human sciences are involved in the justification of moral norms and values.

Unlike the natural world, mankind and its activities are characterized by the ought-to-be. Values are projections of the ought-to-be, i.e., "the ideal models which work as regulative parameters for human operations, performances and actions" (Agazzi 1992 [2004], p. 127). In turn, values are justified through an "image of the human nature": this image is based on biology, psychology, sociology, psychoanalysis, etc., and it should offer a plausible model for human behaviour and actions. This is not falling into the naturalistic fallacy of deducing "ought" from "is", but acknowledging that "it is rational to demand that man behave in accordance with his own constitutive conditions, and accepting the contrary, even if it could be done, would not be rational" (Agazzi 1981b, p. 18).

According to Agazzi, a ground for ethical norms shared by different cultures, religions or philosophies can be provided by the already mentioned principle of human dignity. Consciousness is the *proprium* for human persons, which people can and ordinarily do have; but since it is not a substantial feature, it can be acquired or lost, but persons do not cease to be human when it attenuates or vanishes. Hence, the deprivation of this property cannot become a reason for discrimination (see Agazzi 1992, pp. 28–39, 1992 [2004], Chap. 10). These and other issues concerning the relationships between ethics and the biological sciences are discussed in Gonzalo Miranda's essay "Bioethics".

Finally, any philosophical discussion of science must be set on the background of the most general and encompassing attempts to understand reality, i.e. metaphysics and religion. These are the subject of Paolo Musso's "Metaphysics and Ontology" and Juan José Sanguineti's "Religious Faith, Natural Science, and Metaphysics". In particular, for Agazzi, although metaphysics is a "brief" discourse, it is a necessary presupposition of scientific knowledge. Besides, science cannot substitute nor limit the autonomy of religious faith, for the latter answers questions which science, by its own nature, cannot address.

As mentioned at the beginning, the best homage we can pay to an authentic master of thought, besides knowing and interpreting his or her ideas, is establishing a critical dialogue with them. This too is something we learned from Agazzi, from his attitude towards his own teachers, Gustavo Bontadini in the first place: for he was able to learn from them, with humility, deep respect and gratitude, and at the same time to become an original and intellectually independent thinker.

True dialogue may become easier, *as a matter of fact*, when some basic philosophical assumptions are shared; but *in principle,* it requires above all keenness in the quest for truth and disposition to critical (and self-critical) thinking. The two must always go together, but can be declined in the most different ways, depending on one's personality, background and attitudes. They can elicit objections on particular claims or doubts on basic assumptions; suggest alternative but complementary perspectives; allow to develop cues in the master's ideas, or draw from his or her teaching original and autonomous research strategies. In all of these senses, each of the contributors to this volume can be considered a pupil of Evandro Agazzi and indebted to his research in philosophy.

While this work originates from the desire to acknowledge these debts, and to honour so many years of academic activity and philosophical *paideia* by Evandro Agazzi, it could not have been thought of and published without the patronage and financial help offered by various institutions: thus, we gratefully thank the University of Urbino "Carlo Bo" (*Centro Interuniversitario di Ricerca in Filosofia e Fondamenti della Fisica*—CIRFIS), the Department of Basic Sciences and Foundations of the University of Urbino—*DiSBeF*; the University of Insubria, Varese (*Centro Insubrico "Carlo Cattaneo" e "Giulio Preti"*); the University of Macerata (*Dipartimento di Studi Umanistici*); the University of Messina (*Centro Studi di Filosofia della Complessità "Edgar Morin"*); the National Academy of Sciences, Literature and Arts of Modena; the City of Cesena; and the International Academy of Philosophy of Sciences.

Cesena, Genova, Bologna
December 2014

Mario Alai
Marco Buzzoni
Gino Tarozzi

References

Agazzi, E. 1961. *Introduzione ai problemi dell'assiomatica*. Milan: Vita e Pensiero.

Agazzi, E. 1964. *La logica simbolica*. Brescia: La Scuola (5th ed. 1990).

Agazzi, E. 1967. Alcune osservazioni sul problema dell'intelligenza artificiale. *Rivista di filosofia neoscolastica* 59 (1): 1–34.

Agazzi, E. 1969. *Temi e problemi di filosofia della fisica*. Milano: Manfredi (2nd ed. Rome: Abete 1974).

Agazzi, E. 1979. Analogicità del concetto di scienza. Il problema del rigore e dell'oggettività nelle scienze umane. *Epistemologia*, 2: 39–66.

Agazzi, E. 1981a. Intentionality and Artificial Intelligence. *Epistemologia.* (4, Special Issue, *The Mind-Body Problem*): 195–227.
Agazzi, E. 1981b. Sulla possibilità di una fondazione razionale delle norme. In *Miscellanea Filosofica 1980*. Firenze: Le Monnier: 7–23.
Agazzi, E. 1992. L'essere umano come persona, *Per la filosofia* (Special Issue: "Bioetica e persona"), 9: 28–39.
Agazzi, E. 1992 [2004]. *Il bene, il male e la scienza.* Milan: Rusconi. English ed.: Agazzi, E. 2004. *Right, Wrong and Science: The Ethical Dimensions of the Techno-scientific*, ed. Craig Dilworth. Amsterdam/New York: Rodopi.
Agazzi, E. 2012. *Ragioni e limiti del formalismo. Saggi di filosofia della logica e della matematica* Milan: Angeli.
Agazzi, E. 2014. *Scientific Objectivity and Its Contexts.* Heidelberg/New York/Dordrecht/London: Springer.

Contents

Evandro Agazzi Philosopher

An Overview of His Thought

Fabio Minazzi

Abstract The paper outlines an overall picture of Agazzi's philosophical thought and of its coherent development. In fact, in the beginning Agazzi worked on epistemology, and in particular on the philosophy of mathematics. Then he was led to consider science from a more general and systematic point of view. Thus, reflecting on science and its cultural value, Agazzi's research broadened again its scope, taking into account the links between science, technology and society. But this brought him to analyze the relationships between scientific knowledge and moral reflection. In turn, such a critical survey of the interplay of science and morality required a study of the possible connections between human knowledge and a properly metaphysical dimension. In this last field Agazzi has always referred to the tradition of Western thought, using the method of *analogical discourse* to construct his metaphysical discourse. Finally, Agazzi has just published a book devoted to the problem of the *objectivity* of science, but more in general of human knowledge: a work which rounds up his reflection, highlighting its internal coherence and critical potentialities.

1 From the Aristotelian Categories to the Kantian Critical Trichotomy?

On the occasion of the presentation, at the University of Genoa, of the challenging and complex "encyclopaedic" volume *Filosofia, Scienza e Bioetica nel dibattito contemporaneo. Studi internazionali in onore di Evandro Agazzi* (which I organized, edited and published in Rome, 2007, in the prestigious series of the Presidency of the Italian Council of Ministers, see Minazzi 2007a), Carlo Penco shrewdly observed

---

F. Minazzi (✉)
Centro Internazionale Insubrico, University of Insubria, Varese, Italy
e-mail: fabio.minazzi@uninsubria.it

M. Alai et al. (eds.), *Science Between Truth and Ethical Responsibility*,
DOI 10.1007/978-3-319-16369-7_1

that, to speak comprehensively and coherently of Agazzi's work and thought, one might properly invoke the celebrated Aristotelian categories of *quantity*, *quality*, *relation* and *modality*. As analytically shown by the first systematic and chronological bibliography of Agazzi's output (which significantly concluded the volume mentioned above[1]), the "quantity" of works produced by him, in a life of intense labour, is truly extraordinary. In fact, by exploring Agazzi's *works and days* one discovers that his life has been configured, first and foremost, as a constant effort and unflagging commitment to research, study and reflection, which has left a highly significant mark in his publications (more than a thousand in the period between 1955 and 2006, though his activity has also been pursued steadily until today). The arithmetic and aseptic average of some dozen publications a year records this methodical work unfolding from the years of his university education to those of his maturity. But in addition to the *quantity* of this production we have also to bear in mind, naturally and most importantly, its intrinsic *quality*, with the output over time of works that have had a profound impact on the Italian epistemological research as well as in the context of the international debate. Indeed the many languages which Agazzi has used in the course of his life to present his thoughts are always closely intertwined, giving rise to a very significant international presence. This is, of course, hardly the place to retrace Agazzi's extraordinary career in teaching and research, his work as a visiting professor (which has taken him to a number of European and American universities), or his intense activity as a lecturer (which has taken him to almost every continent and many countries in Europe, Asia, the Middle East, Africa, North America and Latin America). Not to mention the numerous academic appointments and positions in Italian and foreign scientific associations, or his widely varied editorial responsibilities, the numbers of awards, accolades and honorary degrees he has received and much else. However, precisely the admirable sum of these appointments and all these different cultural and editorial responsibilities helps us to better understand not only the specific *quality* of his scientific work, but also the distinctive nature of the multiple international relations that Agazzi has succeeded in forming and building in the course of his scientific and philosophical research. Therefore, speaking of Agazzi is speaking of a philosopher who has always placed himself in an *international* context, up to the point that his cultural presence abroad has come to be even more significant than his presence in the Italian context, where he has worked well for an entire academic life. Last but not least, precisely the consideration of the *intrinsic quality* of his multiple international relations enables us to finally clarify Agazzi's *modality* of "doing philosophy". This modality has always been that of a dialogue and debate at the highest international level, in exchange with the chief interlocutors of the different traditions of thought in the different continents. Penco was thus right to evoke the celebrated Aristotelian categories in presenting a concise and unified overview of Agazzi's work, because the *quantity* of his publications, their intrinsic *quality*, as well as the many international *relations* and the specific *modalities* with which Agazzi has made, step by step, his intellectual journey, in their

[1]See Minazzi (2007b: 1351–1402).

problematic and dynamic whole, constitute a sparkling jewel and an extraordinary cultural polyhedron that is noteworthy for its intrinsic theoretical, purposeful and organizational texture.

However, without denying the heuristic interest of this Aristotelian reference, I think that to grasp dynamically the same qualitative as well as quantitative growth in Agazzi's work, it is perhaps appropriate to follow another suggestion, that which Immanuel Kant delineated in the final pages of his *Critique of Pure Reason*, where he notes how, in the last analysis, the fundamental questions that man must seek to answer by philosophical reflection are basically only three: *What can we know? What ought we to do? What may be hoped?* Three seemingly "simple", and perhaps even "banal", questions but from which arises a powerful *critical trichotomy* that Kant develops within his innovative and fruitful perspective of transcendentalism. For this particular reason the domain of *knowledge* concerns, according to Kant, the descriptive and prescriptive order of scientific objectivity and cognitive truth (that is, the nature of human knowledge), while the question concerning *duty* makes reference to the prescriptive and legal order of ethical correctness, and the normative rules which assume, in society, the form of our particular moral duty; finally *hope* refers to the self-reflexive order of emancipation and authenticity, precisely the teleological order inserted and rooted in the world of praxis. As I have illustrated in my book devoted to the *Teleology of Knowledge and Eschatology of Hope*,[2] this fruitful Kantian critical trichotomy configures, for the whole of Europe and in general for human civilization, a complex and highly articulated philosophical and civil project, in which *knowledge* and *freedom* constitute two different yet intertwined terms in a *single movement of social self-emancipation*. Its driving force is precisely that historical utopia that, in the form of hope, is the keystone of this dynamic and always open-ended tension, between the critical increase of knowledge and the gradual broadening of the civil legacy of freedoms and human rights. This Kantian critical trichotomy, however, went into crisis with the Hegelian and Romantic turning-point that split the ties between knowledge and freedom, delineating a grotesque dichotomy in which the three Kantian transcendental orders were inevitably reduced to three specific historical spheres of pragmatic activities. In this way knowledge was reduced to mere technical instrumentality at the service of economic labour, while the moral plane was dissolved in the dimension of inter-subjective communicational language, and hope was reduced to the need for a liberating mythical praxis, self-reflexive and symbolic, which finally found its emblematic historical expression in the Frankfurt School's critique of ideology.[3]

However, precisely the heuristic strength of the Kantian critical trichotomy offers, at least in my opinion, an adequate hermeneutical instrument to better understand not only our own time but also Agazzi's path of philosophical

[2] See Minazzi (2004).

[3] For a discussion and clarification of this grotesque degeneration of the Kantian critical trichotomy the reader is also necessarily referred to the observations made by Petitot (2009).

reflection and its progressive critical and speculative expansion. Agazzi, in fact, made his scholarly debut with the volume *Introduzione ai problemi dell'assiomatica* (1961), through which he became known as a fine philosopher of science who mastered, with high expertise, the debate related to mathematical logic and the tradition of Hilbert's formalism, in particular. His first studies were mainly (but not exclusively) devoted to logical issues (for example, with *Logica simbolica* of 1964 and with various other entries on mathematical logic appeared in 1967 in the *Enciclopedia filosofica* of the Centro di Studi Filosofici di Gallarate), but Agazzi then published a significant and important epistemological work, *Temi e problemi di filosofia della fisica* (1969) in which his theoretical interests take into more direct consideration the problem of physical knowledge in all its paradigmatic value. Agazzi has never abandoned his initial interest in logical-mathematical thinking, as is attested not only by the book *La simmetria* which he edited in 1973, and above all by his many studies of mathematical logic which have more recently been collected in his volume *Ragioni e limiti del formalismo* (2012), which republished essays in the philosophy of logic and mathematics that appeared along almost the whole of Agazzi's scientific work. It is a fact, however, that this constant interest for reflection within the logical-mathematical framework is then woven with a progressive and significant expansion of his speculative interests, to include the philosophy of physics and the epistemological significance of the history of science and the notion of progress (to which he devoted a volume on *Il concetto di progresso nella scienza*, published in 1976). Let us simply remember the demanding publication of the chief work of a classic of science such as Maxwell: Agazzi made the annotated Italian translation of his celebrated *Treatise on Electricity and Magnetism*, (published in two volumes in 1973 in the "Classici della scienza" series founded and directed by Geymonat for UTET in Turin). So if in 1978 Agazzi published *Le geometrie non euclidee e i fondamenti della geometria* (written in collaboration with Dario Palladino), in the same years he also embarked on a reflection on the problem of meaning, to which he devoted a collective volume of *Studi sul problema del significato*, which he edited in 1979; it brought together the results of a celebrated major seminar he had organized and directed for many years at the University of Genoa. Also in these years he published as editor *Modern Logic. A Survey* (1981) and authored several papers dealing with the relations between science and religion (*Science et foi. Perspectives nouvelles sur un vieux problème*, 1983, and *Il pensiero cristiano nella filosofia italiana del Novecento*, 1980). He also promoted and edited a collective *Storia delle scienze* (1984, in two volumes), a reflection on *Weissheit im Technischen* (1986) as well as a comprehensive stocktaking of *La filosofia della scienza in Italia nel* '900 (1986), without failing to investigate the links between *Philosophie, Sciences, Métaphysique* (1987), the relations between *Linguaggio comune e linguaggio scientifico* (1987), the problem of *L'objectivité dans les différentes sciences* (1988), *Probability in the Sciences* (1988), a theoretical debate (conducted in open and sincere dialogue with Geymonat and the present writer) on the relations between *Filosofia, Scienza e Verità* (1989), the analysis of the relations between *Logica matematica e logica filosofica* (1990), as well as a question on *Quale etica per la bioetica?* (1990), *La comparabilité des théories scientifiques* (1990), *The Problem of Reductionism in Science* (1991), the study of the link between *Science et sagesse* (1991), and the investigation of *Philosophy and the Origin and Evolution of the Universe* (edited in collaboration with Alberto Cordero in 1991).

This variety of investigations found a kind of emblematic and programmatic outcome in his ambitious monograph *Il bene, il male e la scienza* (1992, promptly translated into several languages: German, French, Spanish, Hungarian, Polish, Russian, and English), with which Agazzi's reflection, after having dealt with many aspects of the broad field of objective knowledge as it is configured in many scientific disciplines, felt the need to also address certain ethical issues directly relevant to the ambit of duty that oversteps the horizon of knowledge as such. To this perspective also belong, moreover, other publications in the field of moral philosophy, such as the collective volumes which he promoted and edited on topics such as *Bioetica e persona* (1993), *Philosophy and Cultural Development* (edited with Ioanna Kuçuradi in 1993). This astonishing plurality of interests and energy of scientific production in Agazzi's intellectual life and profession is testified by the simple listing of some of his publications in the last two decades: *Cultura scientifica e interdisciplinarità* (1994), *Filosofia della natura. Scienza e cosmologia* (1995), *Le techno-science et l'identité de l'homme contemporain* (1997), *Realism and Quantum Physics* (ed. 1997), *Philosophy of Mathematics Today* (edited with György Darvas in 1997), *Advances in the Philosophy of Technology* (edited with Hans Lenk, in 1999), *The Problem of the Unity of Science* (edited with Jan Faye, in 2001), *Complexity and Emergence* (edited with Luisa Montecucco, in 2002), *Valori e limiti del senso comune* (2004), *Operations and Constructions in Science* (edited with Christian Tiel, in 2006), *Science and Ethics. The Axiologic Contexts of Science* (edited with Minazzi, in 2008), *Le rivoluzione scientifiche e il mondo moderno* (2008), *Relations Between Human Sciences and Natural Sciences* (edited with Giuliano Di Bernardo in 2010), *Evolutionism and Religion* (edited with Minazzi in 2011), *La ciencia y el alma de Occidente* (2011), *Ragioni e limiti del formalismo, Saggi di filosofia della logica e della matematica* (edited with a Foreword by Minazzi in 2012), *Representation and Explanation in the Sciences* (ed. 2013), *The Legacy of A.M. Turing* (ed. 2013).

This already intense and exceptionally fruitful program of research, study and reflection then finds a significant culmination and theoretical crowning in the publication of his most recent work, looked forward to for some twenty years, his systematic study *Scientific Objectivity and its Contexts*, published by Springer in 2014.

As shown even by this concise and elliptical overview of the most important volumes published and edited by Agazzi in over fifty years of scientific and academic research, it can certainly be said that this thinker, building initially on a strictly epistemological study, has gradually expanded his research program, taking into consideration many problems and issues that have ultimately led him to develop a broader, more systematic and "complete" philosophical reflection, capable of dealing with moral philosophy and the very significance of the presence of man in history. Precisely for this reason I evoked the Kantian critical trichotomy above. Not so much to place Agazzi's thought forcibly under the "Kantian bushel" (the reasons for his evident critical distance from the horizon of the Kantian critique are too many and also extremely profound, see below), but rather to highlight how the overall articulation of his scientific-philosophical research enables us to grasp, thanks to the Kantian critical trichotomy, the whole openness and the rhythm of the theoretical breadth of his original proposals. This is also because, as Kant himself well knew, these three different questions concerning knowledge, duty, and hope, are reduced, ultimately, to a single, strategically decisive question: *what is humanity?* And Agazzi himself arrived rather early at such a reflection on humanity, considered from multiple points of view, and he significantly argued that, in his view, the main problem of contemporary culture (to put it in his own paradoxical words) is trying to "*prove the existence of man*" with the same commitment that in other times philosophy has devoted to the task of proving the existence of God.

2 *Philosophy as Rational Understanding of the* Lebenswelt

But how has Agazzi sought to conceive philosophy? And how does he understand the precise meaning of the *conceptual work* peculiar to philosophical reflection?

> I conceived it - recently replied Agazzi himself - as the effort to rationally understand the complex 'world of Life' in order to find a rationally justified solution to the 'problem of Life.' By the world of life I mean the totality of whatever falls within experience and surrounds us, namely the set of material, natural, historical, social and cultural conditions in which we are immersed and conduct our lives. By the problem of life I mean the need to find the 'right' way to conduct one's existence in order to 'save the value of Life", i.e. to give it a positive *sense.* In both cases philosophy is characterized as a *rational* inquiry that arises 'from the point of view of the Whole' (or, in other terms, of the *Absolute*), investing the totality of experience to ask oneself if, from the comprehension of it, arises a solution to the problem of life. That is tantamount to saying an answer which, in particular, is capable of attributing a sense within the totality of experience itself, or requiring a dimension of the Whole that goes beyond the totality of experience (solutions of the problem of the Absolute of the immanent or trascendentist type respectively). In the case that this undertaking fails, we will have an *irrationalist* outcome, or if it arrives at conclusions that are neither positive nor negative, we will arrive at an *agnostic* position (Agazzi 2013: 8, italics in the text).

This quotation enables us to immediately attain a detailed framework of reference within which the conceptual work of philosophy (at least according to Agazzi's approach of overt rationalist aspiration) is confronted with a highly articulated complexity of problems and open questions directly connected with life and its pragmatic problematicity. It is not very difficult to discern an underlying affinity between this position of Agazzi and the tradition of phenomenological research which has always related the *experience* of the world of praxis to a need for *rational understanding* of life and its problems, a critical understanding that is then capable of establishing a relationship of authentic "critical suspension" (*epoché*) of that experience, so as not to be a victim of the most uncritical and pervasive immediacy and pragmatism of life itself. But in Agazzi there is also the emergence of a different *critical* and even *metaphysical* curvature.

Critical, because our philosopher relies, in the first place, on the intrinsic *reasonableness* of the solutions gradually developed and applied in different historical and theoretical situations. In other words, Agazzi sees human reason as certainly, to put it again in Kantian and Husserlian terms, a precious and irreplaceable *function of the critical integration of experience*, but Agazzi then adds to this heuristic, *Aristotelian* function, the ability to always develop open and dynamic critical solutions, capable of finding his own strategic Archimedean point of reference precisely in the *reasonableness* of the solutions adopted. In other words, for Agazzi critical rationality is configured as a balanced heuristic instrument for the conceptual understanding of the complex articulation of reality. A "reasonable" heuristic instrument which, in each specific case, identifies a possible emergent solution as the most suitable and, indeed, the most "reasonable", namely as the solution most capable of understanding the rich articulation of the real, without however ever slipping into prejudicial, rigid or abstractly dogmatic positions. For this reason the critical rationality to which Agazzi increasingly appeals is always configured, in all his works, as a patient art of knowing how to unravel problems, weaving rational arguments that always analyse the whole of reality, seeking to offer the light of rational understanding as a dynamic and plastic reason that illustrates the complex aspects (*phenomena*) of reality.

Metaphysical, because Agazzi does not neglect to deal also with "the point of view of the Whole". In his philosophical argument we can in fact see that, within this specific critical perimeter of conceptual understanding, the reason which Agazzi addresses constitutes, at the same time, a peculiar practice (an argumentative praxis) which does not ignore the "point of view of the Whole" or of the Absolute. Precisely on this ground is then delineated the second component of this rationality, namely the explicitly *metaphysical* component, whose Aristotelian root, however, is critically mediated through the whole history of Western thought, without of course neglecting the specific formation of Agazzi himself at the school of Bontadini which, on this specific point, emerges very clearly (because on this point Agazzi agrees with Bontadini's critique, 1947, 1952, 1996 of the so-called "dualistic metaphysical realism", namely the "naturalistic assumption of the transcendence of thought", introduced by Descartes which is seen as surviving even in Kant, see Agazzi 1996). In fact, as has been noted, in Agazzi the requirement of the Whole (with a capital initial, just as the word "Life" is also capitalized in his text) is one with the "Absolute" and in this same ambit significantly emerges the aspiration to a "totality of experience" that constitutes, in fact, an explicit and systematic metaphysical requirement. In fact, if one bears in mind what Kant writes about the *Transcendental Dialectic,* in his first *Critique*, in which the philosopher of Königsberg brings out just the fact that the "aspiration to the totality of the requirements for the single reality", "opens, inevitably, to the metaphysical dimension" (in the worst sense of the term, meaning the illusion of knowledge) one can perceive how on this particular point a significant divergence exists between

Agazzi's program of philosophical research and that opened up by the revolutionary Kantian transcendentalist breakthrough. Agazzi certainly does not in the least defend metaphysics in its traditional strictly ontological approach, precisely because his thinking is constantly interwoven and nurtured by a continuous critical comparison, moreover one that is extremely sophisticated, with the latest critical reflections conducted in different fields of philosophic and scientific knowledge. Precisely for this reason Agazzi always has the critical sagacity to re-propose the requirement of placing himself "from the point of view of the Whole", as a heuristic point of view, capable of recovering, by using the method of "analogy", the prospect of the "Absolute" *within* and *beyond* the more limited and circumscribed ambit of human experience. But precisely this strategic point reveals his distance from a qualifying component of modernity, namely that conceptual tradition which—with Kant, but not only with Kant, of course—holds that it is not critically legitimate to go *beyond* the ambit of any possible experience. This theoretical approach stresses, in fact, the constantly circumscribed, limited and always finite character of possible human knowledge. Agazzi does not, however, endorse this need for critical caution and being at the same time aware of all the problematicity of the traditional metaphysical ontologism, raises the need to be able to satisfy a "point of view of the Whole", appealing in particular to the use of *analogy* as a privileged and fruitful instrument in order to defend positively the theoretical *possibility* of being able to construct this particular path of metaphysical inquiry (see Agazzi 2014: 437–455).

On the other hand, this twofold rational need, at the same time *critical* and *metaphysical*, enables him also to avoid two opposite uncritical dogmatisms that often occurred in the history of Western thought: namely the dogmatism of *scientism* (which transforms science itself into an absolute and a sort of taboo, above all possible criticism) and conversely the dogmatism of *fideism,* (which is opposed in an abstract and prejudicial attitude to scientific knowledge and pursues an alleged absolute symbolic knowledge of reality). Once more, against these two uncritical unilateral approaches, which result in unique forms of irrationalism, Agazzi maintains the *sense of critical measure* of his sophisticated rationalism, directed towards the identification of multiple *rational arguments* capable of better illuminating the complex nature of human knowledge, always studied and grasped in its intrinsic historical and conceptual determinacy.[4]

This enables us to better understand the original epistemological approach with which Agazzi has always analysed the peculiar nature of scientific knowledge. Our philosopher, in fact, has not only always defended the precise cultural value of the scientific tradition (see Agazzi 2008a, b), but has always grasped the nature of scientific knowledge, highlighting both its *criticality* and the nature of its *rigour*, and its intrinsically *objective* scope (Agazzi 2014). In other words, for Agazzi science

[4]From this point of view his annotated edition of the writings of Maxwell (1973) remains emblematic, as well as his own *Storia della scienza* see Agazzi (1984), to be compared with that of Geymonat (1970–1976).

constitutes *objective, critical* and *rigorous* knowledge that is such precisely because it delineates, at the highest possible level (albeit always within a certain technical-cognitive patrimony historically configured) an objectivity, a criticality and a rigour which are conceptual and dynamic paradigms of reference. However, Agazzi, while recognizing the fundamental role of this threefold characterization of modern scientific knowledge, at the same time points out its insufficiency in providing a critical understanding of the very patrimony of knowledge available to us from the history of scientific thought. In other words, in his view, objectivity, criticality and rigour are *necessary* components but certainly not *sufficient* to characterize the entire nature of scientific knowledge as a whole. Agazzi feels, in short, the need to supplement these characteristics with the consideration of the *foundation* and *sense* of these same kinds of scientific knowledge. Again Agazzi feels, in short, the requirement that an adequate understanding of the philosophical critique of science entails, in turn, a recognition of how much lies "outside" science itself, because, in his view, *value judgments* themselves cannot find their adequate justification *within* science (for the critical analysis of this complex problem of contemporary philosophical and epistemological reflection, I allow myself to make reference to the collective work that we jointly edited, see Agazzi and Minazzi 2008).

In any case, precisely this particular epistemological-critical approach has enabled Agazzi, from his earliest studies on the philosophy of physics, to avoid, critically, both every possible *phenomenalist* outcome and any drift that has led many epistemologists to talk about a hypothetical science of the *unobservable*. Even in the case of the philosophy of quantum physics Agazzi has instead qualified science as *objective knowledge*, distinguishing, however, two different meanings of *objectivity* itself, i.e. a *weak* objectivity from a *strong* objectivity (Agazzi 1974: 339–357, 2014: 51–57). In fact, if we limit ourselves to defending the *weak* sense of objectivity, science is inevitably reduced (and returned) to a dimension of mere *public intersubjectivity* that is rooted, ultimately, in the linguistic consensus of a given community of scientists. But Agazzi holds that in science there also exists another component, equally fundamental and indispensable, that goes well beyond mere consensual public intersubjectivity, and is rooted precisely in the actual cognitive capacity of scientific thought, which enables us to know the world, revealing some significant aspects of its material and real configuration. Therefore, on this level of *strong* objectivity Agazzi defends the full and legitimate *realist* scope of scientific knowledge, in complete harmony with the classic lesson of thinkers such as Aristotle, Galileo, Newton, Darwin, Maxwell and Einstein.

If the epistemological position of Agazzi is set in relation to the, albeit prominent and complex, traditions of the conventionalist phenomenalism (from Duhem to Poincare, to give just two emblematic names), of the logical empiricism that grew out of the *Wiener Kreis* (which then went through various epistemological phases and seasons, in which Carnap has, however, always been a key point of reference, and so coming down to Hempel's most mature reflections) and Popperian falsificationism itself (not to mention the outcomes of his school, from Lakatos to Feyerabend), it is now easy to understand the originality and uniqueness of the realist position supported by Agazzi. In the first place, because in his reflection on science Agazzi has always defended *the reasons of realism,* so finding himself in a position of substantial isolation and originality. In fact, much of the epistemological debate of the twentieth century has been decidedly anti-realist. And even when it has defended the reasons of realism—as, for example, a philosopher like Popper did, throughout his life—it was a minimal realism, closer to that typical of common sense. In short, it was configured as an uncritical realism that failed to develop a philosophical vision, critical and fully articulated, of its own perspective. To clearly grasp all the reasons and also the theoretical and philosophical features of Agazzi's epistemological realism, it would suffice to bear in mind the intense and memorable discussion, theoretical and dialogic, that he (and the present writer) conducted with a long-standing realist and acknowledged father of the Italian philosophy of science like Geymonat (with whom Agazzi himself had studied, immediately after his early training under Bontadini: see Agazzi et al. 1989, but see also Agazzi 1985, 2001, 2009, as well as Geymonat 1977, Mangione 1985, Minazzi 2001, 2009, 2010). In this regard we should not overlook the influence on Agazzi's thought exerted by an original thinker like Mathieu, in particular by the valuable study that Mathieu devoted in the sixties to the problem of objectivity in science and modern and contemporary philosophy (Mathieu 1960).

In any case, in relation to the different positions that interpret objectivity as mere intersubjectivity, as invariance, or, again, as correspondence to the objects dealt with in a scientific theory, Agazzi, ever since *Temi e problemi di filosofia della fisica*, has had no doubts in stating that, in his view, "the right position of correct realism is rather that which, between objective and real, sees a relationship of *inclusion*: all that is objective is real, even though not all that is real is objective" (Agazzi 1974: 365). Agazzi is thus induced to support a position of original *critical realism* (see Agazzi 2014: 243–312), precisely because he has clearly in mind an observation that has instead often been overlooked or removed from the philosophical debate of the twentieth century, namely the critical awareness that

> The concept of truth is never, in practice, absolute but relative, in this precise sense: a proposition (or set of propositions) is almost never true or false *simpliciter*, but true or false *of* a certain universe of objects, so that the question itself concerning its truth is not formulated completely until one says *of what objects* it must be true. In practice, therefore, the truth is always a truth within a theory, because only within it are objects, as we know, given (Agazzi 1974: 369, italics in the text).

But then how can we qualify the "objects" of scientific knowledge? For Agazzi, the best solution to this challenging question lies in recognizing that "the object is *nothing more* than the sum of *all* its determinations" (ibid.: 370, italics in the text), with the result, then, that if we agree to grasp the determinations of objects as real and existing, consequently the objects must also be thought of as real and existing. The realism proposed by Agazzi, however, is "critical" precisely because it never overlooks the fact that, in the history of thought, the twilight of a determined and genuine scientific theory does not mean recognizing that it was *false* (as would claim the Popperian falsificationism which is thus forced to offer a cemeterial vision of the history of science[5]), but rather that it was *partial*. As a result, its replacement by a new theory always involves the development of a new approach that will be *better* than the previous one, precisely because it will enable us to seize a larger number of determinations of the reality that is the object of our study.

On this plane we therefore see how Agazzi agrees with the Kantian approach, according to which human knowledge is always circumscribed and delimited, because, to quote again Agazzi,

> an absolute truth could not be anything but a truth that applies *to all possible objects*, that is, a truth that by holding true for all possible objectivities, focuses on reality no longer as objectified, but *as such*, which, therefore [...] goes beyond the ambit of consideration of science and rather concerns philosophy (which, characteristically, when it wishes to give itself a cognitive task, proposes the study of *reality as such* and is configured as *metaphysics*) (Agazzi 1974: 369-370, italics in the text).

With this we can clearly see that Agazzi's significant proximity to the epistemological horizon of Kantian transcendentalism is characterized, however, by a specific and wholly decisive difference. Indeed, though admitting (with Kant) that all human knowledge is always confined to certain specific objects within precise cognitive boundaries, Agazzi yet seeks also to recover the "point of view of the whole", as a characteristic and specific investigation of philosophical inquiry which, in his view, leads to that *metaphysics* which Kant instead intended to definitely banish from the epistemological plane (reserving it only a different function within the world of practice and our ethical choices).

In any case, for Agazzi it is the *predicates* that define *operationally* the object of scientific knowledge, precisely because the object, by its intrinsic epistemological nature, is configured as "a structure of relations, most of which can be the result of operations but whose 'being together' is not justified by any operation, despite having to be objectively verifiable" (Agazzi 1974: 374). The very presence of this conceptual framework (as rightly pointed out, among others, by Weyl (2009), explicitly mentioned by Agazzi) stresses how the nature of the objects studied by science cannot be deduced solely from the experimental dimension. Indeed the *conceptual determination* of the said structure depends on a theoretical component that is not reducible, *without residuals*, to the plane of experimental experience (*pace* all the systematically reductivist dreams variously cultivated and replicated, by the tradition of classical empiricism, and Viennese logical empiricism). Also Geymonat in *Filosofia e filosofia della scienza* states that "the history of science shows us that in many cases progress was achieved by the replacement of principles, immediately suggested by observation, with others, seemingly much more contrived and more distant from the facts" (Geymonat 1960: 60). For this reason Agazzi concludes by observing that

> experience, in other words, by itself 'does not speak'; it is rather like the oracle of Delphi, of which Heraclitus said that it 'neither speaks nor conceals, but gives signs',

[5]On this point, however, I may be permitted to refer the reader to Minazzi (1990, 1994).

> i.e. it provides the basis for the constitution of the semantic logos, but does not explicitly indicate an apophantic logos. Just like the response of the oracle, experience has to be "interpreted" and this interpretation is primarily an intuitive act: 'In science,' Goethe wrote, "everything depends on what can be called an *aperçu*, on a recognition of what underlies phenomena. And this recognition is infinitely fruitful (Agazzi 1974:376).

In other words, the real world that we want to know is always, in Galileo's words, "deaf and inexorable": experiences become significant not so much thanks to experience as such, but thanks to that particular "point of view" (the *aperçu* Goethe speaks of) by virtue of which we can construct a theory with which, in the words of Kant, we interrogate nature in the same way as a judge examines a defendant or a witness, forcing nature to answer *our* questions, though we know that nature's answers are also decisive for our own theories since they can, in fact, verify or falsify the predictions derived from our particular theoretical framework.

3 From Intensional Semantics to a New Conceptual Image of Knowledge

The mention of the relation between the semantic logos and the apophantic logos, which concluded the previous section, not only explicitly brings out again the link that connects Agazzi to the classical and fundamental Aristotelian lesson of the *Organon,* but once again indicates his proximity to (and at the same time also his critical distance from) the lesson of Kant, with particular reference to the breakthrough connected with transcendentalism. To the extent that Agazzi emphasizes and highlights the irreplaceable role of theory in the constitution of experimental experience it is evident his similarity with a classical Kantian problem. Actually he notes that "even outside a Kantian discourse, one cannot help but recognize the authenticity of this fact and draw precisely the consequence that, without a minimum of theory, one cannot even *begin* to do science" (Agazzi 1974:377). However, in Agazzi's case this very recognition has led to a progressively comprehensive rethinking of the philosophical problem of meaning (Agazzi 1979), also by advancing a complex examination of the philosophical roots of the different senses of meaning. Indeed, faced with the so-called "linguistic turn", and also against the related "relativistic turn" which has variously characterized the post-neo-positivist philosophy of science, Agazzi was gradually induced to develop an original and detailed analysis of the *intensional semantics* of empirical theories, explicitly raising the problem of the impact of semiotics on the philosophy of science, according to a research program and reflection currently consigned, in its most significant achievements, to the pages of his book *Ragioni e limiti del formalismo* (Agazzi 2012). A critical reflection on Hilbert's formalism and also on the heuristics developed by the axiomatization of scientific theories was the theoretical ambit in which Agazzi's philosophical exordium was already delineated since the publication of his *opera prima* (Agazzi 1961) and constitutes a fruitful and ever-present thread running through nearly all his highly articulated program of philosophical research. Indeed Agazzi has gradually and increasingly specified how the *scientific object* cannot fail to emerge as a peculiar *intellectual construct.* But the very recognition of the existence of this intellectual construct has since led him not only to clarify the eminently relational nature of truth, but also the reasons for a critical realism that cannot but accept a perspective aimed at safeguarding a valid epistemological pluralism capable of identifying the multiplicity of different *levels of reality* investigated and studied by different scientific disciplines. Within this specific dilatation of his program of philosophical research, Agazzi has reconsidered the link that can be established, even within a strictly axiomatized theory, between the *syntactic* component (related with the linguistic and conceptual plane), with its precise *meaning*, as well as its relation to the horizon of *referents*. As is well known, in philosophy and methodology of science it is usually held (think of Morris, Carnap or even Tarski, to suggest only a few exemplary names) that the task of semantics is to assign a specific "interpretation" to a set of syntactic symbols that are held to be "devoid of meaning". In this perspective, the attribution of meaning to a theory is interpreted, *à la* Russell, as the assignment of certain referents (individuals or groups of individuals *et similia*) that appear to be appropriate to the theory to be interpreted. Now Agazzi's perspective, in an attempt to develop a three-level semantics, opposes this conceptual and logical approach, which is widespread and shared by both epistemologists and mathematical logicians. In his view, in fact, the task of semantics is certainly to assign a sense or a meaning to linguistic expressions, but Agazzi also believes that this task is quite different from (and independent of) that of associating referents to syntactic symbols.

This critical perspective draws, in particular, on Gottlob Frege's logical reflection, but it is also conscious of an older tradition of thought that goes back directly to scholastic logic. According to this approach it is necessary to distinguish between meanings and referents, both because meanings, by themselves, do not constitute referents, and also because neither do referents, by themselves, constitute meanings. Agazzi writes in this respect:

> This distinction was clearly developed by scholastic logic as a distinction between *intentio* and *suppositio* and was recovered by Frege in the distinction between *Sinn* and *Bedeutung* [...]. Hence it is far from obvious that when we offer an "interpretation" of a formal system, associating its expressions with certain referents, we give a meaning or sense to these expressions. Naturally, we can offer them meanings (senses), but this requires us to associate with them certain conceptual entities and not referents [...]. This resistance to merging meaning with reference has a long tradition in the history of philosophy. It is implicit, for example, in all the criticisms of the so-called ontological argument for the existence of God, and is at the root of the Kantian demand that some "synthetic" (i.e. empirical) condition must be present in order to be able to attribute the character of knowledge to a statement" (Agazzi 2012: 249).

Frege's semantics lays particular stress on the objective contents of thought [the *Gedanken*], by means of which the *conceptual plane* of scientific thinking is rightly brought out fully, and is recognized as a precondition for the determination of the referents:

This is all the more true if we reflect on the fact that, according to him [Frege, *ed.*] referents can be reached only through the sense and for this reason he attributed a sense even to proper names, which are the typical *linguistic signs* which have individuals as their referents. But this three-level semantics lost its intermediate level already with Russell and the meaning of *linguistic signs* was reduced to their referents or denotations, although Russell remained Fregean in some respects. This trend was reinforced in the extensional semantics for formal systems introduced by Tarski and developed in model theory in mathematical logic" (Agazzi 2012: 250-151, italics in the text).

But, in this way, the paradoxical exit was the losing sight of the specific and autonomous (though relative) *conceptual plane* that always qualifies the scientific enterprise, hence precisely that component of *the conceptual framework* through which we can develop the very notion of the "scientific object", as we have seen. This, however, confirms, from the point of view of Agazzi's espistemology, the close link that always exists between the objective knowledge brought into being by scientific theories and the defence of a critical realism. For what reason? Precisely because, to quote Agazzi again,

the realist position argues that the scientific discourse has a real referent. As is well known, at least since Frege's famous essay on *Sinn* and *Bedeutung*, which recovered distinctions and concepts already widely present in the scholastic treatments of the *suppositio* and the *intentio* of terms, a difference exists between the *meaning* of a term (Frege's *Sinn*), which is a content of thought expressing "what is meant" by that term, and its *referent* or *denotation* (the Fregean *Bedeutung*), which is an object constituting "that about which" that meaning is thought or expressed. Unfortunately, such a distinction has been left unproductive by those who, for a fairly long period of time, have occupied a prominent position in developing theories of meaning, that is by mathematical logicians, who have quickly embraced, with regard to the interpretation of formal calculi, an extensionalist semantics according to which the meaning of a term is the set of its referents (Agazzi 1985: 175-176).

The needs of "practicality" adopted by mathematical logicians to justify abandoning Frege's distinction (an abandonment reinforced by the hegemony exercised by the Hilbertian formalism according to which a formal set of symbols does not possess any meaning, except that of so-called "implicit definitions"), produced an increasing separation between *meaning* and *referent,* leading to the curious (epistemological) paradox of legitimating both a discourse devoid of meaning (which then would not say anything) and a discourse devoid of referents (which then would not speak of anything). The aim of science, however, is very different; it is

to be a referential discourse, since it cannot be affirmed that a statement is true without admitting that it is true of something. [...] The empirical sciences make use of non-linguistic operational criteria of reference in order to grasp the referents of many of their propositions (those that directly describe experimental results), but now we can also add that the same theoretical concepts of a theory must have a 'real' referent (Agazzi 1985:180).

Precisely this recognition enables one to understand the specific function of the apophantic logos which is different from the semantic, because

the institution of the apophantic logos is characterized by the fact that, in addition to the meaning, there emerges the referent and, moreover, in such a way as not being independent of meaning. In fact the search for the referent requires a non-linguistic activity [...] which is in many cases (especially in the case of science) actually of a markedly "practical" kind, such as operatively manipulating by means of instruments, observing in appropriately created conditions, and so forth. This activity therefore consists in *exploring the world* and not in *exploring language*. [...]. The apophantic logos is therefore one which institutes the *notion of truth* directly related to that of reference" (Agazzi 1985: 182, italics in the text).

This then enables one to better understand why

> Each scientific discipline is presented as a discourse that intentions reality from a certain 'point of view', namely proposing to investigate only certain aspects or qualities of it; for that reason it selects a limited number of "predicates" and, in order to be successful in its referential effort, it associates them with some standardized *operations*, which we can call 'criteria of objectification' or 'protocollar criteria' or 'criteria of referentiality'. It is these operations that 'cut out' the specific objects of a given science from the vast ambit of reality and, precisely because they are transactions that do not apply to anything, but to referents already identified (the 'stuff' of everyday experience which is practised within a certain historically determined community) and moreover subject to empirical, and not purely linguistic or intellectual, manipulations, single out specific referents that are necessarily *real* (Agazzi 1985: 188, italics in the text).

A much more complete and elaborated presentation of the theses presented here regarding the peculiar semantic and operational foundation of Agazzi's realism is offered in Agazzi (2014), the life-work in which he has presented the global portrayal of his epistemology. We have preferred to give a documentation of these positions with reference to older publications, in order to show the continuity of the maturation of these ideas.

This overall outcome of Agazzi's critical realism thus proves particularly attuned to other very different programmes of philosophical inquiry—for example with that of a highly original Italian philosopher like the critical empiricist Preti (on whom see Preti 2011 and Minazzi 2011), or with that of the "regional ontologies" of the phenomenology outlined by Husserl in *Ideen zu einer reinen Phänomenologie und phänomenologischen Philosophie* (1913)—which, however, also insisted on both the specific and fundamental *conceptual dimension* of the scientific process as well as the desirability of recovering, heuristically, but also phenomenologically, the fruitful scholastic doctrine of *intentionality* and *suppositio*, in order to develop a richer, more articulated, appropriate and plastic critical image of scientific knowledge.[6]

This specific philosophical approach in Agazzi's reflection also explains the original way in which our philosopher has always been able to engage discussions with some of the principal positions of his time, highlighting their inherent one-sidedness and also their dogmatisms. Take, for example, the problem of the historical determinacy of scientific theories or, again, the no less extensive and profound debate about the alleged "neutrality" (or non-neutrality) of scientific knowledge or, again, the debate concerning the relation between science and

---

[6]On this point, however, I may be permitted to refer the reader to Minazzi (2011).

ethics, or also the relation between science, evolution and religion (for which, in this context, I refer only to Agazzi 1992 and Agazzi and Minazzi 2011). In all these cases, by using an approach based on general systems theory (see Agazzi 1978), Agazzi identifies the privileged comprehensive system of reference, and then takes into consideration the multiple subsystems, *open* and *adaptive*, in accordance with the systems-theoretic methodology inaugurated by Bertalanffy (1968), which he has, however, reworked in a fruitful way within his epistemological and even philosophical reflection. Therefore, while many interlocutors in these debates insist on contrapositions that constitutes a drastic and unilateral "*aut/aut*", Agazzi, on the contrary, has always endeavoured to investigate critically the links of connection, relationship and coordination (i.e. "*et/et*") that can (and must) be identified, always considering them as flexible and complex, mirroring in such a way the actual articulation of a real world, the tangled skein which must always be unravelled, *à la* Leonardo, with critical intelligence, taking into account its multiple, varied and even conflicting actual components, so as to be able to hope to grasp, to again quote the genius of Vinci, any possible "threads of truth".

References

Agazzi, Evandro. 1961. *Introduzione ai problemi dell'assiomatica*. Milano: Vita e Pensiero.

Agazzi, Evandro. 1974. *Temi e problemi di filosofia della fisica*. Roma: Edizioni Abete.

Agazzi, Evandro. 1978. *Le geometrie non euclidee e i fondamenti della geometria* (con Dario Palladino), Est Mondadori, Milano (new edition La Scuola, Brescia, 1998)

Agazzi, Evandro.1979. *Studi sul problema del significato*, ed. E. Agazzi. Firenze: Felice Le Monnier.

Agazzi, Evandro. 1984. (a cura di), *Storia delle scienze*, Roma, Città Nuova, 2 voll.

Agazzi, Evandro. 1985. La questione del realismo scientifico. In Mangione 1985, 171-192.

Agazzi, Evandro, Minazzi, Fabio, Geymonat, Ludovico. 1989. *Filosofia, scienza e verità*. Milano: Rusconi Editore.

Agazzi, Evandro.1992. *Il bene il male e la scienza*. Rusconi: Milano (English translation *Right, Wrong and Science*, ed. Craig Dilworth. New York-Amsterdam: Rodopi 2004.

Agazzi, Evandro.1996. Introduzione. In Bontadini 1996, IX-XXII.

Agazzi, Evandro. 2001. Prefazione. In Minazzi 2001, 9-15.

Agazzi, Evandro. 2008a. *Le rivoluzioni scientifiche e il mondo moderno*. Milano: Fondazione Achille e Giulia Boroli [edition not for sale].

Agazzi, Evandro. 2008b. *Scienza* (interview by Giuseppe Bertagna). Brescia: Editrice La Scuola.

Agazzi, Evandro, Minazzi, Fabio. 2008c. *Science and Ethics. The Axiological Contexts of Science*. Eds. Evandro Agazzi, Fabio Minazzi. Bruxelles-Bern-Berlin-Frankfurt am Main-New York-Oxford-Wien: P. I. E. Peter Lang.

Agazzi, Evandro. 2009. Geymonat o della sincerità. In Minazzi 2009, 65-75.

Agazzi, Evandro, Minazzi Fabio. 2011. *Evolutionism and religion* (Proceedings of the joint meeting of the International Academy of Philosophy of Science and the International Academy of Religious Sciences - Florence, 19-21 November 2009). Eds. Evandro Agazzi, Fabio Minazzi. Milano: Mimesis.

Agazzi, Evandro. 2012. *Ragioni e limiti del formalismo. Saggi di filosofia della logica e della matematica*. Edited and with a *Preface* by Fabio Minazzi. Milano: Franco Angeli.

Agazzi, Evandro. 2013. Che cosa è dentro e che cosa è fuori dalla scienza. Un riflessione filosofica. in *Filosofi italiani contemporanei*. Eds. Giuseppe Riconda, Claudio Ciancio, 7-42. Milano: Mursia.

Agazzi, Evandro. 2014. *Scientific Objectivity and its Contexts*. Cham-Heidelberg-New York-Dordrecht-London: Springer.

Bertalanffy, Ludwig von. 1968. *General Systems Theory: Foundations, Development, Applications*. New York: George Braziller (revised edition 1976).

Bontadini, Gustavo. 1947. *Studi sulla filosofia dell'età cartesiana*. Brescia: La Scuola Editrice.

Bontadini, Gustavo. 1952. *Indagini di struttura sul gnoseologismo moderno*. Brescia: La Scuola editrice.

Bontadini, Gustavo. 1996. *Studi di filosofia moderna*. Milano: Vita e Pensiero.

Geymonat, Ludovico. 1960. *Filosofia e filosofia della scienza*. Milano: Feltrinelli.

Geymonat, Ludovico. 1970-1976. *Storia del pensiero filosofico e scientifico*. Milano: Garzanti, 7 voll.

Geymonat, Ludovico. 1977. *Scienza e realismo*. Milano: Feltrinelli.

Mangione, Corrado. 1985. *Scienza e filosofia. Saggi in onore di Ludovico Geymonat*. Milano: Garzanti.

Mathieu, Vittorio. 1960. *L'oggettività nella scienza e nella filosofia moderna e contemporanea*. Torino: Accademia delle Scienze.

Maxwell, James Clerk. 1973. *Trattato di elettricità e magnetismo*, ed. Evandro Agazzi. Torino: Utet, 2 voll.

Minazzi, Fabio. 1990. Riflessioni critiche sulla filosofia di Popper. *Epistemologia* XIII: 221-36.

Minazzi, Fabio. 1994. *Il Flauto di Popper. Saggio critico sulla "new philosophy of science" e la sua interpretazione di Galileo*. Milano: Franco Angeli.

Minazzi, Fabio. 2001. *La passione della ragione. Studi sul pensiero di Ludovico Geymonat. Prefazione* di Evandro Agazzi. Milano-Mendrisio: Edizioni Thèlema-Accademia di architettura dell'Università della Svizzera Italiana.

Minazzi, Fabio. 2004. *Teleologia della conoscenza ed escatologia della speranza. Per un nuovo illuminismo*. Napoli: La Città del Sole.

Minazzi, Fabio. 2007a. *Filosofia, Scienza e Bioetica nel dibattito contemporaneo. Studi internazionali in onore di Evandro Agazzi*, ed. Fabio Minazzi. Roma: Presidenza del Consiglio dei Ministri, Dipartimento per l'Informazione e l'Editoria, Istituto Poligrafico e Zecca dello Stato.

Minazzi, Fabio. 2007b. *Bio-bibliografia di Evandro Agazzi* in Minazzi 2007a, 1351-1402.

Minazzi, Fabio. 2008. See Agazzi, Minazzi 2008.

Minazzi, Fabio. 2009. *Ludovico Geymonat un Maestro del Novecento. Il filosofo, il partigiano e il docente,* ed. Fabio Minazzi. Milano: Edizioni Unicopli.

Minazzi, Fabio. 2010. *Ludovico Geymonat epistemologo. Con documenti inediti e rari (un inedito del 1936, il carteggio con Moritz Schlick, lettere con Antonio Banfi e Mario Dal Pra)*. Milano: Mimesis.

Minazzi, Fabio. 2011. *Suppositio pro significato non ultimato. Giulio Preti neorealista logico studiato nei suoi scritti inediti*. Milano: Mimesis.

Petitot, Jean. 2009. *Per un nuovo illuminismo. La conoscenza scientifica come valore culturale e civile*. Prefaced, edited and translated from the French by Fabio Minazzi. Milano: Bompiani.

Preti, Giulio. 2011. *Philosophical Essays. Critical Rationalism as Historical-objective Transcendentalism,* ed. Fabio Minazzi. Translation from Italian by Richard Sadleir. Bruxelles - Bern - Berlin - Frankfurt am Main - New York - Oxford - Wien: P. I. E. Peter Lang.

Weyl, Hermann. 2009. *Philosophy of Mathematics and Natural Science*. Translated by Olaf Helmer, with a new Introduction by Frank Wilczek. Princeton and Oxford: Princeton University Press.

Part I
General Philosophy of Science

The Perspectivist Conception of Science

Craig Dilworth

Abstract The paper compares Evandro Agazzi's "*Gestalt* view" with the Perspectivist conception of science developed by the author. What Agazzi means by *Gestalt* or "point of view", as well as what the author means by "perspective", is not something of the subjective sort. It is rather a particular way of *conceiving* of reality—a way that can be *shared*. Agazzi's *Gestalt* view, however, differs in certain respects from the Perspectivist conception. First, on the *Gestalt* view, since theories concretely express their *Gestalt* in declarative sentences, the theories themselves must be true or false; on the Perspectivist view, the paradigm of a scientific thought would not be a true-or-false statement, but a *more or less applicable concept*. Second, Agazzi's legitimation of *truth* on the *Gestalt* view entails realism, while Perspectivism is neutral as regards the empiricism/realism issue. Finally, some differences between both conceptions of science are mentioned as far as the solution of the incommensurability problem is concerned.

Keywords Perspectivism · Agazzi · Realism/antirealism · Idealisations · Incommensurability

Evandro Agazzi's long-awaited *Scientific Objectivity and Its Contexts*, as well as his earlier work, support an aspect of a complete new approach in the philosophy of science, an aspect which he terms "the *Gestalt* view." Particularly noteworthy in this regard is Agazzi's presentation of the *Gestalt* view as leading to his ideas of *the empirical-scientific 'clipping out' of objects*, *idealisation*, and the analogy of scientific theories to *maps*.

C. Dilworth (✉)
Uppsala University, Uppsala, Sweden
e-mail: craig.dilworth@gmail.com

M. Alai et al. (eds.), *Science Between Truth and Ethical Responsibility*,
DOI 10.1007/978-3-319-16369-7_2

According to the first idea, of 'clipping out' objects, we can see scientific objects as 'clipped out' by their disciplines through the performance of particular empirical operations, each discipline containing just those objects clipped out by its operations. We might

> simply consider some 'thing' and ask what science is competent to deal with it. For instance, if we take a watch and ask what the area of its face is, we are considering it as an object of topology; if we ask what its mass is, or what the laws are that regulate the motion of its balance wheel, or what its influence would be on the magnetic field inside the room where it is located, we are considering it as an object of physics; if we ask what the composition of the alloy is out of which its case is made, or what the degree of purity is of the rubies that are inside it, we are considering it as an object of chemistry.... (p. 83).

On the contemporary, formalist, view in the philosophy of science, an object is what it is, and different scientific disciplines are thought to provide various sorts of information concerning it. The difference between Agazzi's approach with his 'clipping out' of objects and the contemporary view might be thought to be insignificant, but if looked at more closely his 'clipping out' idea reveals an aspect of a conception of science fundamentally different from the contemporary view (the Perspectivist conception). On Agazzi's view, the operations performed in the 'clipping out' of objects are performed according to concepts derived from the categories of the discipline, the discipline itself constituting a *Gestalt*.

I have myself worked extensively with the *Perspectivist* conception of science, which is in many respects similar to Agazzi's *Gestalt* view. In fact, Agazzi's and my common holding of this insight with regard to the nature of knowledge lay the ground for a friendship between us that has lasted from 1977 to the present.

What Agazzi means by *Gestalt* or *point of view*, and what I mean by *perspective*, is not something of the subjective sort, as when one *sees* a certain gestalt. It is rather a particular way of *conceiving* of reality—a way that can be *shared*. For Agazzi and me *Gestalts* and perspectives are structured by various *principles* and the *categories* they relate.

On the contemporary view, on the other hand, scientists begin with the assumption-free basis of *experience*, and inductively build conceptual systems (theories) from there. As distinct from this, Agazzi and I say that science begins with certain a priori preconceptions as to the nature of reality, and that these preconceptions—or *principles*—themselves *constitute* the scientific perspective.

Agazzi's *Gestalt* view, however, differs in certain respects from the Perspectivist conception. On the *Gestalt* view, but not the Pespectivist conception, since theories concretely express their *Gestalt* in declarative sentences, the theories themselves must be true or false. When a theory is 'falsified,' what is falsified is actually some particular sentence (statement) of the theory, a sentence that turns out to be false according to the referential operational criteria of the theory's discipline. According to Agazzi, this entails a re-adjustment of the gestalt going from partial retouches to a genuine gestalt switch (Cf. Agazzi, pp. 367ff).

But, given this, one wonders how *idealisation* is to fit into the *Gestalt* view, all idealisations, if taken to be statements, being false. As is in keeping with the Perspectivist conception, the value of idealisation in modern science lies rather in its ability conceptually to capture the physical essence of a particular situation, such an essence paradigmatically taking the form of a *cause*.

Also according to Agazzi, his legitimation of *truth* on the *Gestalt* view entails *realism*, since a sentence is always true or false 'of' something, and in the case of science this something is the sentence's operationally accessible referents. As a consequence, if a theory is supposed true, it must also be supposed that its intended referents exist.

But as regards *truth*, while it is true that Pegasus has two wings, this does not entail Pegasus' real existence. Similarly, true or false statements may be made regarding an intended scientific object, without their being meant to imply the object's existence.

Perspectivism, on the other hand, is neutral as regards the empiricism/realism issue. On the Perspectivist view, while true or false statements may be made with regard to a theory and its application, this does not imply that the theory itself is a entity that may be considered to be true or false. In its application to modern science, however, Perspectivism has shown the endeavour to be thoroughly realist in nature.

For the last hundred years, the philosophy of science has been almost wholly confined to the (formal) *logic* of science, i.e. to thinking of science in terms of the Deductive Model. On this view, the paradigm of a scientific thought is an Aristotelian *statement*, which is either *true* or *false*. On the Perspectivist view, on the other hand, the paradigm of a scientific thought would not be a true-or-false statement, but a *more or less applicable concept*.

On the Perspectivist view, the primary aim of science is not to *know*, but to *understand*. Though the notions of truth and knowledge do belong to science, they belong to its *empirical* aspect. We acquire knowledge of the empirical facts, i.e. of laws determined by measurement. But our higher aim is to *understand* these laws. And this we do using theories to link the laws to the principles of the discipline. The Perspectivist view has many advantages not only over philosophies of science based on the Deductive Model, but over those based on set theory as well.[1]

As shown in *Scientific Progress*, the logical empiricists' and Popper's attempts to depict science in terms of the Deductive Model fail, the empiricists being unable to account for theory conflict, and Popper having no conception of progress. Further, neither the empiricists nor Popper can capture the notion of incommensurability introduced by Kuhn and Feyerabend. One cannot provide a purely formal account (depiction) of incommensurability. The Perspectivist conception, on the other hand, solves the incommensurability problem by showing incommensurability in the case of scientific theories to be a relation between conceptual perspectives (applied concepts) sharing the same intended domain. And it shows how in science one can have a form of *subsumption* that differs from empiricist *deductive* subsumption, and a form of *theory conflict* that differs from Popperian *contradiction*. And the Perspectivist view shows quite generally how one theory may be

[1]As is treated in Dilworth (2008, Chap. 11).

considered to be scientifically more acceptable than another, thereby a consistent conception of scientific progress.[2]

On the Perspectivist view, the principles underlying science determine its *ontology*, and in so doing also its *epistemology*, the latter including its *methodology*. We can also say that the principles of science determine the *perspective* we call science, which includes science's conceptual *paradigm*.[3]

Each individual *scientific discipline*, such as physics, chemistry or biology, *refines* the basic principles of science in its own way. These refined principles determine the particular aspect of the scientific ontology (reality) the discipline investigates. In this way, the refined principles set limits on the discipline itself. So, for example, considering reality from the point of view of the categories of matter, motion and force determines the objects of mechanics rather than those of biology, while both disciplines accept the deeper metaphysical principle that no physical entity comes from nothing. In agreement with Agazzi, each discipline also contains more specific concepts, at least some of which have to be *operationalised* for the discipline to make contact with physical reality (ontology). And one and the same part of reality can become the object of a new and different discipline every time a new perspective (epistemology) is taken on it. Thus different disciplines study different aspects of reality, and are incommensurable in this way. Further, each scientific discipline's refined principles serve to distinguish it from other disciplines, and provide it with its own conceptual paradigm.

Each scientific discipline constitutes a specific perspective on precisely the objects picked out by its operations, which are its *intended object*. The perspective constituted by the discipline includes the categories in terms of which reality is to be conceived. Thus, following Agazzi, the concept of area belongs to the topology perspective, the concept of mass to the physics perspective, and so on. Every scientific discipline has its own domain of intended objects; (p. 64); and "one and the same 'thing' can become the object of a new and different science every time a new specific point of view or viewpoint is taken on it." (p. 84).

It is important to note that for Agazzi the same operations (using e.g. measuring instruments) by means of which the objects of a given science are 'clipped out' of reality are those by means of which it is possible to reach empirical-scientific agreement, due to the intersubjectivity of the operations. This provides commonality of *reference* and *intention*, *within* each discipline. On this basis, on the Perspectivist conception, various theories are advanced in each discipline concerning how best to conceive of the discipline's subject-matter so as to *understand* how particular empirical phenomena can be or are manifestations of the refined principles. This is how scientific theories *explain* the phenomena. And not only disciplines but also theories may be incommensurable.

---

[2]Cf. Dilworth (2008, esp. pp. 85–88).

[3]As I try to show in Dilworth (2007), modern science is based on a conceptual paradigm consisting of three particular physically-interpreted principles relating the categories of uniformity, substance and cause.

It is important to note the two senses of *incommensurability* used here. Incommensurable disciplines have different subject-matters; incommensurable theories can have the same subject-matter. As regards the latter, the situation is more that of a gestalt switch, where each conception constitutes an alternative depiction of a common reality.[4]

Thus, on the Perspectivist view, different theories within a discipline can the *same* objects but say different things about them. And if the theories do differ in how they characterise a common reality, there are a number of ways their superiority in this regard might be determined.

This is essentially the view I present in *Scientific Progress*, and develop in *The Metaphysics of Science*. Both books presuppose my latest book, *Simplicity*.

References

Agazzi, E. 2014. *Scientific Objectivity and Its Contexts*. Cham, etc.: Springer.
Dilworth, C. 2007. *The Metaphysics of Science*, second edition. Dordrecht: Springer.
Dilworth, C. 2008. *Scientific Progress*, fourth edition, Dordrecht: Springer.
Dilworth, C. 2013. *Simplicity*, Lanham: Lexington Books.

[4]Cf. the Gestalt Model in Dilworth (2008).

Science and Operationality

Marco Buzzoni

Abstract One of the most important aspects of Evandro Agazzi's operationalism lies in his attempt to wed the main idea of operationalism with a perspectival view of scientific knowledge. In the Sect. 1 of this paper I argue that this connection is essential to understanding Agazzi's substantial contribution to the philosophy of science. In the Sect. 2, I briefly compare Agazzi's and Searle's treatment of Turing's test, to show how important the notion of perspectival knowledge is for Agazzi. In the last section of my paper, even though I essentially agree with Agazzi's operationalism, I raise some doubts concerning the relationship between theory and experiment and the connection between science and technique, and I propose the modifications that I believe are needed to make Agazzi's operationalism more consistent.

Keywords Bridgman · Searle · Hacking · New experimentalism · Operationalism · Perspectivism · Technique

1 Bridgman's and Agazzi's Operationalism

According to the main principle of Bridgman's "operational analysis", empirical concepts generally mean nothing more than a set of operations: a scientific concept "*is synonymous with the corresponding set of operations*".[1] A body has a position only in so far as this position can be measured; if a body's position cannot be measured in principle, then that body's position does not exist. Operations

I wish to thank Dr. Paolo Baracchi for valuable stylistic suggestions.

[1]Bridgman (1927: 5).

M. Buzzoni (✉)
Department of Humanistic Studies, University of Macerata, Macerata, Italy
e-mail: marco.buzzoni@unimc.it

M. Alai et al. (eds.), *Science Between Truth and Ethical Responsibility*,
DOI 10.1007/978-3-319-16369-7_3

should be "uniquely" defined and intersubjectively repeatable; that is, different scientists must be able to perform the same operations with reasonable agreement in their results. As Bridgman says, operations should be such that they "can be repeated by the same person or different persons under the same or different conditions without hesitation and with the accompaniment of no phenomena which demand the assertion that there has been failure to repeat".[2]

Operationalism in the form it takes in Bridgman's writings, incurs serious difficulties; this fact has induced many authors to dismiss operationalism per se. Since the meaning of a physical term is nothing but the operations leading to its measurement, different measuring operations define different physical magnitudes, so that we end up with an implausible proliferation of physical magnitudes. According to Bridgman, the operations by which length is measured should be uniquely specified. Lengths measured by a ruler and lengths inferred from the time that light takes to travel a given distance and return are to be considered as different physical magnitudes and, strictly speaking, they should have different names: since we have more than one set of operations, and therefore we have more than one concept of length.[3]

Bridgman was well aware of this consequence,[4] but the identity of the measurement results of lengths that are 'different' because they are measured by different instruments remains for him an unexplained accident. Why there is no one-to-one correspondence between the homogeneity of measurement results concerning for example 'length' and the operations leading to these results? This lack of correspondence highlights a fundamental limitation of his approach, which is unable to explain this 'empirical' aspect of the scientific endeavour.

Agazzi's philosophy of science is decidedly operational in character yet avoids such difficulties. Like Bridgman, Agazzi holds that scientific concepts are intimately connected with instrumental operations, but he strongly disagrees with Bridgman as to how experience and theory are to be understood.

On the one hand, for Agazzi it is doing, not sense-data, that is the basis of experience. He insists on the fact that there are meaningful statements that are accepted or rejected on the basis of non-linguistic conditions, that is, of conditions which concern the sphere of "doing something" rather than that of "saying something".[5] Building on Poincaré and the later Wittgenstein, Agazzi claims that people's agreement about cognitive content does not hinge on their 'private' data but on determinate actions that they perform:

---

[2]Bridgman (1950: 29).

[3]Cf. for example Hempel (1954, 1966, Chap. 7), Carnap (1966, Chap. 10). Moreover, as Hempel (1966) rightly observed, there is no way of setting limits to the proliferation of concepts corresponding to the same physical magnitude (such as length), since the development of measurement instruments only slightly different from each other would lead, strictly speaking, to new and different magnitudes. The same objections have been, and still are raised by many authors.

[4]Cf. Bridgman (1927: 10).

[5]Cf. Agazzi (1989: 87).

> Is such an agreement possible? It is, through operations. This fact is very general and is not limited to scientific practice: When we wish to test whether we agree with someone else about a certain notion (that is, about any content of knowledge) the only means at our disposal is to see whether we both make the same *use* of that notion. It is not *apprehending the same thing* in applying the notion that can demonstrate agreement about the notion, but *applying the notion in the same way* in what are otherwise the same circumstances. [...] If I have certain reasons to be doubtful about my interlocutor's having the same notion of red as mine, I could, for example, invite her to select from a bundle of pencils a red one. If the person's way of *operating* is the same as that which I should have adopted in all circumstances of this kind, I am fully justified in concluding that 'red' is an intersubjective notion for us.[6]

On the other hand, Agazzi differs from Bridgman not only in his view of experience, but also in his view of theory. For him theory plays a very different role in shaping the cognitive object. Beside objectivity as intersubjective agreement (which Agazzi calls "weak" objectivity), there is a stronger sense of objectivity as reference to "objects". According to Agazzi, "objects" are constituted by bundles of attributes which we single out from the specific view point of any particular science. It is in connection with this sense of objectivity that Agazzi insists upon the fact that the sciences do not investigate 'things' as ultimate primitive entities, but consider them under different points of view. One and the same 'thing' or entity can become the object of a new and different science if it is considered from a new specific 'viewpoint'. In other words, the object is the result of the application of certain "criteria of protocollarity": by considering reality from the point of view of matter, motion and force, for example, we constitute the "objects" of mechanics rather than those of biology. As Agazzi writes:

> if we take a watch and ask what the area of its face is, we are considering it as an object of topology; if we ask what its mass is, or what the laws are that regulate the motion of its balance wheel, or what its influence would be on the magnetic field inside the room where it is located, we are considering it as an object of physics; if we ask what the composition of the alloy is out of which its case is made, or what the degree of purity is of the rubies that are inside it, we are considering it as an object of chemistry; if we ask its price relative to other watches and in relation to the present conditions of world watch production, we are considering it as an object of economics; if we ask whether wearing a watch of a certain kind might be an indication of its owner's having a certain sort of temperament, we are considering it as an object of psychology; or if our watch is rather old and we ask whether it once belonged to a certain prime minister whose biography we are writing, we are considering it as an historical object.[7]

The two characterizations of objectivity are intimately connected with each other because the operations by means of which the objects of a given science are "extracted" from reality are the same as those by means of which it is possible to reach an intersubjective agreement among researchers. Thus, the strong (ontological) and the weak (epistemic) sense of objectivity are two different sides of the

[6] Agazzi (2014: 76); cf. also Agazzi (1969: 346).

[7] Agazzi (2014: 83); cf. also Agazzi (1976: 12–13). For the change in the meaning of the word "object" in Agazzi, see Agazzi (2009: 171).

same coin: the conditions according to which the objects of a science are given are at the same time the conditions for knowing them objectively.[8]

In general, we may say that Agazzi's view of operations, in comparison with Bridgman's, is closer to that of Hugo Dingler and of German "methodical constructivism" in maintaining that doing is the basic condition for us to have an epistemic access and theoretical-perspectival reference to reality.[9] Together with German "methodical constructivism" Agazzi anticipated the "new experimentalist" turn, which in the 1980s stressed the importance of experimenting and, more in general, of acting and operating in science. A comparison with Hacking's entities realism will enable us to better understand Agazzi's operationalism.

2 Hacking's and Agazzi's Realism

According to Hacking experimentation "has a life of its own".[10] This statement, which has become the motto of the new experimentalism, means that experiments, unlike theories, "are organic, develop, change, and yet retain a certain long-term development which makes us talk about repeating and replicating experiments".[11] This entails a sharp distinction between high-level theories and the phenomenological laws and models, endowed with a low level of generality, which provide the basis for the stability of experimental results. Low-level laws and models, being common to many general theories, are the touchstone for inter-theoretical comparisons. In fact, low-level theories change much more slowly than general theories, even though this fact has been overlooked because of the disproportionate attention paid to the supposed revolutionary changes at the higher theoretical level.[12]

---

[8]Cf. Agazzi (1969, Chap. 10).

[9]Dingler's operationalism was taken up by the 'methodical constructivism' that rose at Erlangen and developed especially at Konstanz and Marburg. Among the authors inspired by Dingler, see e.g. Lorenzen (1987); Holzkamp (1967), Mittelstraß (1974), Tetens (1987), Janich (1993). Special mention must be made of Janich, who has renamed his own account "methodical culturalism" in order to distinguish it from other more "naturalistic" versions. In general, however, Agazzi's version of operationalism, compared to that of the German "methodical constructivists", is much closer to scientific realism. For a comparison between Evandro Agazzi and Peter Janich, see Buzzoni (1997). For an analytical reconstruction of Agazzi's scientific realism, cf. Alai (2009).

[10]Hacking (1983: xiii).

[11]Hacking (1993: 208). For some examples of works inspired by the new experimentalism, cf. also: Hacking (1992, 1999), Cartwright (1983, 1989), Ackermann (1985), Franklin (1986, 1999), Harré (1986, 1998), Galison (1987, 1988), Giere (1988, 1999, 2006), Gooding (1990, 1998), Radder (1988, 1996).

[12]Most 'new experimentalists' subscribe to this thesis: cf. e.g. Cartwright (1989: 352), Galison (1987: 211, 261), Gooding (1990: 88, 190) (where he explicitly accepts Hacking's thesis, even though elsewhere he is much more cautious), Franklin (1999: 273).

In this connection, Hacking bases the reality of theoretical entities such as electrons on the fact that we can manipulate them in the same way as we manipulate everyday objects. We can put electrons to various uses—for example, we can use them to alter the charge on a niobium ball.[13] Theoretical entities are real chiefly because they can be used as "instruments" capable of interacting causally so as "to manipulate other parts of nature in a systematic way". By the time we succeed in doing this, the electron "has ceased to be something hypothetical, something inferred. It has ceased to be theoretical and has become experimental".[14]

By embracing these theses, Hacking tries to prevent experimentation from being swallowed up by the theoretical aspect. However, by so doing he ends up endorsing, in an inverted form, a dichotomy of theory and experiment similar to the one he criticises in Popper and in relativistic philosophies of science.

Despite insisting on a technical-operational criterion for the reality of theoretical entities, Hacking stresses the contrast between experiment and theory rather than their connection. This contrast underpins his claim that one can defend realism "about entities" but not "about theories". In this respect Hacking runs into two main difficulties. In the first place, it is unclear what more might a theory say about empirical reality other than what it says about it through its theoretical entities. It is impossible to distinguish the evidence for the existence of certain particles from the evidence for the theory that talks about them, *since the content of the theory is just the claim that these particles and their properties exist*. As Franklin noted, when experimenters find a particle with a charge, mass and lifetime equal to those ascribed by the theory to K mesons, they rightly assert the existence of K-mesons—just as we would assert the existence of the philosopher Bas van Fraassen if we found an entity with height, weight, gender, hair colour, eye colour, date of birth and home address which were exactly those listed on his driver's license. The point is that the successful performance of operations and measurements on certain particles warrants both the existence of those particles and the truth of the laws about them. The use of the laws that allow us to find out the properties of elementary particles gives these laws the same epistemic status as the particles and their properties.[15]

Secondly, to maintain that experiments have a life of their own, thus separating theory and experiment, would make action punctual, limited to the entities involved in a particular action here and now. Action would lose its generality, a generality which depends on the encounter between the linguistic-representational sphere and empirical reality, and is expressed by natural laws in the form of theories. Without theory, action could only afford access to a reality which would be indeterminate and formless, an unknown "*x*" that would be scientifically irrelevant and should be discarded in accordance with the principle of economy since, being compatible with any theory, it is incapable of "taking sides" in disputes between rival theories.

[13]Cf. Hacking (1983: 23).

[14]Hacking (1983: 262).

[15]Cf. Franklin (1999: 151–155).

Although Hacking is quite right in rejecting theory-ladenness in its most common interpretation, he neglects its element of truth, which consists in the theoretical-perspectival character of scientific knowledge.[16] By changing the theoretical framework, two instances of the "same" experiment can become two different experiments. Two experiments, identical *as to the experimenter's actions and the experimental apparatus*, can stand for two distinct experiments, or even two experiments in distinct scientific disciplines, if performed in order to answer distinct theoretical questions. Before 1905, experiments on the composition of velocities were considered the "same" irrespective of how close the velocities involved were to the velocity of light, since Newtonian mechanics does not distinguish on this basis.[17] After that date, in the light of the special theory of relativity, these experiments take on entirely different meanings according to how close the velocities considered are to the velocity of light; therefore, they are to be considered different experiments. Otto von Guericke showed that sound travels through water by means of an experiment in which he regularly rang a bell before he fed fish in a pond; but if we consider that the hungry fish arrived at the ringing of the bell, this experiment in physics could perfectly well count as a psychological experiment in animal conditioning.

Therefore, there is an intrinsic connection between theory and experiment. The Neopositivist and Popperian thesis of the independence of theory from experiment is as flawed as its mainly new experimentalist converse, the independence of experiment from theory. Both theses presuppose a dichotomy between theory and experiment that makes both concepts incomprehensible.

Returning to our comparison between Hacking's and Agazzi's realism, we can say that (despite some inconsistencies which we shall discuss later on) in Agazzi the problems arising from the dichotomy between theory and experiment are, at least in principle, solved. What may be called the "perspectival character" of scientific knowledge[18] is one of the pillars of Agazzi's philosophy of science, namely the notion of objectivity as reference to "objects". As we saw, it is in connection with this sense of objectivity that Agazzi insists upon the fact that the different sciences investigate "things" from different points of view, so that the same "thing" can become the object of different sciences when considered from different "viewpoints". Basically, for Agazzi theoretical entities cannot have the status of a *quid incognitum* because the perspectival character of empirical knowledge demands that scientific objects or entities do not *have*, but *are* their properties, which are selected from the point of view of determinate theoretical concepts: scientific entities are not an unknown *quid* beyond their properties, but are exhausted in them.[19]

---

[16]From this point of view, Giere's perspectival approach to scientific knowledge (Giere 2006) illustrates the need to accept the element of truth contained in theory-ladenness. Whether Giere'sperspectivism remains realistic enough when separated from any version of the correspondence theory of truth is another question, which need not be discussed here.

[17]For this example, cf. Franklin (1989: 438–439).

[18]Cf. Buzzoni (1995, Chap. 1, Sect. 3.)

[19]Cf. e.g. Agazzi (1978: 36); cf. also Agazzi (1994: 95, 2014).

Later on we shall touch on some aspects of Agazzi's thought in which the operational and theoretical aspects are not perfectly harmonized. However, before we do that I wish to say something about the central role played by the perspectival character of scientific knowledge in Agazzi's thought. For this purpose it will be interesting to compare Agazzi's and Searle's views on the limits of AI. At first sight, this might seem a strange comparison, but a re-examination of Searle's point of view in the light of the perspectival character of scientific knowledge will prove to be instructive.

3 Operationalism and the Perspectival Character of Scientific Knowledge

The Chinese Room thought experiment sums up Searle's view about the limits of AI. Here it is in a very concise version:

> Imagine a native English speaker [...] who knows no Chinese locked in a room full of boxes of Chinese symbols (a data base) together with a book of instructions for manipulating the symbols (the program). Imagine that people outside the room send in other Chinese symbols which, unknown to the person in the room, are questions in Chinese (the input). And imagine that by following the instructions in the program the man in the room is able to pass out Chinese symbols that are correct answers to the questions (the output). The program enables the person in the room to pass the Turing test for understanding Chinese, but he does not understand a word of Chinese.[20]

According to the prevailing interpretation, the main point of this thought experiment is that the purely formal, abstract, syntactical processes of an implemented computer program "could not by themselves be sufficient to guarantee the presence of mental content or semantic content of the sort that is essential to human cognition."[21]

It is interesting to note that in 1967 Agazzi devised a thought experiment in some way similar to Searle's. If we exclude "Leibniz's Mill",[22] Agazzi's thought experiment is the first antecedent of Searle's Chinese Room argument, anticipating for example Ned Block's "Chinese Gym".[23] Agazzi too argues against the Turing Test by pointing out that it is intrinsically impossible to determine intentionality through behaviour. He claims that a computer or a robot can imitate a human being only so far as syntax is concerned. Now, syntax is neither identical with nor

---

[20]Searle (1999: 115). For the first formulation of this thought experiment, cf. Searle (1980: 417–418).

[21]Searle (2002: 51–59). Searle put his thought experiment also in the form of a logical argument: 1. Programs are formal (syntactical). 2. Minds have contents (semantic contents). 3. Syntax is neither identical with nor sufficient for semantics. From these premises we can derive: Programs are neither sufficient for nor identical with minds; i.e. strong AI is false (Searle 1991: 526).

[22]Cf. Leibniz ([1714] 1902: 254). Among the authors who have pointed out the similarity of Searle's thought experiment with Leibniz's, cf. Sharvy (1983), Cole (1984), Rapaport (1986), and Jacquette (1989: 616–617).

[23]Cf. Block (1978).

sufficient for semantics; but minds have semantic contents and it is impossible to understand the meaning of a symbol without human intentionality:

> it is quite reasonable to suppose that a person who does see may simulate the behaviour of a blind person (i.e. there is no apparent behavioural experiment to discriminate a genuinely blind man from one who is not). On the other hand, if we accept for a moment that it would really be possible to equip a robot with such electronic devices as would allow it to simulate perfectly the behaviour of a seeing person (i.e. to avoid obstacles, to read, to recognize colours and figures, etc.), we may also imagine equipping a blind person (or simply somebody whose eyes were adequately sealed) with similar electronic devices, and teaching him/her to interpret the stimuli received through these electronic "sense organs" (stimuli that might be for example of a tactile character). In this condition our person would be able *ex hypotesi* to *behave* as if he/she were seeing, without seeing at all.[24]

I think that Agazzi, at least to some extent and in certain respects, clearly anticipated Searle's thought experiment. In my opinion, this anticipation is no accident, but depends on (at least) one important assumption shared by the two authors, namely the perspectival view of empirical knowledge.

Why, Agazzi asks, do we not say that a camera "saw" the object that is now on the photographic film? The difference is in "the way in which the image is present":

> when we feel compelled to state that a camera, differently from animals, does not "see" an object, we are referring to a "something more" which accompanies the pure recording of an image in the case of the animal, that is, to intentionality [...] The difference between recording an image on a photographic plate and seeing it does not consist in the fact that there is an image, but *in the way in which the image is present*; and it should be no wonder that such a way is not subject to further analysis by means of facts, for there could be a risk of falling into a *regressum* ad infinitum, as one might always formulate a question about the way the subsequent facts are given.[25]

Searle is not only clearly aware of the perspectival character of scientific knowledge; he also realizes that it is an intrinsic ingredient of human intentionality. It is in this sense that intentionality is, in Searle's words, "the feature of certain mental states by which they are directed at or about objects and states of affairs in the world".[26] In particular, the Chinese Room thought experiment presupposes that intentionality has a necessary "aspectual" (that is, perspectival) character which would be very hard to ascribe to a computer or to a robot. Searle uses this claim against the thesis that intentionality is reducible to computability, when he argues that every intentional state has an "aspectual shape", in the sense that it is directed at an object only "under an aspect".[27]

---

[24]Agazzi (1967: 17–18). This example is also in Agazzi (1981: 211–212).

[25]Agazzi (1981: 210); italics added.

[26]Searle (1980: 424), Footnote 3.

[27]Searle (1992: 159–160). "Aspects" or "aspectual shapes" here refer, roughly, to what is involved in seeing-as phenomena: e.g., seeing things as alike or seeing ambiguous pictures as one thing or the other (e.g., Necker cubes, duck-rabbits). Seeing the duck-rabbit picture as a duck would (in these terms) be seeing its duck aspect or seeing it "under" its duck aspect. This point is made in Husserl's concept of "Abschattungen" and Wittgenstein's analysis of seeing-as phenomena (cf. Wittgenstein 1958[1980], II, xi) and, more recently, especially by Haldane (1989), Putnam (1992), McGinn(2004), and Giere (2006).

Given that they share the important thesis of the intentional-perspectival view of empirical knowledge, it is unlikely that it was a coincidence that in 1967 Agazzi relied on a thought experiment that was very similar to Searle's in order to show the limits of AI. Indeed, the intentional-perspectival view of empirical knowledge is a strong reason for rejecting any strong reductionism. The fact that reality must be investigated from particular points of view which cast light on particular aspects of it, is sufficient to dissolve the reductionist illusion that there is a single privileged and fundamental level of description. Conversely, to assume a single privileged and fundamental level of description (which is the very heart of radical reductionism, whether idealistic or naturalistic) would be in stark contradiction with the abstractive, selective, partial, idealizational and theoretical character of any empirical investigation of reality, which requires thought-projects which are "subjective" or, more precisely, always guided by our purposes and values.

There is an obvious difference between Searle and Agazzi. Searle does not exclude the possibility in general of duplicating human intelligence but only the possibility of duplicating it *by computational means*. Agazzi, on the contrary, maintains that, from a scientific point of view, the question in general as to whether an automaton, besides being able to behave like a seeing entity, is or is not actually able to see, must remain open, even though "the *onus probandi* must be taken on by those who maintain this fact, for it is a matter of asserting "something more" than is necessarily implied by the robot's behaviour."[28]

However, this is a minor point, in the light of what we have been saying. The most important difference between the two authors lies elsewhere. Agazzi's solution of the problem of intentionality is consistent with the main tenets of his philosophy of science, and especially with the perspectival sense of objectivity. Searle's position, on the other hand, is not internally consistent, since the perspectival (or "aspectual") account which he rightly gives of the problem of intentionality is clearly incompatible with the main tenets of his philosophy.

As is well known, Searle insists that intentionality is produced by the brain. Intentionality is for Searle a "biological phenomenon, and it is as likely to be as causally dependent on the specific biochemistry of its origins as lactation, photosynthesis or any other biological phenomena."[29]

This claim, which sees the brain or some of its properties as an ultimate constituent of matter or reality, is inconsistent with the aspectual or perspectival character of intentionality. If an ultimate biochemical phenomenon existed independently of any particular point of view, if it existed in itself apart from our knowledge interests, it would be a kind of atomic, self-enclosed reality. But what we call the brain and its properties (including intentionality as a biological phenomenon) cannot exist or be understood apart from theoretical constructions of some type. The brain and its properties appear as biochemical realities only through the concepts, terms and technical apparatuses that define the viewpoint from which biochemistry investigates reality.

---

[28]Agazzi (1981: 212).

[29]Searle (1980: 424).

Searle's claim that intentionality is produced by the brain is inconsistent with the perspectival character of scientific knowledge according to which nature can be known scientifically only from a potentially infinite number of perspectives or theoretical points of view. If a "part" of reality, such as a table, can be examined as a physical phenomenon as well as a chemical or a commercial one, it turns out to be a mechanical phenomenon only when it is considered from a point of view that takes into account only some of its properties such as force, mass and certain spatial and temporal relations. If this is so, whenever physics and chemistry grasp certain law-like connections inherent to a phenomenon, it would be arbitrary to assume that the phenomenon is thereby fully explained, perhaps apart from a few minor details. It would also be arbitrary to assume that any other kind of explanation would necessarily clash with the one provided, for example, by biology. Recognition of this fact is sufficient to undermine all reductionist approaches, whether in terms of old-fashioned symbols and rules systems *à la* Turing, or in terms of distributed processing connectionist systems, or in terms of chemical-biological duplication *à la* Searle.

On the contrary, consistently with this point of view, Agazzi writes that reductionism

> always consists in giving to assertions, which are true in their own specific domain of scientific reference, a general scope which they do not possess. For instance, the sentence "the brain is a very powerful computer" is true, provided that it is meant that this is the case if the brain is considered from the point of view of processing speed and storage capacity, but it is clear that this sentence is not the whole truth about the brain, and it would therefore be incorrect to claim (as often happens) that the different brain functions can be reduced to algorithmic processes, even though they also undoubtedly have algorithmic aspects.[30]

To sum up, our brief comparison between Agazzi's and Searle's treatment of Turing's test confirms that one of the most important aspects of Agazzi's operationalism lies in his attempt to connect intimately the main idea of operationalism with a theoretical-perspectival view of scientific knowledge.

In the rest of this paper I shall suggest some modifications which are needed in order to make Agazzi's overall account more consistent.

4 Some Critical Observations

Although by and large I agree with Agazzi's philosophy of science, I have some doubts concerning his account of the relationship between theory and experiment and of the connection between science and technique.

Concerning the relationship between theory and experiment, Agazzi defends a variant of Kant's thesis that an experiment is a question put to nature and rightly argues that, while the purpose of theory is to supply meaning and intension,

[30] Agazzi (2009: 172).

experiments provide theoretical statements with reference.[31] On the other hand, he seems to share the new experimentalist view that experiments can lead a life of their own. Sometimes he goes so far as to say that an experiment, regarded as a question put to nature, is understandable in itself, apart from any theory: in order to be performed, "an experimental test must be entirely describable in terms of operational concepts and their meanings, which [...] do not depend on the theory."[32] The same point has been reiterated by Agazzi more recently:

> the theory provides the significance of the question, its point, its purpose and reason, besides providing the question with a global meaning resulting from its position in the general context of the theory. But it is not as though the question would be meaningless, that is, not understandable without the theory. *This cannot be the case simply because, in order to be performed, an experimental test must be entirely describable in terms of operational concepts and their meanings which, as we have already remarked several times, do not depend on the theory.* This is why we need the theory in order arrive at the experiment, in order to design the experiment (hence the experiment depends "genetically" on the theory, as we have already said) but not in order to justify it. Once the experiment has been performed, it assumes an independent existence and is in no need of help, simply because it has the same character as the data which are the indisputable basis a theory is challenged to account for, and which it cannot modify or dispense with.[33]

This connection between theory and experiment, by itself, comes too close to the thesis of the new experimentalism that experiments can lead a life of their own. As far as this point is concerned, in order to avoid the relativist consequences of theory-ladenness, Agazzi almost restores pure experimental data. True enough, the performance of an experiment can always be described with concepts related to another particular theory or to common sense. But where could the question *with respect to which the outcome of the experiment is relevant* get its significance, if not from the theory to be tested empirically? The assumption that data are "an indisputable basis a theory is challenged to account for" would seem vulnerable to objections similar to those which I raised against Hacking.

If in this respect Agazzi concedes too much to new experimentalism, on the other hand (especially as to the relation between theoretical and observational, or operational, terms) he inclines to the opposite view, vindicating the independence of theoretical interpretation. Theoretical concepts are necessary because they subsume different characteristics under the unity of an object:

> every operational procedure reveals one *single* feature to be attributed to the object so that, after performing all the operations we need, we have a set of such features. But no object of any science is represented by a *pure* collection of features; it is always a *structured* collection, in the sense that all these features are mutually connected by certain mathematical and/or logical relations, which are not obtained directly from any instrument, but must be arrived at through the intellectual activity of the researcher.[34]

---

[31]Cf. for example Agazzi (1988: 6–9).

[32]Agazzi (1988: 9).

[33]Agazzi (2014: 376); italics added.

[34]Agazzi (2014: 100). For previous formulation of this thesis, cf. for example Agazzi (1967: 374) and (1978: 35).

On closer scrutiny, this argument presupposes that operations can show single determinations of reality but not the *link* that connects them within one scientific object: this link would be in principle irreducible to operations. If this were the case, such a component of the scientific object, rather than *something by means of which* we seek access to reality, would be *itself* a term and independent object of the cognitive act, capable as such of existing on its own with no function of connection to the real world.

Moreover, the claim that theoretical terms are needed to express the connection between the properties of an object clashes with another basic thesis of Agazzi's realism, which I have already mentioned, according to which a scientific object does not have, but is its properties. If this is true, it seems to be unnecessary to assume terms that both designate the way in which predicates are connected in an object, and are irreducible as to their content to operational terms. If we reject an unknown x beyond the phenomena, we must regard an object's properties not as atomistic, isolated determinations of reality but as internally connected with the other properties that constitute the object (where this connection too must be established operatively by means of experimental testing).[35]

The second main point on which, in my opinion, Agazzi's view requires some correction is the connection between science and technique. Agazzi's account of this connection changed considerably over his philosophical career. In Agazzi (1969) we find a decided predominance of the theoretical and rational element over the technical one: the nature of scientific theory is discussed without mentioning the importance of its connection with technical applications.[36] This remained unchanged until Agazzi (1985b), which contains, so far as I know, the first recognition of the crucial importance of technique for science.[37]

In my opinion, this was an inconsistency in Agazzi's operationalism. If one recognizes that scientific hypotheses have cognitive value only through some connection with our operational interventions on reality, then one must admit that technical reproducibility is not a criterion among many but *the distinctive criterion of the truth of scientific propositions*. For in experimentation the performance of actions which are in principle intersubjectively repeatable in order to test a hypothesis is a technical intervention in reality. From this point of view, strictly speaking, technical applicability does not depend on theoretical truth ascertained by other means: even though truth is not identical *in every sense* with technical applicability, reproducible technical applications are, *in the experimental natural sciences*, the only way of ascertaining and justifying the truth of a theory. Thus, from a strictly epistemological point of view, the use of radio waves for practical purposes was a decisive reason for the truth of the electromagnetic theory. Similarly, the explosion of the first atomic bomb provided a terrible confirmation of Einstein's equation expressing the convertibility of matter and energy.

---

[35]For more details, cf. Buzzoni (1986).

[36]Cf. Agazzi(1969: 36–37, 155–168, 372).

[37]That Agazzi (1985b) is the first recognition of the crucial importance of technique for scienceisalso indirectly suggested by the fact that in Agazzi(1985a) the technical applications of the Newtonian physics are still regarded as a simple consequence of a theoretical truth ascertained by other means (cf. Agazzi 1985a: 69).

To be more precise, truth is plainly not the same thing as technical usability, just as a theory is not merely an instrument. However, a theory, in so far as it is true and affords us knowledge of the way things actually are, is potentially also useful. On the other hand, the means by which we ascertain the truth of a theory (the *ratio cognoscendi*, one might say) proceeds in the opposite direction: we can prove that a theory says something true about the world only by showing that it can be translated into operationally, technically reproducible results. The theoretical and technical aspects within experimental natural science can be separated only by methodical abstraction: in the concreteness of doing science, they are inextricably connected. The theoretical aspect is defined as the condition of the possibility of the knowledge of determinate aspects of reality in so far as it allows one to envisage *as possible* causal connections that must be translatable, in principle without residue, into successful technical applications. Conversely, the technical aspect has alethic relevance only in so far as it translates into actions a conceptual mediation without which the technical aspect would appear isolated from any causal context—that is, as a mere coincidence, a chance event not reproducible outside the precise and punctual situation in which it occurs (this is probably the case when animals use tools).[38]

In *Il bene, il male e la scienza* (Agazzi 1992[2004]), we find a detailed discussion of the connection of science with technique. In one passage Agazzi goes so far as to admit that, if we take into account the fact that the collection of operations that "cut out" a given field of objects from reality "constitutes a network of *techniques* (that is, a knowledge of how to do or to work) whose goal is to make pure research possible",[39] then "technique is "consubstantial" with science itself [*la tecnica è consustanziale alla stessa scienza*]".[40] The last quoted sentence is left out in the English edition, and this omission is no accident. The claim of an intimate connection between science and technique was not consistent with the rest of the book. Indeed, Agazzi insists that science and technique are different in principle. Technique consists only in "a *knowing how* (one does certain things), without necessarily implying a *knowing why* (they are done that way)": the efficacy and success of those actions emerge "empirically, that is in the concreteness of practice, without one being able (or at least without having to be able) to give the *reasons* or the explanation of their success". Unlike technique, science is different from other kinds of knowledge "precisely insofar as it proposes to *explain* empirical facts, suggesting reasons that tell us *why* these are in a certain way".[41]

[38]Cf. Buzzoni (1982, Chap. 3, Sect. 4, above all pp. 190–192), (1995: 85–99), and (2008, Chap. 1).

[39]Agazzi (1992[2004]: 184, Engl. transl.: 135).

[40]Agazzi (1992[2004]: 185, my translation, since there is no corresponding passage in the English edition).

[41]Agazzi (1992[2004], 75–76, my transl.). Here too I could not find any exactly corresponding passage in the English edition; as, for example, the following quotation from the same book clearly shows, this is not due to a change of mind: "technique is essentially the competent application of a certain *know-how* attained through the accumulation and transmission of concrete *experience* (which also entails a careful performance of acts), without necessarily being accompanied or supported by a knowing *why* such concrete procedures are especially efficacious." (Agazzi 1992[2004], Engl. transl.: 56).

The distinction between knowing "how" and knowing "why" is acceptable only functionally, as it were, as a distinction *concerning both the theoretical and technical aspects*. What counts as "knowing why" at a certain cognitive level appears as a given, as "knowing how" at a further level where deeper questions arise; this deeper questioning changes the previous "knowing why" into a given (a "knowing how") in need of further explanation. For example, one could think that we only have a "knowing that" about the functioning of the more common household appliances; but what is *prima facie* a "knowing that" (say, that the dishwasher is turned on by pushing a certain button, with no deeper knowledge of its functioning) for a child may well be a "knowing why". To the child's question *why* the dishwasher has started making that noise, we may reply, for instance, that this happens just "because" we pushed a certain button which turns it on. Likewise, we can distinguish between our knowledge "that" the dishwasher does not work and the technician's knowledge of "why" that is the case (say, "because the condenser is broken"). However, the technician's "knowing why" is, from the point of view, say, of an electrical engineer, a "knowing how" which in turn calls for an explanation as to "why" the condenser is broken—and so on without end, at least in the sense that it is not possible to establish a frontier beyond which science can progress no more.

One could respond by adducing *prima facie* more convincing examples supposedly demonstrating the possibility of a mere "knowing how". Many technical improvements proceed from chance discoveries and can further improve without probing the reasons behind this improvement. For example, if an angler all of a sudden caught many more fish than usual and noticed that the hook had been accidentally bent for some unknown reason, from then on that angler may always use that hook and may also bend it more or in different ways actually producing more efficient hooks.

At first it would appear that the angler has no insight into the reasons of his undeniably technical behaviour. But if we look at the example more closely, it soon becomes evident that this is not the case. The angler would have never embarked on the search for more efficient hooks had he not noticed that the hook worked better *because it had been bent*; and this is a knowing "why", it does not matter at how elementary or low a degree. Without this *explanatory* hypothesis, the angler would not have progressed to using the bent hook systematically, let alone to improving on it technically. In the course of the historical development of scientific knowledge, a split developed between those who operate in the field of basic science and those who operate in the applied sector; but this does not call into question the fact that science can know only by acting and intervening technically in reality, and that this intervention, in so far as it is not blind but has some access to its reasons, is from the very beginning to some degree scientific.

From this point of view, also the distinction between "technique" and "technology" is legitimate in only one of its meanings, namely as the distinction between technique and discourse on technique: it is clear that the concept of technique cannot be defined by technical means. However, it is in principle impossible to distinguish between technique and technology by alleging that the former is a pure "knowing how to do" lacking knowledge of the reasons of this doing, while the latter is an "efficient" operating which "is conscious of the reasons for its efficacy and is based upon them, that is, where operation is nourished by its grounding in *theoretical knowledge*".[42] This distinction, intended to separate technology from technique, in actual fact only separates human from animal technique. In the human sphere all "knowing how to do", even in the weak forms of habit and/or compulsion to repeat, *qua* knowing, involves at no matter how infinitesimal a level a noetic aspect of critical awareness. Certainly animals too interpret their environment and thereby use something similar to our concepts, but (with all the caution due when talking about animal capacities) these: "concepts" probably lack the human prerogative of criticality, that essential openness that lets them be freely modified according to the changing of situations.

It is certainly legitimate to distinguish between a scientific-theoretical attitude that concerns itself with the way things are (the search for truth) and a practical-technical attitude that aims at transforming things according to certain concepts or values. It is also correct to distinguish between pure and applied science on the basis of the different practical intentionality of the scientists involved.[43] However, all of this is quite irrelevant to our problem, namely the epistemological relationship between science and technique. In particular, none of this excludes the fact that one can know empirical reality only by acting and intervening in nature, and that one can act on nature only by means of meanings or concepts without which acts would be no more than chance events.

References

Ackermann, R.J. 1985. *Data, Instruments, and Theory: A Dialectical Approach to Understanding Science*. Princeton: Princeton University Press.

Agazzi, E. 1967. Alcune osservazioni sul problema dell'intelligenza artificiale. *Rivista di filosofia neoscolastica* 59(1): 1-34.

Agazzi, E. 1969. *Temi e problemi di filosofia della fisica*. Rome: Abete.

Agazzi, E. 1976. Criteri epistemologici fondamentali delle discipline psicologiche. In *Problemi epistemologici della psicologia*, ed. G. Siri: 3-35. Milan: Vita e Pensiero.

Agazzi, E. 1978. Eine Deutung der wissenschaftlichen Objektivität. *Allgemeine Zeitschrift für Philosophie* 3: 20-47.

Agazzi, E. 1981. Intentionality and Artificial Intelligence. *Epistemologia. An Italian Journal for the Philosophy of Science* (IV, Special Issue, The Mind-Body Problem): 195-227.

Agazzi, E. 1985a. Commensurability, Incommensurability, and Cumulativity in Scientific Knowledge. *Erkenntnis* 22: 51-77.

---

[42] Agazzi (1992[2004]: 77, Engl. transl.: 57).

[43] Among the many other authors who hold similar views, cf. e.g.: Skolimowski (1966: 374) (science concerns what is, technique what ought to be); Lenk (1982: 48–50) (technique concerns what may or must be produced); Bunge (1985: 234); Sachsse (1989: 364) (emphasises technique's practical and economic criteria); Granger (1989: 58) ('scientific intentionality' [*visée scientifique*] 'intends to *constitute objects*', while 'technical intentionality' [*visée téchnique*] 'intends to concatenate *acts* in order to obtain a result, and is not interested in the knowledge of *objects* except insofar as this knowledge may serve this end').

Agazzi, E. 1985b. Gli strumenti e l'oggettività scientifica. *Epistemologia. An Italian Journal for the Philosophy of Science* 8: 3-14.
Agazzi, E. 1988. Do Experiments Depend on Theories or Theories on Experiments? In Batens and van Bendegem (eds.)1988: 3-13.
Agazzi, E. 1989. Naive Realism and Naive Antirealism. *Dialectica* 43: 83-98.
Agazzi, E. 1992[2004]. *Il bene, il male e la scienza*. Milan: Rusconi. English edition: Agazzi, E. 2004. *Right, Wrong and Science: The Ethical Dimensions of the Techno-scientific*, ed. C. Dilworth. Amsterdam/New York: Rodopi.
Agazzi, E. 1994. Raisons et formes du réalisme scientifique. *Revue de Métaphysique et de Morale*, n. 1: 89-103.
Agazzi, E. 2009. Commenti. In *Il realismo scientifico di Evandro Agazzi*, ed. M. Alai, 167-172. *Isonomia* (Special Issue).
Agazzi, E. 2014. *Scientific Objectivity and Its Contexts*. Heidelberg/New York/Dordrecht/ London: Springer.
Alai, M. 2009. Il realismo di Evandro Agazzi. In *Il realismo scientifico di Evandro Agazzi*, ed. M. Alai, 77-96. *Isonomia* (Special Issue).
Batens, D. and van Bendegem J.P. (eds.) 1988. *Theory and Experiment: Recent insights and new perspectives on their relation*. Dordrecht: Reidel.
Block N. 1978. Troubles with Functionalism. In *Perception and Cognition: Issues in the Foundations of Psychology*. Ed. C.W. Savage, 261-325. Minneapolis: University of Minnesota Press.
Bridgman, P.W. 1927. *The Logic of Modern Physics*. New York: MacMillan.
Bridgman, P.W. 1950. Reflections of a Physicist. New York: Philosophical Library.
Bunge, M. 1985. *Treatise on Basic Philosophy*, vol. 7, part II Life Science, Social Science and Technology. Dordrecht: Reidel.
Buzzoni, M. 1982. *Conoscenza e realtà in Karl R. Popper*. Milan: Angeli.
Buzzoni, M. 1986. L'operazionismo e il problema dei termini teorici. *Epistemologia. An Italian Journal for the Philosophy of Science* 9: 39-76.
Buzzoni, M. 1995. *Scienza e tecnica. Teoria ed esperienza nelle scienze della natura*. Roma: Studium.
Buzzoni, M. 1997. Operazionismo, realismo e tecnica. Tre forme di operazionismo Ian Hacking, Peter Janich ed Evandro Agazzi, 179-195. In *Contesti filosofici della scienza*, ed. L. Montecucco. Brescia: La Scuola.
Buzzoni, M. 2008. *Thought Experiment in the Natural Sciences*. Würzburg: Königshausen& Neumann.
Carnap, R. 1966. *Philosophical Foundations of Physics*. New York: Basic Books.
Cartwright, N. 1983. *How the Laws of Physics Lie*. Oxford/New York: Oxford University Press.
Cartwright, N. 1989. *Nature's Capacities and Their Measurement*. Oxford/New York: Oxford University Press.
Cole, D. 1984. Thought and Thought Experiments. *Philosophical Studies* 45: 431-444.
Franklin, A. 1986. *The Neglect of Experiment*. Cambridge: Cambridge University Press.
Franklin, A. 1989. The Epistemology of Experiment. In *The Uses of Experiment. Studies in the Natural Sciences*, ed. Gooding D., Pinch T.J., and Schaffer S. 437–460. Cambridge: Cambridge University Press.
Franklin, A. 1999. *Can That be Right. Essays on Experiment, Evidence, and Science*. Dordrecht: Kluwer.
Galison, P. 1987. *How Experiments End*. Chicago: University of Chicago Press.
Galison, P. 1988. History, Philosophy, and the Central Metaphor. *Science in Context* 2: 197-212.
Giere, R.N. 1988. *Explaining Science: A Cognitive Approach*. Chicago: University of Chicago Press.
Giere, R.N. 1999. *Science Without Laws*. Chicago: University of Chicago Press.
Giere, R.N. 2006. *Scientific Perspectivism*. Chicago: University of Chicago Press.
Gooding, D. 1990. *Experiment and the Making of Meaning*. Dordrecht: Kluwer.
Gooding, D. 1998. Picturing Experimental Practice. In Heidelberger und Steinle (hrsg.)1998: 298-322.

Granger, G.G. 1989. Peut-on assigner des frontières à la connaissance scientifique? In *Karl Popper et la science d'aujourd'hui*, ed. R. Bouveresse, 47–61. Paris: Aubier.
Hacking, I. 1983. *Representing and Intervening*, Cambridge: Cambridge University Press.
Hacking, I. 1992. The Self-Vindication of the Laboratory Sciences. In *Science as Practice and Culture*, ed. A. Pickering, 29-64.Chicago: University of Chicago Press.
Hacking, I. 1993. Do Thought Experiments Have a Life of Their Own? Comments on James Brown, Nancy Nersessian and David Gooding. In *PSA 1992*, 2: 302–308, ed. D .Hull, M. Forbes and K. Okruhlik. East Lansing (MI): Philosophy of Science Association.
Hacking, I. 1999. *The Social Construction of What?* Cambridge/London: Harvard University Press.
Haldane, J. 1989. Naturalism and lntentionality. *Inquiry* 32, 305-322.
Harré, R. 1986. *Varieties of Realism*. Oxford/New York: Blackwell.
Harré, R. 1998. Recovering Experiment. *Philosophy* 73: 353-377.
Heidelberger, M. und Steinle F. (hrsg.) 1998. *Experimental Essays – Versuche zum Experiment*. Baden-Baden: Nomos.
Hempel, C.G. 1954. A Logical Appraisal of Operationalism. *Scientific Monthly* 1 (No. 79): 215-220.
Hempel, C.G. 1966. *Philosophy of Natural Science*. Englewood Cliffs (N.J.): Prentice-Hall.
Holzkamp, K. 1967. *Wissenschaft als Handlung. Versuch einer neuen Grundlegung der Wissenschaftslehre*. Berlin: De Gruyter.
Jacquette, D. 1989. Adventures in the Chinese Room. *Philosophy and Phenomenological Research* 494, 605-623.
Janich, P. 1993. Die methodische Konstruktion der Wirklichkeit durch die Wis-senschaften. In*Neue Realitäten – Herausforde-rung der Philosophie*, ed. H. Lenk und H. Poser, 460-476, XVI. Deutscher Kongreß für Philosophie (Berlin, 20.-24. September 1993). Berlin: Akademie Verlag.
Leibniz, G.W.v. 1902[1714]. The Monadology. Transl. G. R. Montgomery. La Salle (Ill.): Open Court.
Lenk, H. 1982. *Zur Sozialphilosophie der Technik*. Frankfurt a.M.: Suhrkamp.
Lorenzen, P. 1987. *Lehrbuch der Konstruktiven Wissenschaftstheorie*. Mannheim: Biliographisches Institut.
McGinn, C. 2004. *Consciousness and Its Objects*. Oxford: Clarendon Press.
Mittelstraß, J. 1974. *Die Möglichkeit der Wissenschaft*.Frankfurt a.M.: Suhrkamp.
Preston, J. and Bishop M. (eds.) 2002. *Views into the Chinese Room: Essays on Searle and Artificial Intelligence*. Oxford: Clarendon Press.
Putnam, H. 1992. The Project of Artificial Intelligence. In H. Putnam. *Renewing Philosophy*, 1-18. Cambridge (MA): Harvard College.
Radder, H. 1988. *The Material Realization of Science*, Assen: van Gorckum.
Radder, H. 1996. *In and About the World. Philosophical Studies of Science and Technology*. New York: State University of New York Press.
Rapaport, W. J. 1986. Searle's Experiments with Thought. *Philosophy of Science* 53: 271-279.
Sachsse, H. 1989.*Technologie*. In *Handlexikon zur Wissenschaftstheorie*, ed. H. Seiffert and G. Radnitzky, 361-5. München: Ehrenwirth.
Searle, J. R. 1980. Minds, Brains and Programs. *The Behavioural and Brain Sciences* 3: 417-457.
Searle, J. R. 1991. Yin and Yang Strike Out. In *The Nature of Mind*, ed. D.M. Rosenthal, 525-526. New York: Oxford University Press.
Searle, J. R. 1992. *The Rediscovery of the* Mind. Cambridge (MA): MIT Press.
Searle, J. R. 1999. The Chinese Room. In *The MIT Encyclopedia of the Cognitive Sciences*, ed. R.A. Wilson and F. Keil. 115-116. Cambridge (MA): MIT Press.
Searle, J. R. 2002. Twenty one years in the Chinese room. In Preston and Bishop (eds.) 2002: 51-59.
Sharvy, R. 1983. It Ain't the Meat, It's the Motion. *Inquiry* 26: 125-134.

Skolimowski, H. 1966. The Structure and Thinking in Technology. *Technology and Culture* 7: 371-83.

Tetens, H. 1987. *Experimentelle Erfahrung. Eine wissenschaftstheoretische Studie über die Rolle des Experiments in der Begriffs- und Theoriebildung der Physik*. Hamburg: Meiner.

Wittgenstein, L. 1958[1980]. *Remarks on the Philosophy of Psychology*, Vol. I, ed. G.E. M. Anscombe and G.H. von Wright. Transl. G.E.M. Anscombe. Chicago: University of Chicago Press.

The Issue of Scientific Realism

Mario Alai

Abstract Agazzi stages a complete, very detailed and overall convincing defence of scientific realism, its presuppositions and corollaries (mind-independence of reality, referentiality of theories, truth as correspondence, knowledge as the goal of science, justifiability of beliefs in unobservables through abductive arguments). But his claims that truth is relative to a circumscribed domain, not "pictorial" and not pertaining to theories, and that scientific objects certainly exist because they are nothing but abstract bundles of properties, are potentially ambiguous. Moreover, to the antirealist objections based on radical theory-change he replies that pre- and post-revolutionary theories do not contradict each other and are equally true, because each one deals with a different domain of objects of its own making: but this reply (apparently a legacy of neopositivistic operationalism) risks to make theories analytic, so slipping into conventionalism, or to reduce their content to observable phenomena, thus giving into antirealism.

Keywords Scientific realism · Truth · Scientific theories · Scientific objects · Operationalism · Incommensurability · Pessimistic meta-induction · Deployment realism · Entity realism

1 Introduction

Agazzi's philosophy of science takes its roots in two different traditions of thought: Aristotelian and scholastic realism, which he learned mainly from Gustavo Bontadini; and neopositivistic epistemology, characterized by empiricism and operationalism, which was still the dominant view when he graduated, and with which he later interacted through Ludovico Geymonat and his research

M. Alai (✉)
Department of Basic Sciences and Foundations, University of Urbino "Carlo Bo", via Timoteo Viti 10, 61029 Urbino, Italy
e-mail: mario.alai@libero.it

M. Alai et al. (eds.), *Science Between Truth and Ethical Responsibility*,
DOI 10.1007/978-3-319-16369-7_4

enterprises. What is more important, he has succeeded in synthesizing the best aspects of both approaches.

From the former tradition, he draws the idea that reality is the object of scientific research, and truth is it goal; thus he is able to check both the risk of antirealism inherent in logical empiricism, and the relativism of some early anti-positivistic reactions. From the scholastic approach he also developed his peculiar conception of scientific objectivity, the keystone of his thought, according to which science deals always with *formal* objects, in some measure selected, abstracted and structured by us. This distinguishes him from various forms of naïve or uncritical realism. To be sure, this is also the main tenet of Kantism and subsequent forms of constructivism; but unlike those philosophies, in Agazzi it does not lead to subjectivism and antirealism, thank to the robust objectivist and realist bases of his thought.

From neopositivism Agazzi derives two insights: an uncompromising empiricism, as a safeguard against any rationalist metaphysics; and the awareness that our cognitive approach to reality is at bottom practical or operational, hence direct. Thus he avoids the epistemological dualism and the ensuing antirealism which has characterized much modern philosophy, empiricism in particular. In fact, in spite of his frequent warnings that rational argumentation must complement experience as a source of our knowledge, at some crucial passages of his complex exposition the neopositivistic legacy seems to prevail, and the balance of Agazzi's thought appears to be slightly tilted on the empirical-operational side. But this, I shall point out, may have consequences which contradict his own avowed realism and objectivism.

Thanks to the happy circumstance of the publication, this year, of a long waited-for volume which draws together into a vast and coherent picture Agazzi's teachings in epistemology and philosophy of science over many years, a real *summa* of his philosophy (Agazzi 2014), I shall be able to carry out my exposition and comments exclusively with reference to it.

2 Presuppositions of Scientific Realism

As explained by Agazzi (§ 5.1.4) scientific realism holds that the unobservable entities postulated by the best instances of theoretical science (like atoms, electromagnetic waves, viruses, etc.), (a) exist, (b) are independent of any description we give of them, and (c) can be known. This means that our best scientific theories are true descriptions of the unobservable entities (§ 5.2.3) and can be sufficiently well justified to be rationally believable.

All this presupposes three more claims: that theories are referential (§ 5.3.2), i.e., that they are descriptions, like a story or a map, rather than mere forms of expression, like abstract paintings, or mere instruments, like hammers and nails; that they must be taken literally as such, since their purpose is knowing reality; and that truth is a correspondence between descriptions and their subject matter. In fact, antirealism is often based on the rejection of one of these three further claims (see Alai 2013).

2.1 Independent Reality and Objectification

Both Kant and various recent forms of constructivism have denied that we can refer to reality as independent from thought, and describe it: for them we can at most construct phenomenical entities, which in no way represent reality as it may be "in itself". With the "linguistic turn" of the XX Century, some philosophers have claimed that it is impossible to reach out of language, or refer to an extra-linguistic reality[1]; they held that our expressions don't have referents, but only *senses* (or *intensions*), and that these are purely determined by our epistemic states (verificationism) or by relations to other expressions (contextualism) (§ 5.3.1).

According to Agazzi these are all forms of epistemological dualism, the belief (typical of modern philosophy) that the immediate object of knowledge is not reality, but our own representations, or ideas (§ 4.2), so that "the fundamental question of modern epistemology became that of determining whether or not, starting from our ideas, we can *indirectly* obtain knowledge of reality" (p. 245).

To this conception Agazzi opposes his own idea of objectification: each scientific discipline studies its subject matter exclusively from the point of view of some *basic* attributes, for whose attribution it employs specific *operational* or protocollarity criteria. The latter typically consist of operations (including observations and measurements, but more generally various forms of interaction with things). The propositions which attribute or deny a basic attribute are called "protocollar", as they are immediately judged to be true or false by the application of operational criteria (although Agazzi grants that this immediate judgement of truth or falsity is fallible).

Theoretical attributes are then defined through relations among basic attributes. Theoretical propositions, concerning them, are supported by rational inferences from protocollar propositions (and other theoretical propositions) (§§ 2.5, 2.6). The specific *scientific objects* of each discipline are just ordinary things seen from the point of view of the attributes of that discipline, hence they are *structured bundles of attributes*. As such, they are abstract objects, which however are *exemplified* by the concrete objects (constituted by a complete totality of attributes) which are their concrete referents (§§ 2.7.4, 5.4.1). For instance,

> in physics we define the term "electron" through a structured set of mathematically formulated properties which together constitute a certain abstract object. But this does not entail that these are meant to be properties of the abstract object; they are meant to be properties of the single electrons which are the intended referents of the mathematical model we have constructed (p. 113).

Even *things*, from which scientific objects are "clipped out", are constructs, i.e. concrete objects seen from the point of view of a limited number of predicates; but they can be identified without problems within common sense, independently

[1]This is the doctrine which Hintikka and Hintikka (1986) call "the thesis of language as universal medium".

of any theory: they are the scientific objects of some theory, once the latter has become part of common sense. For instance, electricity existed before being known; later it was introduced by some theories, and now it is a thing of common sense (§§ 4.1.6, 4.3).

Thus, from a cognitive point of view, we never encounter any non-structured material, as we always work from within some subjective point of view: we should not imagine "a reality 'in itself,' which should on the one hand have its intrinsic fixed structure, independent and unaffected by language and thought, while on the other be such as to be mirrored by thought and language" (p. 229). Thus, Agazzi might seem to deny that reality has its own nature and properties independently of thought, for he refuses to "conceive of reality (considered at a given moment) as being absolute and structured *in itself*" (p. 215). But this is not the case, since for him thought and language just select and "clip out" properties which reality has independently of us.[2] He explains that the active role of the subject

> results in the determination of attributes which are known as they are brought to light and, at the same time, are those actual aspects of reality which are effectively known through a particular intervention. Under different conditions, reality would manifest itself under different aspects or in the form of other attributes, but these too would be real" (p. 229).

The operational nature of basic predicates ensures that, against constructivism, objects are the abstract reconstruction of a concrete and independent reality; they are "clipped out" of things (p. 97, passim); and since things are "clipped out" of the concrete reality, "objects are *part of reality* (i.e., that part which has been 'objectified' through the operations), and are not something 'behind' which or 'under' which reality remains hidden" (p. 97); besides, "the process of objectification takes place in a referential situation, and is carried out under strictly referential conditions" (p. 181). Therefore, "(a) science attempts to represent a reality independent of science itself, ...; (b) what science states is an adequate representation of this reality 'as it is'" (p. 263).

Moreover, since protocollarity criteria consist of interactions with concrete reality, it follows that acceptance of propositions as true depends on the way in which reality is independently on us. Agazzi also grants that meanings of both basic and theoretical terms are not determined only by protocollar operations, but also by the inferential relations with other terms; hence, their *sense* is at least partially contextual. But terms have also referents, i.e. the concrete objects to which terms are shown to be applicable (or not) through protocollar operations (the basic terms) or protocollar operations *plus* rational inferences (the theoretical terms) (Agazzi 2014, Chap. 2).

Thus, he largely employs the logical empiricist conception of language and scientific methodology, in particular its distinction between empirical and theoretical terms, whose meanings are basically derived from experience and passed on to theoretical terms via inferential relations. But his stress on the interactive nature of experience provides this picture with solid realistic underpinnings.

---

[2]As I also argued in Chap. 3 of Alai (1994).

Agazzi stresses that the operational nature of experience warrants the existence of scientific objects. But it should be noticed that, while an apple as object of mechanics is clipped out of ordinary apples, nothing like this happens for electrons, photons, etc., which are not introduced through abstraction, but by a creative postulation. Thus, existence is *guaranteed* only for the basic objects of science, the observable ones. Otherwise we should grant that even objects like caloric, phlogiston or ether exist.

2.2 The Goals of Science

For Agazzi "it is uncontentious (a) that science has a referential *intention*" (p. 271), and "there is a general agreement" that "the intrinsic goal of science is to offer reliable means for attaining *truth*" (p. 260). This is perhaps a little too optimistic, for radical contextualism and panlinguism deny the possibility of extra-linguistic reference, and instrumentalism denies that theories should be taken literally as descriptions purporting to be true. In his view, however, the intrinsic goal of science (as opposed to its many possible extrinsic goals, like success, money, career advancements, etc.) "is that of obtaining *reliable knowledge*" (ibid.). Of course, those who doubt the possibility of knowing the unobservables (like for instance van Fraassen (1980)), restrict the goal of science to empirical adequacy, i.e., the truth about observable entities and phenomena. But Agazzi holds the realist position that science endeavours to describe also the unobservable reality (§ 5.5.3). Nonetheless, he grants that beside the primary (intrinsic) goal of science there can be also secondary and subordinate (intrinsic) goals, like the practical utility stressed by pragmatist and instrumentalists (§ 5.2.2).

A controversial problem is also whether science should only describe, synthesize and predict, or also explain. This question is linked to the preceding one, since due to the empirical underdetermination of theories, our main route to the knowledge of unobservables is the inference to the best explanation (see Lipton 1991; Alai 2014b). So, those who deny either the possibility or the usefulness of explanations are usually antirealist. According to Agazzi explanation is a primary aim of science, on a par with description: it is "a component of that process of *unification* that … operates at the level of perception. … knowing is in a broad sense unifying" (pp. 336–337). More precisely, we describe observable phenomena by empirical laws (§ 7.1.1), and we explain them by theoretical hypotheses (§ 7.1.2). Moreover, assuming a statement or a theory as an explanation is assuming it to be true (§ 5.5).

2.3 Referentiality

Science could not pursue truth and knowledge unless it purported to be referential (§ 5.3.2). For Agazzi formal sciences are semantic, since their propositions are definitions, constitutive of the senses of their terms; while empirical sciences are *apophantic*, since their propositions are understood as descriptions. Their task is "*exploring the world*, and not *exploring language*. However … this exploration of the world in search of referents takes place on the basis of sense; otherwise, we would not be able to *recognise* the referent when we meet it" (p. 273). Thus Agazzi answers Plato's problem, how can one look for what one doesn't know, yet: he explains that what we ignore, and try to find, are the referents of our terms, and we can find them because we know their senses (§ 5.3.4). That theories are referential is clear because (a) the propositions contradicted by experience are rejected, although fully endowed with sense (§ 5.3.3); and (b) terms acquire their referents through the operations which define the basic predicates and constitute scientific objects.

2.4 Truth as Correspondence

But referentiality is not enough for realism, a correspondence conception of truth is also required. In fact, after developing a theory of meaning understood as both sense *and* reference (§ 4.1), Agazzi proposes a correspondence theory of truth, answering the main objections raised by its opponents (§ 4.5). For instance, it is often objected that we lack criteria for ascertaining the correspondence of our descriptions to reality conceived as wholly independent of our epistemic states. To this he replies that the operations by which we assign objects and attributes as referents to our terms are the same by which we ascertain the possession of an attribute by an object; thus, they are criteria of truth for observational propositions (§ 4.5.3). Moreover, we have criteria of truth for theoretical propositions, since they are justified by inferences from observational propositions (§ 4.6).

Many objections concern the problem of *truth-makers*, i.e., what do true propositions correspond to. If it is said that they correspond to *facts*, it is objected that since the same state of things can be truly described by infinite propositions with different senses, there should exist an infinite number of facts, one for each true proposition. For Davidson, instead, correspondence would have the opposite, but equally absurd consequence, that there should be just one "Great Fact" (for, given certain logical assumptions, apparently plausible, but correct only within an extensional perspective, there would follow that any true proposition corresponds to any fact).[3]

To the former objection Agazzi replies that just as an object, constituted by infinite attributes, admits of infinite correct descriptions, each of which describes only some of its attributes, so also a state of things is not multiplied when different propositions bring out the different facts about it (4.5.3). To Davidson's objection he answers by his own doctrine of objectification: each scientific discipline deals only with certain attributes and with the objects made up of them: so, it deals only

[3]On this see Volpe (2005: § 3.9).

with partial aspects of concrete things. Therefore statements and theories are not true of reality in general, but only of the restricted domain of their discipline, or of their own specific models: they have no correspondence to a "Great Fact", but only to the limited portion of reality to which they refer.

2.5 Truth as Circumscribed and Relative

This is how Agazzi explains his doctrine that truth is circumscribed and relative:

> a correct realist approach … limits the truth of the propositions of a given science to its specific and *empirically circumscribed* domain of objects. It is incorrect to say that, for a realist, "a theory cannot be true unless it can be *extended* consistently, without correction, to all of nature" (p. 298).

> Scientific truth is always a *relative* truth, in the sense that every scientific sentence is always true (or false) 'of' the specific objects which constitute the particular domain of the theory in which the sentence occurs (p. 402. See also p. 408, §§ 4.5.3, 4.5.3.5, 5.5.2, 5.5.3).

This idea is widely shared, in particular by those who reject the "statement conception" (according to which theories are sets of universal statements), adopting instead a "structuralist" conception (according to which they are rather constituted by classes of models, which become more and more restricted as they are enriched with theoretical predicates: see Suppe (1977: 221–229); Stegmüller (1976)). Now, this way of speaking is correct if it means that each statement or theory describes only certain aspects of reality; but it conflicts with a more established way of speaking of truth, and it might be mistaken with the following relativistic doctrine: any statement or theory is true, since it is true at least of its own objects or models.

We normally use the relative expression 'true of …' only for predicates, while for propositions or statements we use the absolute term 'true': we say that 'black' is true of ravens, but that the statement 'all ravens are black' is true, simply. It would be misleading to say that this statement is true only of ravens, for it actually speaks of all things, saying that each of them either is black or is not a raven.

The same holds even if we describe theories as sets of models (for, as explained by Agazzi (p. 258), sets of models are still translatable into sets of statements): a theory selects a certain class of "empirical" models EM, and one of "theoretical" models TM, and it is true if and only if all the empirical models, once enriched with certain theoretical predicates, become theoretical models (see Stegmüller 1976; Alai 1985). So, in effect, the theory speaks of *all* objects, claiming that if any of them is an empirical model, it is also a theoretical model.

This is not to say that all (universal) statements and theories speak of the one "Great Fact", because although the statement 'all ravens are black' speaks of all objects, it doesn't make about them the same claim as, e.g., the statement 'all swans are white': it describes a different fact. Therefore Davidson's objection, based on logical subtleties and on an unwarranted extensionalist presupposition, can be rejected without relativizing truth to subjects.

2.6 Pictorial Truth

Agazzi also says that correspondence is not "pictorial", in the sense that (1) propositions are not "a kind of reduplication of reality under the form of representation", because language can only describe particular aspects of reality, it cannot offer a complete representation of it; and (2) there is no "point-to-point correspondence" between the elements of a proposition and those of reality, as held by logical atomism and (in part) by Wittgenstein's *Tractatus* (§§ 4.5.5, 5.2.1; p. 258).

I grant that no *reduplication* of reality is possible: we could not represent Saint Peter's Basilica in Rome by making a perfect duplicate of it, for even a full size replica could not occupy the same spatiotemporal position. This is to say that *any* representation is partial and selective. But then also *pictorial* representations are such: they share some traits of their subject (say, shapes, colours, etc.), but not others (say, dimensions, materials, etc.). So, it is not clear in which sense our representations of reality should not be pictorial. Besides, in *Tractatus* Wittgenstein has shown precisely that propositions *are* pictures of reality, in that they reproduce its logical structure. For instance, in 'Romeo loves Juliet' the words do not resemble their referents in any way. But since whenever a person A loves a person B there is a true proposition with the word 'loves' between the names of A and B, and whenever Romeo is involved in a state of things there is a true proposition including the word 'Romeo', etc., between language and reality there exists a structural correspondence by which the former pictures the latter. This is clearly a point-to-point correspondence, even if not *all* "points" of a state of things correspond to "points" of a proposition, and vice versa.

At any rate, Agazzi stresses that truth is a *correspondence* relation, in the sense that "we certainly speak *about* something, that this something consists of substances endowed with attributes, and that these attributes result from the encounter between our way of investigating reality and *what reality is*" (p. 231). Now, this is right, but it wouldn't yield a realist conception of knowledge if the "encounter of our investigations with reality" didn't yield a picture of (some of) the structures which reality has independently of us (see Alai 1994, pp. 94–99).

2.7 The Truth of Theories

Agazzi also argues that theories cannot be true or false. Apparently this claim contradicts a crucial tenet of scientific realism, so it must be carefully examined. There are two reasons why he holds this: first, his propensity for the non-statement view of theories (although he grants that also the statement view is partly correct) (§ 7.2.5). So, he claims that theories are not sets of statements (even if they can be translated into sets of statements), but models, or, metaphorically speaking, maps. As such, they cannot be literally true, but only more or less faithful or accurate. This does not conflict with realism, however, since he also claims that maps, hence also theories, are true in a non-literal sense, in that they contain information which can be translated into propositions, and so become true or false (p. 258).

A similar idea is advanced by Giere (1988, Chap. 4), for whom theories are families of models, hence neither true nor false, but more or less resembling reality. The implicit premise is that only statements can be true or false: for instance, questions, wishes, orders, etc., are neither. However, although a map is not a statement, if proposed as a map of a particular region it is implicitly *asserted*, i.e., understood as a description of that region, hence true or false, even if it is not translated into propositions.

Agazzi's second reason for denying that theories are true or false is they postulate abstract objects like rigid bodies, perfect gases, adiabatic transformations, etc. Hence "the aim of theories is far from that of telling a 'literally true story' concerning the world, but is rather to give the most faithful depiction of a certain (partial) vision of the world under a specific point of view …" (p. 256),[4] and "of *causally explaining* empirical laws" (p. 259).

Now again, this is right, and compatible with realism, if it means that we don't claim that a given concrete marble slate *is* a rigid body, or that a particular concrete body *is* free-falling, etc. But theories don't say this: rather, they claim that *the more* a concrete system approximates one of these idealized models, *the more* its behaviour approximates that described by the theory's laws. Hence, theories can be *literally* true. After all, as Agazzi himself makes clear, a scientific object A is abstract, but it has a concrete referent B, which must possess the attributes characterizing A. Besides, since nothing can be considered as an explanation unless it is taken as true (§ 5.5.3), theories could not be explanations of empirical laws (against Agazzi's own claim) unless they were true.

Agazzi's conclusion is that "while … the problem of realism has significant links with the question of truth, we do not … need to relate this truth to theories in order to investigate this issue" (p. 259).[5] But I have argued that this is correct only in some limited sense. In any case, his considerations should not be understood as a rejection of realism about theories, in the way in which for instance Hacking (1983), Cartwright (1983) or van Fraassen (1980) reject it. For, while refraining from attributing truth values to theories, he attributes them to the statements about unobservables which are derived from theories.

3 The Justifiability of Belief in Theories

As said, Agazzi correctly characterizes the problem of scientific realism as that concerning the reality of the unobservables postulated by theories (§ 5.1.4). But this problem has various aspects, such as whether it is possible to refer to

[4]Actually, to avoid the risk of epistemological dualism, this should be better formulated as "to give the most faithful depiction of *the world* under a specific point of view …".

[5]This reminds of Devitt's aphorism: "What has truth to do with Realism? On the face of it, nothing at all" (Devitt 1984: 34).

unobservables, or whether the terms referring to them are eliminable, or whether discourse about them can be true or false, or whether knowledge of them is the aim of science. These questions were at the centre of the stage until the 1960s and the 1970s, but now the most debated issue is whether we can justifiably believe what theories say about unobservables: this is the question of *epistemic* scientific realism (Alai 2013). The strongest opposition to realism from this point of view has come from Hacking (1983), Cartwright (1983), Laudan (1977), and especially Van Fraassen (1980, §§ 5.1.4, 5.2.1); in fact Agazzi argues at length against them in defence of epistemic realism.

Their position, in his view, is based on the "Radical-empiricism argument" (p. 290), which illicitly infers

> from the easily admissible thesis (a) that every existence claim about the world must explicitly *be linked* with sense experience … to the already more controversial thesis (b) that such claims must *ultimately rest* on sense experience, but even to the extreme thesis (c) that every existing entity must be *directly ascertainable* by sense experience (p. 291).[6]

Van Fraassen, for instance, holds that the only ultimate justification of beliefs can be unaided sense observation. But to this severely restrictive position Agazzi replies that justification also comes from the rational arguments, by which statements about the unobservables can be inferred from statements about the observable reality (§ 5.5.2).

3.1 Explanatory Arguments

The most important of those rational arguments are the abductive ones, based on the need to explain the observable phenomena thorough unobservable causes or structures:

> A *general* characteristic of our knowing activity (that we find also in everyday life) is that *in order to explain … what we 'see', we look for something we do not 'see'* (p. 297).
>
> Occam's razor (*entia non sunt multiplicanda praeter necessitatem*) is certainly a wise intellectual principle, but it also admits of a 'counterpart' which is no less wise (*entia non sunt diminuenda praeter necessitatem*). The conjunction of these two principles says that we must have *good reasons* both for introducing and for denying entities, properties, and so on (p. 299).
>
> We introduce hypotheses (i.e. conjectures) to explain facts (i.e., states of affairs established beyond reasonable doubt, though, like every claim, subject to the possibility of error). … Scientific knowledge, like every form of human knowledge, walks on *two* legs, experience *and* reason! (p. 359).
>
> A theory usually postulates other entities and speaks of their properties and processes in order to explain the laws. But in such a way these entities could hardly be denied a right of citizenship among the referents of the science involved (pp. 361-362).

---

[6] I made a very similar point in Alai (2010, p. 663).

Thus, even without explicitly mentioning it, Agazzi accepts the *inference to the best explanation* (see Lipton 1991) as the backbone of theoretical reasoning in science.

Antirealists typically deny the need for explanation or think, like Van Fraassen, that something can be explained by deducing it from a hypothesis, even without believing in the truth of the latter. In reply Agazzi asks: "would we honestly accept, as an explanation of a fact, a hypothesis which we know to be *false*, but which accidentally happens to be such that we can logically deduce this fact from it?" (p. 296). In fact, mere acceptance of a theory without belief is possible from a pragmatic point of view, not from a cognitive one (§ 5.5.3).

A particularly effective argument is the "no miracles argument" of Smart (1968, p. 150), Sellars (1962, p. 97), Boyd (1983), Putnam (1975a, p. 73; 1978, pp. 19–21), and others: only a miracle could explain the success of science, unless we acknowledge that theories are largely true (§ 5.6). Agazzi notices that even if success were not cumulative (a problem on which we shall come back soon), its systematic nature would show that it cannot be due just to lucky coincidences. The most evident manifestation of success is "the whole world of technology, we must admit that we are in the presence of a gigantic and irrefutable confirmation of the truth of our scientific theories and of the *realist* purport of science" (p. 309).

Hacking (1983) and Giere (1988) supported *entity realism* by arguing that unobservables (like electromagnetic waves or microparticles) certainly exist, since we *produce* and *manipulate* them (e.g., by television sets, microwave ovens, X-rays generators, particle accelerators etc.). Agazzi employs the same argument: "The specific 'criterion of reality' for scientific objects that technology introduces … is the fact that technology *makes use* of such objects, and it is obviously not possible to make use of something that does not *exist*" (p. 309). But in this formulation the argument risks to be understood as a *petitio principii*, since the premise that we manipulate and use waves or microparticles is not based on direct perception, but on the very belief in the truth of theories which this argument is supposed to justify. But the same reasoning becomes fully plausible when phrased as another variant of the "no miracle" argument, as Agazzi does elsewhere: "if we succeed in *operating* on reality, letting ourselves be guided by a science, it follows that this science has picked out some actual properties of reality" (p. 285).

Larry Laudan objected in his famous (1981) that many theories that enjoyed large success in the past, later proved to be wrong, hence success is not evidence of truth. Agazzi replies in two ways: first, by endorsing what is commonly called *deployment realism*,[7] today the most accredited form of scientific realism: as a conjunction of a certain number of statements, a theory can be called "false" when even just one of them is false. But then "false" theories can still include many true statements; in particular, it is reasonable to assume that those statements which

[7]A position advocated by Kitcher (1993) and Psillos (1999).

were *essentially deployed* in deriving the theory's successful predictions are in fact (at least approximately) true. In Agazzi's words,

> It may well happen that a particular theory which turns out not to be adequate from several points of view and is therefore replaced by another, remains *partially* adequate from certain points of view; and this is enough to afford an understanding of its predictive success. This success depends on those parts of the theory which were adequate (p. 301).

> Only when the postulated referents are characterised through properties which actually play a logical (and not just a psychological) role in explanation and, especially, in prediction, can they be credited with a solid ontological status (p. 302).

It might be objected that often a theory succeeds in predicting certain phenomena precisely because they were known in advance, and the theory has been formulated in order to accommodate them. So, there is no compelling evidence of its truth. The reply is that in other cases theories predict phenomena previously unknown, or not used in the construction of the theory, and then the partial truth of the theory is the only plausible explanation (see Alai 2014a). While Agazzi does not explicitly consider this objection, he notices that "We are even more obliged [to believe in an object postulated by the theory] if it is possible for us to derive from the existence of this object, in a logically cogent way, certain previously unobserved features that we actually observe in conformity with our prescriptions" (284).

Agazzi's second reply to Laudan's historical objection is that in some cases a theory replaces another without contradicting it, but by introducing a new domain of objects (p. 302). But on this point, a keynote in Agazzi's epistemology, I shall come back shortly.

3.2 How Do We Know that Unobservable Entities Exist?

A different antirealist objection is reported by Agazzi as follows:

> There are certain measurements that we can perform in order to attribute to [an electron], let us say, a mass, a charge, a spin, and so on. It is also helpful to speak of such measurements as expressing *properties* of 'some object' because it helps our mind to synthesise them. But, as a matter of fact, all we can do is perform these measurements – there is no moment when we are actually acquainted with the object, i.e. the electron. Why should we then be authorised to speak of it as something really existing without a perceptual evidence for its existence? (p. 283).

To this he replies by his doctrine of objectification:

> This seemingly reasonable argument is actually involved in the old superstition of epistemological dualism which, in this case, consists in conceiving of the electron as a kind of 'substance' that lies 'behind' its properties, and which is such that we never encounter *it*, while we are able to encounter its *properties*. If one thinks this way, however, one conceives of the electron as a 'thing' and not as an 'object,' and one has removed oneself from physics by this very fact. If we instead conceive of the electron as an object, it must be conceived of not as something to which properties are attached, but as something which is *constituted* by these properties. An object is to be considered as the 'structured' totality of the objectively affirmable properties and not as a mysterious substratum of

> these properties. This might sound as a Humean positivism, but it is not, since we do not maintain that such properties are exclusively our perceptions: they are ontological aspects of reality and may even be perceptually not attainable (p. 283).

> An object is a complex *structured* reality, as we have pointed out, and there is no *reason* to pretend that the *all* the properties that go into this structure be observationally testable. [Therefore] the distinction between a *realism of properties* and a *realism of entities* ... is useless since, in science, entities (objects) are (as we have maintained) nothing but *structured sets* of properties (p. 284).

But I don't think this reply is successful, because the objection could also be phrased as follows: for Agazzi each abstract object must have a concrete referent. But even granting that the electron exists as an abstract object, i.e. as a set of properties, how do we know that it also exists as a concrete referent? After all, even ether, caloric and phlogiston existed, if conceived as sets of experimentally observable effects. Agazzi might reply that also the concrete referent is just a bundle of properties, but then the objection would be: how do we know that even the properties which are not directly observable or measurable exist, and that they are actually related to each other and to the observable properties just as claimed by the theory? So, also the distinction between realism about entities and realism about theories is not useless, for we might justifiably believe that an entity exists, without believing that it has all the properties attributed to it. Eventually, however, Agazzi comes back to the only effective realist answer, the inference to the best explanation of the novel predictive successes of the theory:

> In the case of some objects it may be that none of the properties attributed to them is empirically testable. In such a case we are nevertheless obliged to admit the existence of such objects for theoretical *reasons* ... We are even more obliged if it is possible for us to derive from the existence of this object, in a logically cogent way, certain previously unobserved features that we actually observe in conformity with our prescriptions ... The 'realist' import of scientific *applications* acquires relevance, though not in the grossly pragmatist sense according to which success is the best guarantee of truth. Rather ... in the more rigorous sense according to which if we succeed in *operating* on reality, letting ourselves be guided by a science, it follows that this science has picked out some actual properties of reality ... (pp. 284–285).

3.3 The Problem of Incommensurability

In the 1960s and 1970s a strong threat to realism came from Kuhn's (1962) and Feyerabend's (1975) claim that the theoretical change brought about by "scientific revolutions" is so radical to make theories "incommensurable": since the meaning of each term is contextual, i.e. determined by its relations to all other terms, a term does not have the same meaning when occurring in different theories. So, theories cannot be rationally compared, and there cannot be any rational justification for believing one theory rather than any other.

Agazzi tackles this problem by two complementary strategies, which now we shall briefly examine: (A) as far as alternative theories share at least some basic predicates, we can identify some common referents, hence rationally compare them; (B) if instead they don't share any referent, they are incommensurable, but not incompatible, since they deal with different objects. In both cases, therefore, there can be progress.

(A) The first strategy was followed in the 1970s and in the 1980s by the best advocates of scientific rationality (Putnam 1975b; Brown 1977, 1988; Shapere 1981, 1984, etc., beside Agazzi himself: 1976a, 1976b): meaning has two components, sense and reference; and while senses are contextual, so varying across theories, as claimed by Kuhn and Feyerabend, referents are not, and can be univocally identified across theories, if these share some basic predicates. For instance, even if a theory classifies whales as mammals and another as fishes, it is clear that they refer to the same animals, since they attribute them many common properties. Equally, it is clear that Avogadro's "molecules" and Dalton's "atoms" were the same objects, as they shared many properties (§§ 3.2, 3.3, 5.3.5).

But while thus attributing a key role to reference, Agazzi keeps a descriptivist theory, according to which reference is fixed by properties, or descriptions. On this account, whenever two entities postulated by different theories share many properties, but not the vast majority of them, it remains undetermined whether they are the same entity differently described, or two distinct entities described in partially similar ways. The problem is even more serious in the case of "theoretical" terms (i.e., terms for unobservable entities), because Agazzi grants that "their entire meaning depends on the context of the theory, being influenced in particular by the logical relations existing with the operational as well as the other theoretical predicates. Therefore, they are endowed with only a 'contextual meaning'" (p. 379). There follows that there is *meaning variance*, just as claimed by contextualists, hence theories cannot be rationally compared.

This problem can be overcome only by (i) embracing a referential theory of meaning also for theoretical terms, and (ii) accepting a "causal" theory of reference, as suggested by Putnam (1975b), according to which reference is directly fixed to what a given entity actually is, not to a (possibly wrong) description of it. This may happen either by ostension (in the case of observation terms, like 'gold'), or through the causal role played by the referent (in the case of theoretical terms). Thus reference, even of theoretical terms, can be identified non-descriptively, hence independently of the theory, so that shared referents can be identified.

As concerns basic observational predicates and the operational procedures, contextualists like Kuhn and Feyerabend held, and Agazzi grants, that "these procedures are certainly bound to some *non-empirical* context (the context which allowed for the design and use of the instruments, the context depending on the 'historical' and the 'hermeneutic' dimensions of science)" (p. 379), hence also their interpretation is *theory-laden*. But like Stegmüller (1976, Chap. 3), Pera (1982, Chap. IV.3), Shapere (1984), Kosso (1992, Chap. IX), and others, Agazzi points out that the context on which the interpretation of basic predicates and operational procedures depends need not be that of the theory under scrutiny: hence, its evaluation can still be objective. In other words, the theoretical/non theoretical distinction is theory-relative, and what matters is that the empirical basis of a given theory be non-theoretical with respect to it (§ 7.2.5).

(B) Agazzi's second strategy in facing the problem of radical theoretical change hinges on his doctrine of objectivity: if two theories share the same operational predicates, and there is an operational sentence implied by one and denied by the other, they are *comparable* and *incompatible* (pp. 380–381). But if

> two theories contain operational predicates which are not *completely* identical … they do not speak about the same objects, and *because* of that they are to be considered as *incomparable* or *incommensurable,* by resorting to our criteria (p. 381).

> It is then clear that two 'rival' theories can both remain true, each obviously *about* its own objects (…), so that they are not really rival (p. 286).

> In the history of science there are several theories which have established a rich set of true sentences about certain specific domains of objects, and this truth is never destroyed by the fact that other theories have proposed new systems of true sentences about new domains of objects (p. 381).

There follows that in any case theories can be true, and there can be progress: when successive theories share referents, progress is linear: in particular, it is cumulative if the superseding theory includes the superseded one, and non-cumulative if the former contradicts the latter.[8] Instead, when there is a complete change of objects we have *non-linear progress*, which is cumulative in the sense that knowledge about the new objects is added to the knowledge of the old ones, even if "this progress cannot be understood as a purely logical fact. … theory change very often means the opening of a new domain of inquiry … and this is by no means a matter of pure logic" (p. 382; §§ 7.2.7, 7.2.8; p. 286).

The idea of a radical change of objects strikes me as crucial in understanding the relations among different *disciplines* dealing with partially overlapping domains: for instance, psychology and neurophysiology both deal with our mental life, and they are not incompatible, but complementary. Perhaps, as suggested by Agazzi (p. 286), in this way one can also plausibly explain at once the radical contradiction and the pacific coexistence we observe between classical and quantum mechanics: for quantization, indeterminacy and casual behaviour pertain to the micro-world, while continuity, determinacy and causality pertain to the macro-world.

In general, however, it seems difficult to apply this idea to the relationships among different *theories* of the same discipline: it is implausible that in order to speak of the same objects two theories must share *all* the operational predicates, as claimed by Agazzi. In fact, as shown by the causal theories of reference, the descriptions given through operational predicates are not *definitions* of the terms, but just ways to fix their referents: as it was rightly objected to Bridgman's operationism, when we measure mass by different methods we are not measuring

---

[8]On the other hand, if two theories share some referents, but "no operational sentence may be found that supports one theory and disproves the other, we should say that, as far as we were able to determine, the two theories are *comparable* and *compatible*" (p. 381). But this does not seem right, for then the two theories would be *empirically* equivalent, but they might well be theoretically incompatible, like for instance quantum mechanics and Bohmian mechanics, or like the Ptolemaic theory and the Copernican theory in the XVI Century.

different properties, but the same property (Suppe 1977, p. 19). Probably, therefore, in typical cases of "scientific revolution" we have theories purporting to deal with identical or largely overlapping domains of observable objects, but postulating different theoretical objects and properties. Moreover, the theoretical objects postulated by a theory are not introduced *in addition* to the different objects postulated by the rival theory, but *in the place* of them. For instance, Newtonian mechanics and general relativity deal with the same observable objects (like planets and stars), and with some common theoretical properties (like mass); but the former claims that there is action at distance, and space has no causal powers, while the latter denies that there is action at distance, and claims that space has causal powers. So, they cannot be both (completely) true.

3.4 The "Pessimistic Meta-Induction"

In recent years, however, an even more serious threat to realism and to the idea of progress has been the so called "pessimistic meta-induction": since no theory older than, say, a hundred years is now believed to be true any more, probably in the next hundred years or so even all presently accepted theories will be rejected. So, how could we ever justifiably believe in the truth of some theory? (§ 8.1.5; Putnam 1978, p. 25).

Agazzi formulates this problem as the question: "How can scientific theories be true if they are usually refuted after a more or less short life?" But literally understood, the question is not difficult: except for verificationists, theories might well be refuted, yet true; in fact, all of them could be both refuted and true. The trouble is, theories are not just refuted, but mutually incompatible; so, they cannot be all true. Of course, we don't need that all of them are true, only that *some* are; above all, we need to be able to (fallibly) distinguish the (probably) true ones from the (probably) false ones, i.e., we want to be justified in believing some of them. But how can we believe in those which now appear as most plausible, if the pessimistic meta-induction convinces us that all of them will be discarded in a hundred years or so?

Popper held that successive theories, though false, approximate truth more and more. But for Agazzi this is not an acceptable solution, for it presupposes a substantialist conception of truth, which in turn entails epistemological dualism, i.e. the idea that we don't know reality, but an intermediate entity called 'truth' (pp. 287, 391, 392).

But I think the mere grammatical practice of treating 'truth' as a substantive term does not, by itself, have this wrong implication: 'knowing the truth' is usually understood as a short for 'knowing of some particular statements that they are true', i.e. 'knowing the truth *about some particular facts*'; and this, by semantic descent, simply means 'knowing some particular facts'. So, 'approximating the (total) truth' is a short for 'knowing an ever larger number of facts'. The real problem with the Popperian idea of an increasing number of known truths (i.e., facts) is rather that if really all theories were shown to be completely false, as claimed by antirealists, later theories would contain the same number of truths (zero!) as the earlier ones.

Agazzi tries to counter the pessimistic induction by the semantic principle of the eternity of truth: "If a sentence does not change, and its referents do not change, then its truth (or falsity) does not change" (p. 402). Now, this is obvious, but the problem is that once a sentence has been refuted, the evidence is that it was (and still is) false, and this may happen to all scientific statements. Granted, in some cases

> the purported falsifications must be interpreted neither as an elimination of their respective referents nor as a discovery of false assertions made about these referents, but as a *change* of referents … and with it so too change the *objects* to which theories refer. … But it is then clear that two 'rival' theories can both remain true, each obviously *about* its own objects (or referents as one may prefer), so that they are not really rival (p. 286; § 7.2.7).

But as I noticed earlier, these are limiting cases, for in general theories of the same discipline make incompatible claims about the same things.

3.5 Are Refuted Theories Still True?

Earlier I noticed that Agazzi's idea that theories apply only to the restricted domain of their own objects might be just a misleading way of putting the obvious point that a theory does not speak of everything, which is compatible with the absoluteness of truth. But here it turns out that he understands it in a more radical way, resembling Kuhn's and Feyerabend's relativism: that *prima facie* rival theories actually speak of different objects is not just an occasional contingency, but it must necessarily be always the case.

In fact, he argues that a theoretical sentence of theory T con be refuted only by endorsing its contradiction, which in turn means accepting a theory T' contradicting T. But two theories contradict each other only if they have the same referents, and because of the theory-ladennes of terms, this happens only very seldom (p. 403). (This is like saying that, e.g., a theory entailing 'phlogiston exists' and one entailing 'phlogiston does not exist' are not contradictory, since 'phlogiston' does not refer to the same entity in those theories). It follows that

> the condition for the eternity of truth is always fulfilled for, two theories being necessarily different, either they give rise to the unproblematic situation comparable to that of two sentences being 'complementarily' true of the same referent, or to the even less problematic situation of two sentences being true of two different referents. Our puzzling conclusion is therefore that no falsification of a theory is properly possible, and in such a way the entire objection is met (p. 404).

But this conclusion is really too embarrassing to be acceptable: it contradicts Agazzi's own attempt to preserve the comparability of theories by showing that meaning variance applies to senses but not to referents, and as a consequence it wrecks realism: if all theories are automatically true, then they are merely analytic, and empirical science is void of any factual content (against Agazzi's own claim that only formal sciences are semantic, while natural sciences are apophantic). Even the weaker thesis that no theory is refutable entails the impossibility of distinguishing truth from falsity, hence of justifying any scientific belief, hence of knowing anything.

The Author fully appreciates the paradoxical character of his claim, and he asks: "Do we really believe that Ptolemaic astronomy *is* still true,[9] that the corpuscular theory of light was not disproved by experimental results on the velocity of light, that Newtonian mechanics was not disproved by relativistic and quantum mechanics, and so on?" (p. 404).

And he answers: "yes!": Ptolemaic astronomy was only about the relative motions of the Sun and the Earth, because at the time they couldn't observe their absolute motions; hence, it is still true (p. 404). But this is to reduce the content of theories to their empirical claims, so giving into antirealism: what scientific realism is all about is that theories have a non-observable content, and we can believe in its truth.

The corpuscular and wave theories of light, says Agazzi, describe different aspects of the same "thing", in fact now they are both employed to account for the optical phenomena known today:

> the corpuscular theory was (and is) true of the corpuscular *aspects* of light … and the wave theory was (and is) true of the undulatory aspects of light, and finally … our present wave-particle theory of light is true of light as we objectify it in present-day physics. [Only,] after a certain time an objectification meets its 'limits,' and *without being proven false*, it is proven *partial* … (p. 405).

But this overlooks that the two theories were incompatible, for claiming that light is a wave implies denying that it is made up of particles, and vice versa. Moreover, while our current theory has some aspects of both, it denies that either one was a correct model of light. Nor these three theories intend to describe different domains, for all of them aim to explain the nature of that radiation by which we see the world. One could deny that they contradict each other only by reducing them, again, to their empirical content, thereby ignoring the explanatory aim of science, and conceiving progress as a mere accumulation of empirical data.

Concerning relativistic and quantum mechanics, Agazzi denies that they represent a progress with respect to classical mechanics, or that they offer a more approximately true description of the same objects. On the contrary, they talk about utterly different objects, because

> objects are 'clipped out' of things by operational predicates which are *defined* on the basis of operational procedures. Every operational procedure is given (or, better, is characterised) by a certain order of approximation or margin of error. … This means that … it makes no sense to carry out calculations leading, for example, to the expression of the length of a body as being equal to 5.00021 cm. if the instrument on which the length calculations are based in that context only admits of a margin of error of one millimetre. The alleged accuracy would simply lead to a *meaningless* statement … because if the meaning of the operational predicate has been introduced by means of a measurement limited by a certain order of approximation, it is clear that we are not using this meaning (or we are misusing it) if we pretend that it is bound to a different order of approximation (pp. 405–406).

---

[9] I.e., given the eternity of truth, that it *was and still is* true.

I think quantum mechanics is where Agazzi's idea of a domain shift applies best, but this interpretation is possible even without embracing the old verificationistic and operationistic view of meaning he presupposes here: the measurements made by a particular instrument do not *define* the meaning of terms, which is constituted by both sense and referent. Measurements only contribute (together with other properties described by the theory and by common sense) to constitute the sense, but reference is a relation to things themselves, independently of our definitions. The statement that the length is 5.00021 cm. may be true or false, but it is certainly meaningful, whether we can control it or not. Measurements are only criteria for checking the truth of such attributions, and two instruments with approximation margins of respectively 1 and 0.01 mm still measure the same property.[10]

Summing up, the pessimistic meta-induction cannot be countered by appeal to the relativity of truth or of objects. But Agazzi could successfully reply by the same strategy he endorsed earlier (pp. 301–302), that of *deployment realism*: rather than claiming that rejected theories remain *totally* true, but only about the objects defined by them (i.e., about mere Kantian *phenomena*), he could argue that they remain *partially* true of *all* reality, understood as independent of the criteria employed to identify their referents. A theory which is overall false may still make some true claims (both empirical *and* theoretical). Since those claims can be identified through the essential role they played in the prediction of novel effects, they can be safely (although fallibly) believed.

Agazzi accounts for the cumulative nature of science by claiming that since a theory is true only relatively to its objects, after a sufficient number of checks it becomes practically certain and immune to falsification (pp. 408–409). But what makes scientific knowledge relatively stable is rather that those claims which are singled out as approximately true by their deployment in successful predictions are typically preserved by later theories, so cumulated over time.

References

Agazzi, Evandro. 1976a. The concept of Empirical Data. Proposals for an Intensional semantics of Empirical Theories. In *Formal Methods in the Methodology of Empirical Sciences*, ed. Przelecki et al., 153-167. Dordrecht: Reidel.

Agazzi, Evandro. 1976b. Analyse des racines philosophiques des différents sens de la signification. In *Actes du Colloque International sur Philosophie et langage, organisé à Bruxelles en 1975 par l'Institut des Hautes Etudes de Belgique*, special issue of *Revue de l'Université de Bruxelles*, 1–2, pp. 4-17.

[10]Agazzi objects that even common sense knows that e.g., attributing the height of a mountain with the approximation of 1 mm. would be meaningless (even if one had instruments for measuring with that accuracy) (p. 406). But the reason why we don't attribute to mountains heights with that approximation is not because this would be *semantically* meaningless for lack of operational criteria; rather, it is because the "theory" itself (i.e., in this case, common sense) tells us that this would be pointless, since the heights of mountains vary continuously of various centimetres or more, due to geological, atmospheric, and animal causes; moreover, such an approximation would be practically irrelevant for geographers, alpinists, etc.

Agazzi, Evandro. 2014. *Scientific objectivity and its Contexts*. Cham, etc.: Springer.
Alai, Mario. 1985. Stegmüller on the Structure of Theories. *Scientia* LXXIX: 105-115.
Alai, Mario. 1994. *Modi di conoscere il mondo. Soggettività, convenzioni e sostenibilità del realismo*. Milano: Franco Angeli.
Alai, Mario. 2010. Van Fraassen, Observability and Belief. In *New Essays in Logic and Philosophy of Science*, ed. Marcello D'Agostino, Giulio Giorello, Federico Laudisa, Telmo Pievani and Corrado Sinigaglia, 663-675. London: College Publications.
Alai, Mario. 2013. Ontologia, conoscenza e signifícato nel realismo scientifico. In *Ontologia, realtà e conoscenza*, ed. Mariano Bianca, Paolo Piccari, 119-144. Milano: Mimesis.
Alai, Mario. 2014a. Novel Predictions and the No Miracle Argument, *Erkenntnis* 79, 2: 297-326.
Alai, Mario. 2014b. Explanatory Realism. In *Science, Metaphysics, Religion*, Proceedings of the AIPS 2013, ed. Evandro Agazzi, 99-116. Milano: Franco Angeli.
Boyd, Richard. 1983. On the Current Status of the Issue of Scientific Realism. *Erkenntnis* 19: 45-90.
Brown, Harold I. 1977. *Perception, Theory and Commitment*. Chicago: University of Chicago.
Brown, Harold I. 1988. *Rationality*. London: Routledge.
Cartwright, Nancy. 1983. *How the Laws of Physics Lie*, Oxford: Oxford University.
Devitt, Michael 1984. *Realism and Truth*. Oxford: Basil Blackwell.
Feyerabend, Paul K. 1975. *Against Method. Outline of an Anarchistic Theory of Knowledge*. London: New Left Books.
Giere, Ron. 1988. *Explaining Science: A Cognitive Approach*. Chicago: University of Chicago.
Hacking, Ian. 1983. *Representing and Intervening*, Cambridge: Cambridge University.
Hintikka, Merrill B., Hintikka, Jaakko. 1986. *Investigating Wittgenstein*. Oxford, New York: Blackwell.
Kitcher, Philip. 1993, *The Advancement of Science*, Oxford: Oxford University.
Kosso, Peter. 1992. *Reading the Book of Nature. An Introduction to the Philosophy of Science*. Cambridge: Cambridge University.
Kuhn, Thomas S. 1962. *The Structure of Scientific Revolutions*. Chicago: The University of Chicago.
Lipton, Peter. 1991. *Inference to the Best Explanation*. London, New York: Routledge.
Laudan, Larry. 1977. *Progress and its Problems*, Berkeley: University of California.
Laudan, Larry. 1981. A Confutation of Convergent Realism. *Philosophy of Science* 48: 19-49.
Pera, Marcello. 1982. *Apologia del metodo*. Roma, Bari: Laterza.
Psillos, Stathis. 1999. *Scientific Realism. How Science Tracks Truth*. London, New York: Routledge.
Putnam, Hilary. 1975a. What is Mathematical Truth? In Putnam, Hilary. *Philosophical Papers*, Vol. I: *Mathematics, Matter and Method*, 60-78. Cambridge: Cambridge University.
Putnam, Hilary. 1975b. The Meaning of 'Meaning'. In Putnam, Hilary *Philosophical Papers*, vol. II: *Mind, Language, and Reality*, 215-271. Cambridge: Cambridge University.
Putnam, Hilary. 1978. *Meaning and the Moral Sciences*, London: Routledge.
Sellars, Wilfrid S. 1962. Philosophy and the Scientific Image of Man". In *Frontiers of Science and Philosophy*, ed. Robert G. Colodny, 37-78. Pittsburg: The University of Pittsburg.
Shapere, Dudley. 1981 Meaning and Scientific Change. In *Scientific Revolutions*, ed. Ian Hacking. Oxford: Oxford University Press.
Shapere, Dudley. 1984. *Reason and the Search for Knowledge*. Boston Studies in the Philosophy of Science, Vol. 78. Dordrecht: Reidel.
Smart, John J.C. 1968. *Between Science and Philosophy*, New York: Random House.
Stegmüller, Wolfgang. 1976. *The Structure and Dynamics of Theories*. New York, etc.: Springer.
Suppe, Frederick. 1977. The Search for Philosophic Understanding of Scientific Theories. In *The Structure of Scientific Theories*, ed. Frederick Suppe, 3-243. Urbana, etc.: University of Illinois.
Van Fraassen, Bas C. 1980. *The Scientific Image*. Oxford: Clarendon Press.
Volpe, Giorgio. 2005. *Teorie della verità*. Milano: Guerini.
Wittgenstein, Ludwig. 1922. *Tractatus Logico-Philosophicus*. London: Kegan Paul, Trench, Trubner & Co.

Scientific Progress

Vincenzo Fano and Giovanni Macchia

Das Wahr ist das Ganze.
(Hegel, *Phänomenologie des Geistes*, Vorrede, 1807)

Abstract We deal with the problem of verisimilitude, a notion which, roughly speaking, tries to capture how close a scientific theory is to the truth. Our starting philosophical basis is Evandro Agazzi's approach and his view on scientific objectivity which relies on his particular meaning of 'partial truth'. By following an epistemological approach to the verisimilitude problem and adopting the semantic view of theories, we develop our epistemological proposal about the comparative evaluation of scientific theories and cognitive situations. Our proposal allows to establish, in a qualitative way, in which sense a theory, or a cognitive situation, is better (more verisimilar) than another.

1 Introduction

Does science progress, and can its progress be estimated? Is our knowledge cumulative, so that our theories progress towards some paramount cognitive goal, such as some sort of truth? Are some of our ideas about the world more rational, more scientific than others? Are our beliefs on nature, on its mechanisms, really worthy of credence?

Questions like these about the nature of scientific knowledge—the kind of knowledge that is shaped into scientific theories and consists of the ascertainment, understanding and explanation of the phenomena of the world—are much less

V. Fano (✉) · G. Macchia
Department of Basic Sciences and Foundations,
University of Urbino 'Carlo Bo', Urbino, Italy
e-mail: vincenzo.fano@uniurb.it

G. Macchia
e-mail: lucbian@hotmail.com

M. Alai et al. (eds.), *Science Between Truth and Ethical Responsibility*,
DOI 10.1007/978-3-319-16369-7_5

naïve, and call for answers much less plain than might appear. And, although the progressiveness of science, its rationality, and its cumulative path, seem in a sense obvious facts, probably taken for granted not only by most laymen, but also by the majority of scientists, actually, these questions still remain unanswered, tricky, and therefore of broad and current interest among philosophers of science.[1]

In this respect, it suffices to loosely consider the progress of knowledge in different ages of our history: it is reasonable to think that today the international scientific community has more knowledge than a hundred years ago, but it is also unquestionable that the scientific (in a broad sense) community in the second century before Christ possessed more knowledge than in the seventh after Christ. So scientific progress is not a necessary path of history,[2] the time arrow of knowledge—so to speak—is not at all unidirectional, so also a scientific *regress* in knowledge is possible.

Consequently, concepts like "theory change", "cumulative growth", "truth", "true theory", "verisimilitude", "truthlikeness", "belief revision", all belonging to that field of research named theory of scientific progress and all trying to capture the elusiveness of our knowledge dynamics, have proved to be particularly slippery and controversial.

In this paper, we will be dealing with one of these elements: the problem of *verisimilitude* (usually also called *truthlikeness* or *approximate truth*).[3] The notion of the verisimilitude of a theory attempts to explicate how close a theory is to the truth. In other, more intuitive, terms, this notion can be expressed by saying that a theory is more verisimilar than another when "the first says more things about the domain under investigation and more of all things said are true" (Kuipers 2014: 64). Therefore, the problem of verisimilitude may be summarized with a question like this: What can there be about a theory that makes it closer to the truth than another one?

On the other hand, we know that every scientific theory, however established and cherished, is presumably false, that is, with at least one false consequence. Any theory, indeed, is an idealization and an oversimplification of the real world, and, sooner or later, it will probably be included in some other more general theory, insofar as a perfect match between theories and the world is actually, and for many reasons, impossible (just think, for instance, about the unobservable and spatiotemporally remote aspects of the universe). However, some (false) theories objectively seem to be better than others. Let us consider, as a very simple example, the theories of motion: starting with Plato and Aristotle, they then developed through Philoponus, Buridan, Oresme, Galileo and Descartes, until arriving at Newton and finally Einstein, so giving the appearance of a real improvement:

[1]Recently (March 2014), the journal *Studies in History and Philosophy of Science* devoted an entire Special Section to the progress of science. For a survey of this topic see Niiniluoto (2011a).

[2]Popper (1963, Sect. 1).

[3]These concepts are often used as synonyms, but not always: for instance, Zwart (2001) uses verisimilitude to define distance to the truth by means of truth-value and logical strength, and truthlikeness to establish distance to the truth by means of similarity between the possible worlds.

Einstein's theory, though presumably false, is more successful than its predecessors essentially because it is *closer* to the truth, that is, nearer to the unknown true theory about motion. But this natural appearance shrinks from any attempt to be, as it were, weighed precisely, that is to say, rationally quantified and explained in all its attributes. So, is this basic intuition of "closeness" deceptive? The problem of verisimilitude consists exactly of bringing off that attempt: establishing what are the objective endowments that make one theory better than another.

It should not be overlooked that the concept of truth is, as Oddie (2014) rightly points out, a rather coarse-grained property of propositions, so a more fine-grained ordering, such as the degree of closeness to the truth, is naturally compelling. After all, such a notion is really basic, given also that its importance follows from two modest and minimal realist assumptions, widely accepted: "The truth doctrine (that the aim of an inquiry, as an inquiry, is the truth of some matter) and the progress doctrine (that one false theory may realise this aim better than another). Together these yield the conclusion that a false theory may be more truthlike, or closer to the truth, than another" (Oddie 1986: ix). Furthermore, and still more significantly, Oddie adds: "Truthlikeness is not only a requirement of a particular philosophical outlook, it is as deeply embedded in common sense as the concept of truth. Everyone seems to be capable of grading various propositions, in different (hypothetical) situations, according to their closeness to the truth in those situations" (ibid.).

Notwithstanding, it is important to remark that there are research programmes, such as the belief revision programme, that do not consider truth and falsity as necessary means for an adequate understanding of the methodological rules that guide the change of beliefs and consequently of the nature of science,[4] or also programmes which are against the idea that science progresses as it gains more verisimilitude.[5] Indeed, outstanding philosophers with pragmatist leanings, such as Thomas Kuhn and Larry Laudan, maintain the idea that science is not a truth-seeking enterprise but rather a problem-solving activity.

Our paper proceeds as follows. In Sect. 2 we very briefly hint at Popper's historical proposal on verisimilitude that gave birth to this field of research. In Sect. 3 we introduce Agazzi's approach, which is our starting philosophical basis. After dwelling, in Sect. 4, upon the distinction between logical and epistemological approaches to the verisimilitude problem, specifying our preference for the latter, in Sect. 5 we take into account the traditional syntactic view of theories. Its weak points lead us to adopt the semantic view of theories in Sect. 6, where we will enter into the details of the notions and of the constitutive elements of this view, also developing our epistemological proposal about the comparative evaluation of theories and cognitive situations. Finally, Sect. 7 offers our conclusions.

---

[4]See Niiniluoto (2010, 2011b).

[5]See, for instance, Bird (2007), who evaluates progress in terms of increasing knowledge, and Piscopo and Birattari (2010), who criticize the notion of truthlikeness in a particular approach called the similarity approach.

2 Popper's Proposal

The tough problem of verisimilitude, and its relevance for a realist conception of knowledge, was first seriously discussed by Karl Popper as an attempt to illuminate the rationality of the scientific enterprise.[6] In 1963, he gave the first formal explication of the verisimilitude notion, developing his theory, in *Conjectures and Refutations*, which he combined with his falsificationist epistemology elaborated in 1934 in *Logik der Forschung*. The basic idea of his falsificationism was a Peircean fallibilist view, according to which all kinds of knowledge are uncertain and corrigible, so even our best physical theories are probably false. He thought, indeed, that scientific knowledge consists merely of conjectures, rather than certitudes, which scientists put to severe tests, whose results may or not refute these hypotheses, but never accept as true or as probable. By refuting some conjectures that have proved false, and substituting them with new ones not yet falsified, scientists come closer to the truth about the world insofar as scientific theories have a truth value depending on the relation between the theories and the extra-linguistic world. This realist faith, and his belief in the rationality and cognitively progressive character of the scientific enterprise, naturally necessitated a way to compare a new theory with its predecessor in order to evaluate its degree of improvement, namely its degree of verisimilitude, in this way substantiating the idea that science makes progress by replacing false theories with more truthlike (but probably still false) theories.

Popper's definition of verisimilitude was quite plausible. It stated, in non-technical terms: a theory is more verisimilar than another if the true consequences of the first include those of the second (namely, the former implies more true sentences than the latter), and if the false consequences of the second include those of the first (namely, the former implies fewer false sentences than the latter). The truth approximation theory of Popper was demolished in 1974 by David Miller and Pavel Tichý, who independently demonstrated that, following Popper's definition, a false theory can never be closer to the truth than another true or false theory. This result opened the way to the post-Popperian approaches to verisimilitude, all of which share the idea that the approximation to the truth should be the fundamental aim of scientific research. The concept of verisimilitude, therefore, has since then been developed by other authors, and today there can be distinguished at least five approaches (Kuipers 2000: Chap. 7), but none of them is without its difficulties.[7] So that, despite a good deal of work, such concept—whose importance actually goes well beyond both Popper falsificationism and particular kinds of realism[8]—seems to have eluded a precise, satisfactory characterization.

---

[6]See Popper (1962, 1963: Chap. 10; 1966, 1972: Chaps. 2 and 9).

[7]For a survey see also Niiniluoto (1998) and Zwart (2001).

[8]Such as, for instance, the convergent realism of Putnam (according to which the predictive success of mature theories should be explained by their "approximate truth").

We are here particularly interested in a very influential approach to the problem of progress in science, that of Evandro Agazzi,[9] which we are going to very briefly introduce in order to adopt it as a starting point for our reflections on the notion of verisimilitude.

3 Agazzi's Approach

We start from the following statement by Agazzi, published in 1969, in his first important book where he presented his theory of scientific objectivity:

> It is necessary to understand that the decline of a genuine theory does not mean you have recognized that it is *false*, but that it was *partial*; and replacing it with a new one is not a mere substitution, but its being replaced by a *better* theory, which is able to capture a greater number of reality determinations. Our willingness, in principle, to drop this new theory as well is just a consequence of the fact that at any moment we may not believe to have brought the objectivity horizon to coincide without residues with the reality horizon. (Agazzi 1969: 368; our translation)

Agazzi's very recent latest book, *Scientific Objectivity and Its Contexts*, is not only a further development of his view on scientific objectivity proposed in 1969, but an analysis so long, wide and deep as to constitute, as he himself admits, a whole life's work. Here, he puts forward similar theses, obviously enriched by years of elaboration, such as the following:

> It is not fully appropriate to consider theories as true or false, but rather as more or less adequate. Now, it may well happen that a particular theory which turns out not to be adequate from several points of view and is therefore replaced by another, remains *partially* adequate from certain points of view; and this is enough to afford an understanding of its predictive success. This success depends on those parts of the theory which are adequate. (Agazzi 2014: 301)

One of the central ideas of this discourse is that "after a certain time an objectification meets its 'limits,' and *without being proven false*, it is proven *partial*, that is, not such as to exhaust reality" (p. 405). This does not mean that "each different theory has to do with a *different reality* […], but that they have to do with *different aspects* or attributes of reality, which we express more precisely by saying that each theory has to do with different *referents* that result from different *objectifications* of reality" (ibid.).[10] However, such a "partial truth", according to Agazzi, has

---

[9]We started to analyse his approach in Fano and Macchia (2009).

[10]Objectifications, indeed, "result from isolating only certain features of reality included in a given point of view and disregarding all the rest, which amounts to considering only certain of reality's attributes and accepting only the respective predicates in the scientific language" (Agazzi 2014: 104).

nothing to do with an "approximation to truth" or with an "approximate truth", but it has the meaning of a "complete truth on a partial domain" (ibid.).[11]

As far as the knowledge expressed in scientific theories is concerned, a change in theory is one thing, actual progress is another. Agazzi clearly underlines this point by saying:

> What is really evident is that in every discipline there is a succession of theories in time—that a change occurs—but one which does not by itself imply progress. In order for change to be considered as progress, the factual ascertaining of its having occurred must be accompanied by a value-judgment of some sort, enabling us to claim that the new situation is better than the old. The difficulty lies precisely in the determination of this 'better'. (Agazzi 2014: 370).

Before proposing our own approach, we have to dwell upon a couple of important preliminary specifications.

4 Logical and Epistemological Approaches

A fundamental distinction in the approaches to verisimilitude one can choose, or better in the problems one wishes to resolve, needs to be underlined. Such distinction was already clearly posed by Popper in his seminal work of 1963.

A preliminary definition of an appropriate notion of verisimilitude gives an answer to the *logical problem*, namely to the question about what *we mean* if we claim that one theory is more verisimilar than another. This problem tries to give an appropriate account of the concept, also determining its logical properties. The attempted solutions are various, but can be gathered together under three broad "families"; however, none of these approaches is unanimously considered as the best or most promising one (see Oddie 2014).

The *epistemological problem* of verisimilitude, instead, reads: *how do we know* that one theory is more verisimilar than another; or on what evidential grounds are we to rationally and conjecturally claim that one theory is closer to the truth than another.[12]

According to Zwart (2001: 121), defining the meaning of "approach to the truth" is not only an analytical affair, but is also subject to what Niiniluoto (1987: 265)

[11]One of the examples given by Agazzi concerns the corpuscular and wave theories of light. Both theories are true: the former is true of the corpuscular aspects of light (namely, of light objectified by means of corpuscular predicates), the latter is true of the undulatory aspects of light. It goes without saying that our current wave-particle theory is true of light as we objectify it in present-day physics.

[12]Popper (1963: 234) expressed these two approaches posing, respectively, the following questions: "What do you intend to say if you say that the theory t_2 has a higher degree of verisimilitude than the theory t_1?", and: "How do you know that the theory t_2 has a higher degree of verisimilitude than the theory t_1?". According to Popper, the answer to the latter question (the epistemic one), which depends on the answer to the former, is: "I do *not* know—I only guess. But I can examine my guess critically, and if it withstands severe criticism, then this fact may be taken as a good critical reason in favour of it" (ibid.). So he maintained that the claim "t_2 is more verisimilar than t_1" is falsifiable.

calls *Augustine's objection*: "To judge that a son resembles his father presupposes acquaintance with the father; similarly, to judge that a theory resembles the truth presupposes that the truth is already known". This point embodies the important difference between the logical and the epistemological problem: in the former, in order to make a comparison of any two theories with regard to their closeness to the truth, it is assumed that the truth (or, in a sense, the true theory, supposed as existing) is known, whereas in the latter the relative verisimilitude is estimated without knowing the truth, namely one needs an appropriate notion of *estimated verisimilitude* in order to compare two theories with regard to their closeness to the unknown truth.

Therefore, any answer to the first kind of question, explaining under what circumstances one theory is closer to the truth than another, presupposes an *aprioristic* complete knowledge of the truth. But, *actually*, in most interesting cases, we simply do not know it: scientists do not know the true theory when carrying out their empirical investigations, but they have to choose, on the basis of limited information about the truth, between rival theories. Consequently, an answer to the second kind of question is a rule of theory-choice, so that it is the epistemological problem, and its basic notion of estimated verisimilitude of competing theories, that, in a real account of scientific progress, should be taken into account.[13]

For these reasons we are more interested in, and we will try to give a qualitative answer of, the *epistemological* problem of estimating a possible scientific progress, and not in the *logical* one of establishing, in a state of omniscience, which of two theories is the more verisimilar. It is probable that in our solution of the epistemological problem a solution of the logical one is implicit. Anyway the important point is that we have to evaluate the truthlikeness of a theory without knowing the truth.

5 The Traditional Syntactic View

In the discussions of verisimilitude the so-called *syntactic*, or *logico-linguistic*, *view* to theories has been prevalent. This approach conceives the structure of empirical theories primarily as sets of statements in a formalized language (statements that, in the case of axiomatized theories, are logical consequences of a subset of axioms). Therefore, most, if not all, of the attempts to estimate scientific progress in a realistic perspective move from a set of possible theories $X_1, \ldots, X_n$ exhaustive and disjoint, only one of which X_i, expressed in a given language L, is true. The idea is to apply the notion of truth for formalized languages proposed by Tarski, that is, one should not compare the best theory we have with reality, but with the true theory expressed in the language L (Zwart, 2001: 4). This whole approach, in our opinion, places too much emphasis on language. In fact, scientific progress almost always involves a change in language, so that such comparisons are often impossible, as noted by Kuhn.

[13]There are different proposals to deal with the estimated verisimilitude. For example, Niiniluoto introduces a quantitative notion of *expected verisimilitude* in Bayesian terms. See Niiniluoto (1987: Chap. 7, 1998: Sect. 8), and, more in general, Zwart (2001: Chap. 4), on the solutions to the epistemic problem given by Popper, Kuipers and Niiniluoto.

Furthermore, all traditional approaches undergo the objection (Miller 1974) that their verisimilitude definitions are not linguistically invariant. You can see this quite easily. Consider a very simple language with three sentences A, B, C. Truth is $A \wedge B \wedge C$. Let us consider the theories: $T_1 = \sim A \wedge B \wedge C$, $T_2 = \sim A \wedge \sim B \wedge C$, $T_3 = \sim A \wedge \sim B \wedge \sim C$. Therefore T_1 has one mistake, T_2 two mistakes and T_3 three mistakes. Hence it is reasonable to say that, although all three theories are false, T_1 is more verisimilar than T_2, and T_2 more than T_3. Let us introduce new sentences $D =_{df} A \equiv B$ and $E =_{df} A \equiv C$. It is easy to show that T_1, T_2 and T_3 could be expressed through the three sentences A, D, E as well. In particular, we have $T_1 = \sim A \wedge \sim D \wedge \sim E$, $T_2 = \sim A \wedge D \wedge \sim E$, $T_3 = \sim A \wedge D \wedge E$. But now T_1 has three mistakes, T_2 two mistakes and T_3 one mistake. This means that the order of verisimilitude in the new language is inverted.

There are two possible answers to this objection (Oddie, 2014): either arguing that predicates A, B and C are in some ways the *right* ones, while D and E are wrong, or accepting that the definition of verisimilitude is language-dependent. The former (defined by Kuipers 2000 as "essentialistic realism") is implausible, especially if we are dealing with the epistemological problem of the estimation of verisimilitude without knowing the truth[14]; the latter is the one followed by the majority of scholars. As stated by Niiniluoto (1998), the request that the notion of verisimilitude be independent of a change of language, though inter-translatable, is too strong. In fact, the whole approach to the problem of verisimilitude is built on Popper's attempt to implement Tarski's notion of truth in the analysis of truth in scientific theories. So, from the outset, a given language is assumed as well as the existence of a true theory in that language.[15] These approaches, therefore, are interesting from the logical point of view, but actually they are only partly appropriate responses to the problem of establishing a way to compare the verisimilitude of two theories which do not share the same language; the latter is a very common situation in science.

6 The Semantic View

We instead prefer, and in the rest of our paper will follow, the *semantic view*, which conceives theories as sets of "logico-mathematical" models. We prefer this view because, by using directly the structures picked out by scientific theories, it is much closer to the actual practice of scientists than the syntactic approach, given that for most scientific theories there exist no axiomatic systems able to

[14]Indeed, even if we accept, as many scholars do, that there are natural predicates, we cannot be sure that our best scientific theories have already captured them.

[15]See, for instance, Cevolani (2009: 90–91), a book which is, to our knowledge, the most pedagogical introduction to the verisimilitude problem.

reconstruct their structures.[16] Scientific theories are not in fact sets of sentences, as has been assumed in many preceding approaches to the notion of verisimilitude.[17]

Here we use Patrick Suppes' point of view, which applies model theory and set theory to the analysis of scientific theories and to intertheory relations, and according to which a theory is a set-predicate. He never actually gives a true definition of the notion of theory, but proposes many examples.[18] Let us introduce some important concepts of his approach.

6.1 Theory

To avoid confusion, note that in what follows, in order to explicate the concept of scientific theory, we use a set-theoretic formalism, whereas to explicate the notion of concrete objects we use mereology, a significantly weaker theory. This choice depends on the fact that we attempt to introduce the least possible number of abstract entities.

In order to define what is a theory we must preliminarily define its constitutive elements. First of all, a theory relates primarily to a *domain of objects* O constituted by individual entities. For example, the gravity law concerns any rigid body falling in a gravitational field. With O we refer to the mereological sum of such objects. Therefore, we assume that, for O, GEM (general extensional mereology) holds.[19] In practice GEM states that the relation "to be a part of" (this is not a set of independent axioms) is:

1. a partial order, that is, it is anti-symmetric, transitive and irreflexive;
2. such that if o_1 is a proper part of o_2, then there is another part of o_2 that does not overlap with o_1 (supplementation);
3. such that there is always an object that is exactly the sum of two objects (unrestricted composition).

Concerning the domain of objects O we define a set of types of *experimental operations* $S = s_1, \ldots, s_m$. The operations of S produce *data sets* $d_1, \ldots, d_m$. For example, in the fall of bodies we measure positions and times. The data appropriately processed by a statistical point of view can produce trends of *observable parameters* $F = f_1, \ldots, f_n$. In our example, the positions as a function of time.

However, this is usually not enough. To explain what is going to be, it is necessary to introduce also *unobservable parameters* $G = g_1, \ldots, g_k$ calculated on the basis of F. In our case, gravitational forces.

[16]For a wider comparative analysis of these two approaches see Kuipers (2007: Sect. 2.1).

[17]Not by Kuipers (2000).

[18]Suppes (1962) is perhaps the most interesting of all the papers that the great philosopher devoted to this topic. See also the very clear Suppes (1957: Sect. 12.2).

[19]A good introduction to mereology is Varzi (2014).

Furthermore, parameters G and F are bound by *laws* $L = l_1, \ldots, l_j$, namely sentences, universally closed for all variables F and G, which are at the same time informative and simple (as Mill-Ramsey-Lewis-Earman's approach claims; see Earman, 1986: Chap. 5).

Therefore, putting together all the previous elements, one has that a *theory* X consists of a quadruple $X = \langle S, F, G, L \rangle$, namely it is composed of a set of experimental operations, a set of observable parameters, a set of unobservable parameters, and a set of laws.

6.2 A Model of a Theory

If the operations S can be applied to a domain O_i of objects, obtaining observable and unobservable parameters F and G such that L holds, then one says that O_i is a *model* of the theory X (and X is a (or the) true theory about O_i).

The same domain of objects O_i could be a model of more than one theory. We emphasize that O_i is not a type, but a token, that is a concrete individual part of the world. A part that is the mereological sum of smaller parts.

Notice that the previous definition of what a theory is has no *universality claim*, that is, it does not claim to be true for a certain set (type) of objects. The truth of a theory comes only after we have established that a certain individual domain of objects is a model of it. In this sense, scientific theories are not in fact falsified, but simply they could become useless when they no longer have models in the sense indicated. Recall Agazzi's statement that in general old theories are not false but partial.

Through the notion of model one can also specify the following important characteristic a theory may possess. If a theory X has at least one model O_i, one can say that X is able to produce *knowledge*. This definition is motivated because if O_i is a model of X, then X justifies true statements about O_i, that is the laws L of X applied to O_i. These truths are also justified by the observable parameters F of X. We can therefore say that by endorsing X one produces knowledge.

6.3 Anomalies of Models

Given a domain of objects O_i, being a model[20] of a theory X, there almost always exists a set S_a of operations (belonging to X) that produce data and parameters not fitting, for the objects O_i, with the laws L of X. These operations are the *anomalies* of the model O_i with respect to the theory X. In order for O_i to be a model of X, it is necessary to eliminate the anomalous data d_a. It is not possible to drop S_a, because they probably produce good data as well, so we must in a certain sense ignore d_a. It is well known that all scientific theories, as emphasized by historians of science, live in an empirical environment of anomalies.

[20]Perhaps, keeping in mind the existence of anomalies, one should speak of a "quasi-model".

6.4 Confronting Theories

A theory $X_1 = \langle S_1, F_1, G_1, L_1 \rangle$ is *deeper*[21] than a theory $X_2 = \langle S_2, F_2, G_2, L_2 \rangle$ with respect to the domain O if, at time *t*:

(a) O_{12} is the mereological sum without gaps and without overlaps of all models of both X_1 and X_2, which we know at time *t*;
(b) at least one of these conditions holds: i. $S_2 \subset S_1$ and $\sim(G_1 \subset G_2)$; ii. $G_2 \subset G_1$ and $\sim(S_1 \subset S_2)$;[22]
(c) the difference between S_2 and S_1 (G_2 and G_1) is relevant for O_{12}, that is, it concerns a part of O_{12}.[23]

In more intuitive but rough terms, one theory is deeper than another if the former answers more why questions, about a given area of investigation, than the latter. In this way, a deeper theory increases the explanations of the details of the phenomena at stake, thus allowing our knowledge to become more exact.

We can present this situation in a graphic way:

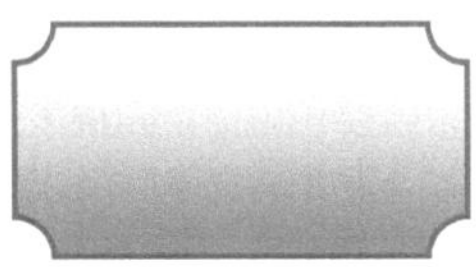

The mereological sum O_{12} seen from X_1

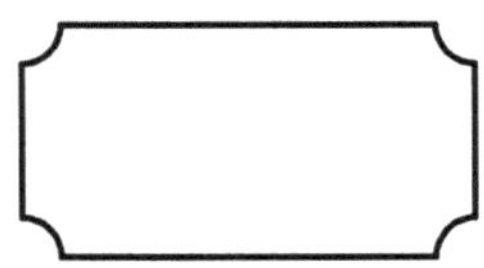

The mereological sum O_{12} seen from X_2

The same part of the world O_{12} is seen from different theories. The deeper perspective from theory X_1 has been colored gray.

Returning to the example of gravitation, a series of data on the freefall in the presence of friction cannot be understood solely through the gravitational force. A theory which also includes friction forces, therefore, is deeper than one that does not. And the introduction of friction forces enlarges G. A similar argument applies to the solar system with respect to classical mechanics and general relativity. Only the latter is able to account for the anomalies of Mercury's perihelion. So the understanding of the Mercury perihelion anomaly enlarges S. General relativity is obviously not only deeper than classical mechanics, but also wider, in the following sense.

A theory X_1 is *wider* than a theory X_2 at time *t*, if, given O_1 the mereological sum of all models of X_1 known by us at time *t*, and O_2 the mereological sum of all models of X_2 known by us at time *t*, O_2 proves to be a proper part of O_1.

---

[21]The distinction between progress in depth and progress in breadth has been proposed by Zwart (2001: 2).

[22]A difference in the Fs is irrelevant, since they are individuated by the Ss. On the contrary, though Gs are calculated on the basis of Fs, the latter do not determine which are the former.

[23]This condition is important to avoid the meaningless cases caused by the irrelevant addition of S and G. Indeed, it would be easy to build an artificial example of a theory deeper than a given theory X by adding to the latter observational and/or theoretical terms irrelevant for O_{12}.

Consequently, one theory is wider than another if it increases the number of objects, and hence of natural phenomena, about which scientists can make warranted claims.

Here too the graphic representation is very intuitive:

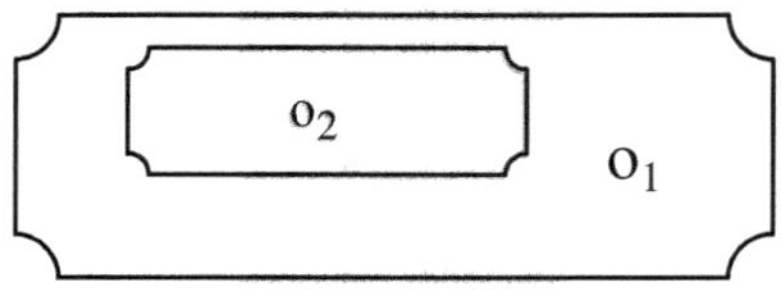

The smaller shape is O_2, that is the mereological sum of X_2 models, whereas the bigger one is O_1, that is the mereological sum of X_1 models.

Let us call $S_\mathbf{x}$ the union of all sets of experimental operations of a set of theories $\mathbf{X} = X_1, \ldots, X_h$ and $G_\mathbf{x}$ the union of all sets of their theoretical terms.

A set of theories $\mathbf{X} = X_1, \ldots, X_h$ is *deeper* at time *t* than a set $\mathbf{Y} = Y_1, \ldots, Y_k$ if:

(a) $O_{\mathbf{XY}}$ is the mereological sum without gaps and without overlaps of all domains of objects we know at time *t* as models of both **X** and **Y**;
(b) at least one of these conditions holds: i. $S_\mathbf{y} \subset S_\mathbf{x}$ and $\sim(G_\mathbf{x} \subset G_\mathbf{y})$; ii. $G_\mathbf{y} \subset G_\mathbf{x}$ and $\sim(S_\mathbf{x} \subset S_\mathbf{y})$;
(c) the difference between $S_\mathbf{y}$ and $S_\mathbf{x}$ ($G_\mathbf{y}$ and $G_\mathbf{x}$) is relevant for $O_{\mathbf{XY}}$, that is, it concerns a part of $O_{\mathbf{XY}}$.

Moreover, a set of theories $\mathbf{X} = X_1, \ldots, X_h$ is *wider* at time *t* than a set $\mathbf{Y} = Y_1, \ldots, Y_k$ if the mereological sum of all models of **Y** known by us at time *t* is a proper part of the mereological sum of all models of **X** known by us at time *t*.

Now we can join the two criteria in the following definition:

A set of theories $\mathbf{X} = X_1, \ldots, X_h$ is *Agazzi-better* ($>_{Ab}$) at time *t* than a set of theories $\mathbf{Y} = Y_1, \ldots, Y_k$ if at least one of the following two conditions holds:

1. **X** is wider at time *t* than **Y** and **Y** is not deeper than **X**;
2. **X** is deeper at time *t* than **Y** and **Y** is not wider than **X**.

We believe that with this definition we have a good explication of Agazzi's perspective on dynamics and comparison of theories. It is, of course, possible that two sets of theories are such that one is deeper than the other whereas the latter is wider than the former. We can represent this probably uncommon situation graphically as follows:

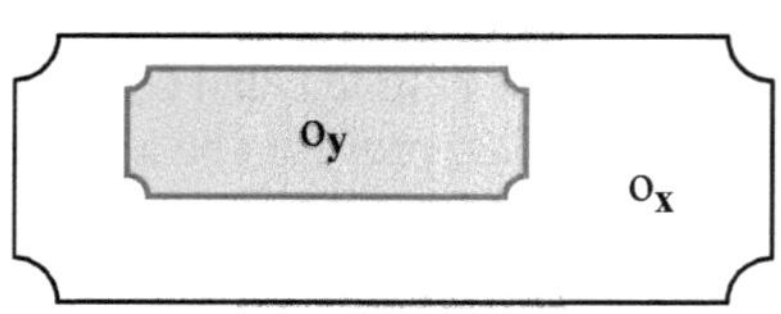

You see that the mereological sum of **Y** models $O_\mathbf{y}$ is a proper part of the mereological sum of **X** models $O_\mathbf{x}$, but the explanation of $O_\mathbf{y}$ given by **Y** is deeper than the explanation of $O_\mathbf{y}$ given by **X**. In this situation it is not possible to determine whether $\mathbf{X} >_{Ab} \mathbf{Y}$ or $\mathbf{Y} >_{Ab} \mathbf{X}$.

A very useful peculiarity of our definition is that a set of theories X is not the conjunction of X_1, ..., X_h. Indeed our approach is model-theoretic and not language-dependent. In the current scientific situation the logical conjunction of accepted theories would often lead to contradictions.[24] In fact, not only does science live in an environment of anomalies, but it is not rare that two accepted theories say something opposite on important topics. See for instance today's situation, in which quantum field theory is background dependent with respect to spacetime, whereas general relativity is not. Yet no one would renounce either of these two theories!

6.5 Confronting Cognitive Situations

The set of theories **X** accepted by a group of scientists U at time *t* is a *U,t- cognitive situation.*

Let us say that a U_1,t_1-cognitive situation is Agazzi-better than a U_2,t_2-cognitive situation if the set of theories **X** accepted by group U_1 at time t_1 is Agazzi-better than the set **Y** of theories accepted by group U_2 at time t_2.

Based on this definition, in many cases we can compare two cognitive situations of different times and determine whether or not there has been an increase in knowledge and hence scientific progress.

7 Conclusions

In general, in the attempt to explicate epistemological notions, there is a sort of trade-off between epistemological and logical values. That is, the more a notion is logically articulated, the less it is epistemologically relevant. This happens for instance to the standard logical treatments of verisimilitude. On the contrary, the more an explication is epistemologically adequate, the less it is logically articulated. This occurs perhaps in our perspective.

Therefore, maybe, we have been able to provide a reasonable definition, quite accurate, substantially language independent, of a better cognitive situation in Agazzi's sense, namely, a definition able to capture a greater number of reality determinations. However, its weak point is that, being in a sense just a conceptual approach providing neither calculations, nor precise estimates, it cannot be concretely used as an effective measure of the verisimilitude of theories.

---

[24]Perhaps it is possible to tackle this same situation through paraconsistent logics, as outlined by Itala M. Loffredo D'Ottaviano during the last congress of the Académie Internationale de Philosophie des Sciences, *Pragmatism and the practical turn in philosophy*, Pont-à-Mousson, 11th–14th September 2014, "On pragmatic truth and quasi-truth".

Acknowledgments We thank Claudio Calosi, Gustavo Cevolani and Theo Kuipers who read attentively drafts of this paper, giving us very useful suggestions and corrections.

References

Agazzi, Evandro. 1969. *Temi e problemi di filosofia della fisica*. Milano: Manfredi.
Agazzi, Evandro. 2014. *Scientific Objectivity and Its Contexts*. Heidelberg: Springer.
Bird, Alexander. 2007. What Is Scientific Progress? *Nous* 41: 64-89.
Cevolani, Gustavo. 2009. *La freccia della conoscenza*. Bologna: Archetipolibri.
Earman, John. 1986. *A primer on determinism*. Dordrecht: Reidel.
Fano, Vincenzo, and Giovanni Macchia. 2009. Realismo dei modelli e progresso scientifico. In *Il realismo scientifico di Evandro Agazzi*, ed. Mario Alai, 156-172. Urbino: Editrice Montefeltro.
Kuipers, Theo A.F. 2000. *From instrumentalism to constructive realism*. Dordrecht: Kluwer.
Kuipers, Theo A.F. 2007. Laws, Theories, and Research Programs. In *General Philosophy of Science. Focal Issues*, ed. Theo A.F. Kuipers, 1-95. Amsterdam: Elsevier.
Kuipers, Theo A.F. 2014. Empirical Progress and Nomic Truth Approximation Revisited. *Studies in History and Philosophy of Science* 46: 64-72.
Miller, David. 1974. Popper's qualitative theory of Verisimilitude. *The British Journal for the Philosophy of Science* 25: 166-177.
Niiniluoto, Ilkka. 1987. *Truthlikeness*. Dordrecht: Reidel.
Niiniluoto, Ilkka. 1998. Verisimilitude: The third period. *The British Journal for the Philosophy of Science* 49: 1-29.
Niiniluoto, Ilkka. 2010. Theory Change, Truthlikeness, and Belief Revision. In *EPSA Epistemology and Methodology of Science*, Vol. I, eds. Mauricio Suárez, Mauro Dorato and Miklós Rédei, 189-199. Dordrecht: Springer.
Niiniluoto, Ilkka. 2011a. Scientific progress. In *Stanford Encyclopedia of Philosophy*, ed. Edward Zalta. http://plato.stanford.edu/entries/scientific-progress/.
Niiniluoto, Ilkka. 2011b. Revising Beliefs Towards the Truth. *Erkenntnis* 75: 165-181.
Oddie, Graham. 1986. *Likeness to Truth*. Dordrecht: Reidel.
Oddie, Graham. 2014. Truthlikeness. In *Stanford Encyclopedia of Philosophy*, ed. Edward Zalta. http://plato.stanford.edu/entries/truthlikeness/.
Piscopo, Carlotta, and Mauro Birattari. 2010. A Critique of the Constitutive Role of Truthlikeness in the Similarity Approach. *Erkenntnis* 72: 379-386.
Popper, Karl R. 1962. Some Comments on Truth and the Growth of Knowledge. In *Logic, Methodology and Philosophy of Science*, eds. Ernest Nagel, Patrick Suppes and Alfred Tarski, 285-292. Stanford: Stanford University Press.
Popper, Karl R. 1963. *Conjectures and refutations*. London: Routledge.
Popper, Karl R. 1966. A Theorem on Truth-Content. In *Mind, Matter and Method*, eds. Paul K. Feyerabend and Grover Maxwell, 343-353. Minneapolis: University of Minnesota Press.
Popper, Karl R. 1972. *Objective Knowledge*. Oxford: Oxford University Press.
Suppes, Patrick. 1957. *Introduction to Logic*. New York: Van Nostrand Reinhold Company.
Suppes, Patrick. 1962. Models of data. In *Studies in the methodology and foundations of science*, ed. Patrick Suppes, 24-35. Dordrecht: North Holland, 1969.
Varzi, Achille. 2014. Formal theories of parthood. In *Mereology and the sciences*, eds. Claudio Calosi and Pierluigi Graziani, 359-370. Heidelberg: Springer.
Zwart, Sjoerd D. 2001. *Refined Verisimilitude*. Dordrecht: Kluwer.

Part II
Philosophy of the Special Sciences

Foundations and Philosophy of Mathematics

Marco Borga

Abstract Philosophy of mathematics does not coincide as such with the research on the foundations of mathematics. This confluence, however, occurred at the beginning of the twentieth century, in the framework of the efforts spent for overcoming the "crisis" produced by the discovery of the antinomies. Hilbert's formalism became soon the dominant view in this connection, that had also its philosophical counterpart in the conception of mathematics as a complex of pure formal systems devoid of specific meanings and referents. Agazzi has constantly opposed formalism, relying especially on philosophical reflections about Gödel's theorems, from which he derived the recognition of meanings and contents of many mathematical theories. This has pushed him to revisit the work of Peano and his school (and to stimulate his pupils to investigate their contributions in depth). It turns out that Peano was a pioneer and a champion of that request of logical rigor that animated much of the mathematical community of his time, so that his defense and practice of axiomatizations and his skillful use of mathematical logic as a tool for this critical analysis remained paradigmatic for the foundational investigations. However he never accepted a formalistic conception of mathematics, and this is why he and his school (after having completed their program) remained outside the main stream formalistic outlook of Hilbert's followers that was dominant in the first half of the twentieth century.

It is a widespread opinion that the field of research called "Foundations of Mathematics" has become a proper field of study at the end of the Nineteenth Century, deeply related with mathematical logic: Frege, Peano, Hilbert and most recently Gödel are just some names among the most representative ones.

Adapted from an earlier Italian version published in Minazzi (2007) and translated by Stefano Calboli, Dept. of Basic Sciences and Foundations, University of Urbino (calbolistefano@gmail.com).

M. Borga (✉)
Department of Mathematics, University of Genoa, Genoa, Italy
e-mail: borga@dima.unige.it

M. Alai et al. (eds.), *Science Between Truth and Ethical Responsibility*,
DOI 10.1007/978-3-319-16369-7_6

The identification of the foundations of mathematics with the philosophy of mathematics is disputable—even if it is widely endorsed. I don't take this identification in a literal sense (indeed it would be obviously false because it would disclaim the existence of philosophy of mathematics before Frege). I rather take it to mean that philosophical thinking on mathematics and foundations of mathematics were identified during the thirty years from Frege to Gödel. Indeed, the so-called "Foundational" schools (Logicism, Intuitionism and Formalism) also focused on philosophical problems, for instance the classic one about the existence of mathematical entities.

Whether or not one agrees with the previous identification, it is unquestionable that it cannot be supported nowadays. There are in fact several recent philosophical views developed in a non-foundationalist or even anti-foundationalist perspective: for instance, extreme forms of Lakatosian mathematical empiricism, which deny that mathematics needs foundations, hence any analysis of them. On the other hand, it is obvious that philosophy of mathematics—as a specific field in philosophy—is influenced by the more general philosophical climate, that in turn can influence foundations in more or less direct way. For instance, if on the one hand the search for rigour and strong foundations for mathematics seems consistent with a Neopositivistic point of view, on the other hand the attention to the fallibility of mathematics, that proceeds through trials and errors, is easily associated with irrationalistic tendencies, especially in some countries.

Under these assumptions, the aim of this contribute is to analyze some topics in the philosophy of mathematics. These topics are chosen especially among those which directly or indirectly concern also the foundations of mathematics, and in particular, those to which Evandro Agazzi has given the most relevant contributions.[1]

Another circumstance is worth mentioning: the research in mathematical logic and in the foundations of mathematics in Italy were brilliantly started by Peano's school, but was rudely interrupted by many causes that I will not expose here totally, but that—summarizing—are only marginally connected with the influences of Croce's and Gentile's Neo-idealism and to the hostility that Peano raised among contemporary mathematicians. In fact, the main cause was that Peano's school had completed his own foundational program and it wasn't interested in joining the new programs that were then starting in other countries.[2] In the Sixties of the last century there was a revival of foundational studies in Italy: the Italian scholars

[1]It is a pleasure for me recall that I was a student of Agazzi in the early 1970s. I have taken part in his courses "Mathematical Logic" (at the Department of Mathematics) and "Philosophy of Science" (at the Department of Philosophy) and he has been my thesis supervisor when I have received my degree in mathematics, in 1973. Of course, these circumstances have influenced this contribution, in which some statements ascribed to Agazzi are not accompanied by accurate bibliographic references because they are the result of my familiarity with Agazzi's scientific approach.

[2]See Borga et al. (1985). I remind that Agazzi promoted these researches on Peano's school in the early 1980s thus filling a cultural gap in Italy. See Borga (2005) for a detailed analysis of the different approaches to foundational analysis of Peano's and Hilbert's school.

had, therefore, "to recover" many years of foundational research during which Italy was absent from the international scene.[3]

The mainstream point of view in those years (among logicians and scholars of foundations) was that mathematical theories were formal systems. This means that the vast majority of scholars believed that both the axiomatization step (of modern type, i.e. that of *hypothetical-deductive* systems, in Pieri's terms) and the subsequent one, the formalization step, had always been accomplished. The latter step was completely unrelated to mathematical practice, and required that any theory made explicit the deductive logical rules used in providing a precise characterization of proofs (*formal* proofs) within that theory. These proofs, in fact, were defined as finite sequences, or finite trees, of formulas linked to each other by logical rules. This latter step appears obvious nowadays, especially in the area of computer science research that strives to attribute demonstration tasks to computers (computers can exclusively make formal proofs!), but in those years it had different reasons: it was related to Hilbert's program, which was then very influential (as it still is in part nowadays). Hilbert's program required a deep analysis of proofs in order to guarantee that no contradiction could be derived: therefore proofs should be rigorously defined (as it doesn't happen in the mathematical practice!), indeed become *formal*, and be studied by what Hilbert calls *Beweistheorie* (proof theory). Let us emphasize—even if I said that above—that this approach affected scholars like logicians and researchers on foundations, but only very little the mathematicians working in the traditional fields of mathematical research. These mathematicians, in fact, considered the use of logic as an obstacle to their research (it was too niggling for them). By the way, this is a respectable approach to logic—perhaps a natural and obvious one—even if sometimes it is referred to by quoting the sarcastic words that Poincaré used with regard to the formal proofs.

In such a context, which saw formalization as the culminating point in the development of a theory (not in order to work within it, but to work *on* the theory), the so-called limitative theorems came in very soon, in spite of Hilbert's optimism (which in hindsight we now see as unjustified). They are theorems in logic (perhaps meta-theorems would be a better world), which show that in looking for rigour at the level of formalization, we incur in irresolvable problems or problems which have unwelcome solutions.[4]

Among these theorems, the second Gödel's Incompleteness Theorem (1931)[5] has a prominent position: for any consistent formalized system that is sufficiently powerful (i.e. which can formalize at least elementary arithmetic) a consistency

[3]See Agazzi (1986), in particular the introduction and Cellucci (1986).

[4]Ladrière (1957) has been a "classic" on this topic; Agazzi (1961)—that we will mention afterwards—is a systematic treatment of the most important steps in the axiomatic method's development. See also Agazzi (1992, 1994).

[5]Gődel (1931), Agazzi (1961) contains the first Italian translation of Gődel (1931). See also the preface by P. Pagli to the Italian edition of Shanker (1988): this preface analyzes Italian circulation of Gődel's Theorems.

proof cannot be carried out by proof techniques belonging to the system in question. It means that this proof cannot be achieved by means of those elementary and reliable methods, i.e. finitary methods, proposed by Hilbert for this purpose. Even if Hilbert has never said exactly what he meant by 'finitary' or 'finististic' methods, it has been immediately evident that they were just a part—actually a very restricted one—of all proof techniques of arithmetic.

Probably, a stronger blow to formalism was dealt by the first Gödel's incompleteness theorem, although this is seldom mentioned. This theorem concerns the syntactic incompleteness of arithmetic: against Hilbert's famous claim "In mathematics there is no *ignorabimus*", the theorem of incompleteness of arithmetic showed that there are mathematical issues that cannot be decided. Indeed, there are closed formulas (i.e. propositions for which, given an interpretation, they can be said to be true or false) of which it is demonstrable that are neither provable nor refutable, and this phenomenon is not due to a deductive weakness of the formal system.[6]

However, a way out (today it seems we should say: an expedient) from Gödel's second incompleteness theorem was proposed shortly thereafter. It was an attempt to extend Hilbert's finitism by carrying out consistency proofs through methods which on the one hand could not be formalized within the theory under scrutiny (so to escape Gödel's theorem), and on the other hand were sufficiently reliable to be used in the research on proof theory. They are the constructive methods, typically used in intuitionistic mathematics, but here employed in meta-mathematics rather than in mathematics. It is remarkable that the consistency proof for arithmetic given by Gentzen in 1936—in the so called 'modified' or 'generalized' Hilbert's program—has been judged "acceptable from an intuitionistic point of view", even if it is surely not finitary. But here a crucial question emerges: who guarantees the reliability of these constructive methods? Of course, no further consistency proof was available, because it would have required further methods to deal with the problem (so generating an infinite regress). So, these constructive methods had to be accepted for their capacity to persuade intuitively; meta-mathematics was by its own nature an informal theory. So, taking the search for rigour to the highest level by formalization, one was eventually obliged to come back to an informal theory, at least at the meta-mathematics level.

More recently, in a deeply changed philosophical context, some have taken a more radical position. Since going back to an informal treatment is unavoidable, sooner or later, why shouldn't we stick to it from the beginning, giving up the formalization step and directing philosophical analysis directly to informal (or pre-formal, not formalized) mathematics? These two approaches are deeply different: one

[6]Gödel's formula expresses within the formal system (through a technical process involving the *arithmetization of syntax*—currently called Gödelization—and the representation of the primitive recursive functions in the formalized arithmetic), the (meta-theoretical) fact of being unprovable. In a similar way—on a semantic level—in the Liar antinomy a formula expresses its falsity. Since this formula is unprovable, by the *adaequatio intellectus et rei* condition it is therefore true on the natural numbers.

thing is resigning oneself to a certain return to the informal in metamathematics; another is to require that the mathematics that has to be studied (not only by mathematicians, but also by philosophers) should be the non-formalized one instead.[7] However, Lakatos—from whom I have taken the above observation[8]– was interested in the philosophical revaluation of informal mathematics. Informal mathematics proceed through trial-and-error processes, by 'proofs and refutations'. Moreover, this is the mathematics practiced daily by mathematicians and it is different from the idealization constituted by formal mathematics. In any case, Lakatos' philosophy of mathematics seems to stall on fundamental questions it posed, particularly dealing with the problem of the potential falsifiers for informal theories.

From his Popperian, quasi-empirical and fallibilist approach, Lakatos rightly analyzes the problem of potential falsifiers for mathematics and correctly distinguishes between the cases of formal and informal theories. The theorems of the informal theories are the potential falsifiers of the formal theories. This is absolutely natural if one believes in the supremacy of the informal theories over the formal ones, although it is acceptable only for well-established informal theories (of which formal systems are intended to be counterparts). As to the informal theories, instead, Lakatos doesn't offer an exhaustive explanation of which their potential falsifiers could be. He only offers unfinished glimpses, that do not seem to have been adequately developed by others, except by proposing again the traditional problem about nature of mathematical entities.[9]

Agazzi's point of view on these matters is not as extreme as Lakatos' one, but there are some common elements: the need for a return to the informal in metamathematics is evident. Agazzi analyzes the question concerning the return to informal approach in mathematics and in meta-mathematics through the distinction between—in Agazzi's terms—the "concrete" theories and the "abstract" ones, which, he stresses, are featured by a very broad and general scope *language*. As a matter of fact some mathematical theories, for instance arithmetic—that deals

[7]There are many doubts concerning the fact that Hilbert considered the formalized mathematic as 'the true' mathematics. Formalization, as we wrote, was a step aimed to the analysis of metatheoretical problems.

[8]Lakatos (1962: 184).

[9]Lakatos marginally analyzed the potential falsifiers problem for the informal theories in his paper titled *A Renaissance of Empiricism in the Recent Philosophy of Mathematics?*, presented at the international conference on philosophy of science which took place in London (1965). Lakatos was one of the organizers of the conference and also editor of the proceedings (1967). This problem is elaborated in the expanded edition of the same paper that has been published posthumous (Lakatos 1976a) with the same title (but it seems to be already completed in 1967). In the last edition Lakatos says: "What is the 'nature' of mathematics, that is, on what basis are truth values injected into its potential falsifiers? This question can be in part reduced to the question: What is the nature of informal theories, that is, what is the nature of the potential falsifiers of *informal* theories? [maybe is it the] *construction* the only source of truth to be injected into a mathematical basic statement? Or *platonistic intuition*? Or *convention*? The answer will scarcely be a monolithic one. Careful historico-critical case-studies will probably lead to a sophisticated and composite solution" (Lakatos 1976a: 40).

with natural numbers—want to describe privileged models. Even if the approach can be syntactic, the guide is always semantic (the intended model). Instead, other theories are "abstract" by their own nature, and to have several models is an advantage in terms of their general application. According to Agazzi, the concrete theories have "a content that doesn't appear far from the content we usually attribute to the empirical sciences".[10] These considerations can be seen as the conclusion of the long course that, starting from the classic axiomatic, arrived at the modern one and was finally crowned by the critical awareness produced by the theorems on the limitations of formalisms.[11] According to the classical perspective, theories were intended to deal with certain mathematical objects (discovering their properties), whereas after the transition to modern axiomatic, theories have—so to speak—emptied of their contents. This means that the syntactic view has become prevailing, if not unique. This fact is well illustrated by the statement (rather unhappy from a terminological point of view) that the axioms implicitly define the primitive concepts. But Gödel's theorem shows that there are true propositions about natural numbers that nevertheless cannot be demonstrated within the formal system for arithmetic. This reveals that *aside from* the formal system (that, by the way, has infinite models, even not isomorphic to each other) the structure of natural numbers exists, regarding which arithmetic's task should be to make true assertions.

This perspective puts forward, however, the problem of what kind of existence should be assigned to the objects of a theory: for instance, Kronecker claims that the numbers are created by God, while for Frege and Russell they are sets. Again, for intuitionistic mathematics numbers are built on the basis of *two-ity*. Who is right? We are dealing with a multiplicity of choices. Agazzi seems to be inclined toward a constructive conception, that also gives him the possibility to treat in a unified manner both mathematics and empirical theories. An empirical theory cuts out its "objects" within a universe of "things" using the operational predicates. They represent the "point of view" of the theory, from which objects are studied.

Similarly, mathematical objects will be identified by the operations that are considered typical of the theory under scrutiny (for instance, let us consider the difference between the arithmetic of natural numbers and that of the integers or of the rational numbers, based on the fact that we want to operate with subtraction and division, without exceptions).

It will not seem inappropriate, I hope, to recall an observation due to Peano, focused on a distinction similar to the one I'm dealing with. In 1906, Peano wrote that a consistency proof is not required for theories such as arithmetic or geometry, while it is appropriate when the postulates are hypothetical and do not correspond to real facts. The context was that of the early meta-theoretical researches at the beginning of the twentieth century. In 1900, in Paris, Hilbert had posed the question concerning the consistency of mathematical analysis. Russell, in 1902, had

[10]Agazzi (1978: 172).

[11]Agazzi (1961).

discovered his antinomy. In 1904, Hilbert had set the ground of what became later his foundational program. Moreover, there had been some "misunderstandings" between Padoa and Hilbert (actually, of Padoa concerning Hilbert's judgment on his work), while Pieri had supported the idea that it was in principle appropriate looking for a consistency proof.[12] In this context, Peano had stayed on the sidelines of the debate: as mentioned above, a consistency proof is not necessary for what Agazzi calls "concrete" theories, since they "speak" about certain real objects. This position could be labeled as a form of Platonism, and one could stress what Pieri pointed out as the difference between his own "abstract" position and Peano's physico-geometrical one. But maybe, this position is more sophisticated. Peano has written: "The axioms of arithmetic, or of geometry, are satisfied by the idea that every writer of arithmetical and geometrical issues has about the number and the point". Moreover, Peano has added: "We think the number, therefore it exists".[13] It's remarkable that right when the formal way to think the mathematical theories has been developed, also a distinction has been made according to which just some axiomatic systems retain the status of theories provided with contents.

If all this seems to undermine the epistemological interest of the attempts to prove the consistency of arithmetic, let us remind that this interest has always been restricted: it was nothing more than a step toward more significant mathematical theories. Moreover, it must never be forgotten that the problem for which the solutions had been sought was the consistency of analysis, and that the first result obtained within Hilbert's program was a proof given by Ackermann in 1924—which later on has been shown to be wrong[14]—whose aim was to prove the consistency of classical analysis.

The above mentioned Lakatos's approach has inspired a few years later the well-known book *Proofs and Refutations: The Logic of Mathematical Discovery*. It is a manifesto of the modern mathematical empiricism (or quasi-empiricism). More recently, this empiricism in mathematics has in turn provided new trends in the philosophy of mathematics. These trends agree on the end of foundationalism and the fallibility of mathematics (often labeled as "loss of certainty"). Again, they agree with the assimilation of mathematics to empirical theories (encouraged by the results concerning computers-aided proofs), and share doubts on the value of

[12]Padoa accused Hilbert to have not given due consideration to his solution of the non-contradiction of arithmetic (of integers), which was based—as we would say nowadays—on the existence of the standard model (Padoa 1903). For his part, Pieri initially seems to put some distance between himself and Padoa, hence to agree with Hilbert on the opportunity of seeking a "direct" proof of the non-contradiction of the arithmetic axioms. But Pieri had the idea—really distant from the approach of the emerging Hilbert's program—of looking for an arithmetic's model within a system that could be considered "of pure logic", that is, without using another auxiliary system (Pieri 1904).

[13]Peano (1906), in *Opere Scelte*, I, p. 343.

[14]Ackermann himself had corrected his mistakes in the preliminary drafts. As a result, his demonstration allows us to prove merely the consistency of a part of arithmetic rather than proving the consistency of mathematical analysis.

traditional proving activities, that sometimes have even been declared "dead". Finally, These trends agree in linking this topic to the chronic troubles affecting the daily teaching of mathematics, and in blaming formalism for them (which, I submit, is at least arguable).[15]

However, on this occasion the target is not merely the formalization, rather mathematical logic itself. Mathematical logic had been the main protagonist in the foundational studies, for instance in Frege and Russell's logicism—which placed it as foundation of mathematics—and in the Hilbertian formalism, in which it was an essential *tool* for the formalization of mathematical theories, as well as for the consistency proofs. It has been considered appropriate to replace—at a methodological level—this kind of logic—that, meanwhile, had become a proper and autonomous mathematical discipline—with a new logic, a logic of discovery, as the subtitle of Lakatos' book points out explicitly. Summarizing, the object of philosophy of mathematics should not be the "justification moment" but the mathematical practice, that is mathematics in its development, that includes, especially, the set of all those procedures followed in the search for proofs, of which there is no trace in the final proof (formal or not). The traditional studies on foundations have disregarded too much this aspect, focusing on the analysis of proofs as finished products. This deficiency has produced a widespread and almost complete lack of interest of mathematicians on the topics related with the philosophy of mathematics. This line of research seems to require our encouragement, but only on condition that—this is my opinion—it is placed side by side with foundational research, that is, it doesn't have to replace it. However, the foundational researches have moved forward in the meantime, even if with purposes dissimilar from the original ones.[16]

Specifically, regarding Agazzi's perspective, it seems his ideas—especially with reference to the limitation theorems and the proposal to treat mathematics and empirical theories unitarily—constitute an authentic and original anticipation

---

[15]See the anthology (Tymoczko 1986) for these new directions in the philosophy of mathematics. Tymoczko's book has been reviewed in *Zentralblatt* by Agazzi. At the end of his review he says: "This book is very stimulating and fulfills its task of providing a philosophical elucidation of several hitherto perhaps too little considered aspects of the nature of mathematics" (*Zbl.Math.*608, p. 6). About the same topic there are other readings, starting with Hersh (1979). They are scientific divulgation books (that have also achieved some success), as Davis and Hersh (1981) whose central chapter is remarkably titled "From Certainty to Fallibility", and Kline (1980), both translated in Italian. To these books we might add Cellucci (2002), that agrees with Tymoczko's conclusion on the end of foundational studies and share with it the hope for a revival of interest for the analytic method and the logic of discovery, and Hersh (2006), that contains the English translation of the introduction by Cellucci's book in the second chapter.

[16]Actually not so different from each other if we think that the recent Simpson's book (Simpson 1999)—that wants to be an important book on the foundations of mathematics—aims to the construction of as much as possible mathematics in "weak" formal systems, for which Hilbert's program becomes achievable. In this case, objects of revision are the formal systems, on which it is conjectured that they have been chosen stronger than what is really needed for the formalization of the current mathematics.

of positions developed in the following years. His latest stances focused on more extreme forms of mathematical empiricism—besides indicating that Agazzi is in line with some basic issues about the cognitive value of mathematics—show a constant interest toward this question and this gives us hope that some pages of his philosophy of mathematics that have yet to be written, will indeed be written in the next few years.

References

Agazzi, E. 1961. *Introduzione ai problemi dell'assiomatica*, Vita e Pensiero, Milano.

Agazzi, E. 1978. Le matematiche come teorie e come linguaggio. *Epistemologia 1*: 165-182.

Agazzi, E. (ed.) 1986. *La filosofia della scienza in Italia nel '900*. Milano: Franco Angeli.

Agazzi, E. 1992. Ragioni e limiti del formalismo. *Epistemologia 15* (special issue: Il problema della conoscenza formale in scienza e filosofia): 9-40.

Agazzi, E. 1994. On formalism. In G. Fløistad (ed.), *Philosophical Problems Today*, vol. I: 75-137. Dordrecht: Kluwer.

Borga, M. 2005. Alle origini delle ricerche metamatematiche: indipendenza e coerenza fra Ottocento e Novecento. *Epistemologia 28*: 3-24.

Borga M., Freguglia, P., Palladino, D. 1985. *I contributi fondazionali della scuola di Peano*. Milano: Franco Angeli.

Cellucci, C. 1986. Logica e filosofia della matematica nella seconda metà del secolo. In Agazzi 1986: 317-336.

Cellucci, C. 2002. *Filosofia e matematica*. Bari: Laterza.

Davis, P.J., Hersh, R. 1981. *The Mathematical Experience*, Birkhäuser, Boston; It. tr. *L'esperienza matematica*. Milano: Edizioni di Comunità 1985.

Gödel, K. 1931. Über formal unentscheidbare Sätze der *Principia Mathematica* und verwandter Systeme I. *Monatshefte für Mathematik und Physik* 38: 173-198. It. tr. in Agazzi 1961: 203-228, also in Shanker 1991: 21-62.

Hersh, R. 1979. Some Proposals for Reviving the Philosophy of Mathematics. *Advances in Mathematics* 31: 31-50; also in Tymoczko 1986: 9-28.

Hersh, R. (ed.) 2006. *18 Unconventional Essays on the Nature of Mathematics*. Berlin: Springer.

Kline, M. 1980. *Mathematics. The Loss of Certainty*. New York: Oxford University Press. It. tr. Matematica. *La perdita della certezza*, Mondadori, Milano, 1985.

Ladrière, J. 1957. *Les limitations internes des formalismes*. Louvain: Nouwelaerts, and Paris: Gauthier Villars.

Lakatos, I. 1962. Infinite Regress and the Foundations of Mathematics. *Aristotelian Society Supplementary Volumes 36*: 155-184; also in Id., *Philosophical Papers*, vol. 2, ed. J. Worral, G. Currie. Cambridge: Cambridge University Press, 1978: 3-23.

Lakatos, I. (ed.). 1967. *Problems in the Philosophy of Science*. Amsterdam: North-Holland.

Lakatos, I. 1976a. A Renaissance of Empiricism in the Recent Philosophy of Mathematics. *The British Journal for the Philosophy of Science 27*: 201-223; also in *Philosophical Paper*, vol. 2, ed. J. Worral, G. Currie, Cambridge: Cambridge University Press, 1978: 24-42 and in Tymoczko 1986: 30-48.

Minazzi F. (ed.). 2007. *Filosofia, scienza e bioetica nel dibattito contemporaneo. Studi internazionali in onore di Evandro Agazzi*. Roma: Istituto Poligrafico e Zecca dello Stato.

Padoa, A. 1903. Le problem n. 2 de M. David Hilbert. *L'Enseignement Mathématique 5*: 85-91.

Peano, G. 1906. Super theorema de Cantor-Bernstein et additione. *Rivista di Matematica 8*: 136-157; also in Id., *Opere Scelte*, vol. I. Roma: Cremonese, 1957: 337-358.

Pieri, M. 1904. Circa il teorema fondamentale di Staudt e i principii della geometria proiettiva. *Atti della R. Accademia delle Scienze di Torino 39*: 313-331; also in *Opere sui fondamenti della matematica*. Roma: Cremonese, 1980: 289-307.

Shanker, S.G. (ed.). 1988. *Gödel's Theorem in Focus*, Croom Helm Ldt., London; It. tr. *Il teorema di Gödel: una messa a fuoco*, Padova: Muzzio, 1991.

Simpson, S.G. 1999. *Subsystems of Second-Order Arithmetic*. Berlin: Springer.

Tymoczko, T. (ed.) 1986. *New Directions in the Philosophy of Mathematics*. Boston: Birkhäuser. New edition, Princeton: Princeton University Press, 1998.

Artificial Intelligence

Mariano L. Bianca

Abstract Before analyzing Agazzi's conception of Artificial Intelligence, an historical overview is offered of the different ways in which the problem of imitating or reproducing human thought has been approached since the Middle Ages till now. My review of Agazzi's position is mainly based on two essays: "Alcune osservazioni sul problema dell'intelligenza artificiale" (Some observation on the problem of the artificial intelligence, 1967) and "Operationality and intentionality: the missing link of the artificial intelligence" (1981). In these Agazzi carries out a deep analysis and criticism of the different approaches to this topic (especially computationalism and functionalism) and outlines his own articulated conception. In his view today's computers (or intelligent artificial systems in general) cannot be said to have thought or feeling analogue to those of human beings because they are not endowed with intentionality. In such a way Agazzi anticipated of about 15 years this celebrated thesis maintained by John Searle. It is possible to build complex machines capable of realizing performances very similar to those of human reasoning or goal-oriented behaviors. This similarity, however, is confined to the capability of performing certain operations, while the difference that still remains is the capability to give meanings to such operations and their results, and to intentionally propose goals to themselves.

1 A Preliminary Note

Evandro Agazzi has dealt with artificial intelligence, both from the critical and the philosophical point of view, in a few essays, the most salient of which are: "Alcune osservazioni sul problema dell'intelligenza artificiale" (*Some observation on the problem of the artificial intelligence*, 1967a) and "Operazionalità e

M.L. Bianca (✉)
University of Siena, Siena, Italy
e-mail: mariano.bianca@unisi.it

M. Alai et al. (eds.), *Science Between Truth and Ethical Responsibility*,
DOI 10.1007/978-3-319-16369-7_7

intenzionalità: l'anello mancante dell'intelligenza artificiale" (*Operationality and intentionality: the missing link of the artificial intelligence*, 1991). We say "salient" because the 1967a paper contains the first extensive critical treatment of the issue of artificial intelligence proposed by Agazzi, with particular insistence on intentionality as the decisive characteristic that distinguishes human intelligence from machine or artificial intelligence. In such a way he was anticipating of 15 years this thesis that is often credited to John Searle. Actually Agazzi had presented his paper in English at the "Wiener Memorial Meeting on the Idea of Control" held in Genoa in 1965, whose proceedings however never appeared. Therefore in 1967 he published an Italian translation of that paper, that was also reprinted in Agazzi (1978), and in addition presented his ideas in a shorter paper at the 21th National Congress of Philosophy (1967b). Agazzi had the opportunity of taking up again and expanding his ideas in the paper "Intentionality and artificial intelligence" presented at a meeting on "The Mind-Body Problem" organized by the International Academy of Philosophy of Science in 1980 and whose proceedings appeared as a special issue of *Epistemologia* (see Agazzi 1981). An Italian translation of this paper was later published (Agazzi 1987).

Agazzi's treatment of artificial intelligence reflected also, from the very beginning, his strong focus on operations in scientific epistemology. Therefore he published a paper in which both intentionality and operationality are stressed as central concepts in the debate on artificial intelligence (Agazzi 1991) closing in such a way the circle of his discussion of this topic. The 1991 paper was newly reprinted as Agazzi (2010).

This brief historical reconstruction explains why, in the present contribution, quotations are taken only from Agazzi (1967a, 1991), that is from the initial and the final points of this trajectory. Indeed the 1967a seminal paper contains the priorities and originality of the whole treatment devoted by Agazzi to artificial intelligence, while the 1991 paper stresses that strict connection between intentionality and operationality that is the backbone of the whole reflection on human knowledge developed by Agazzi in his work, a connection whose significance is well exemplified in the treatment of artificial intelligence. Of course in Agazzi (1981) one finds a more systematic, detailed, analytically deepened and also thematically enriched treatment of the issue, but the real novelties are rather linked with Agazzi's investigations in the domain of logic, semantics, foundations of mathematics to which are specifically devoted other contributions contained in the present volume, so that we did not consider essential to pay them a particular attention.

2 A Few Introductory Remarks

Before examining Evandro Agazzi's theses about artificial intelligence, it is useful to dwell upon the idea of reproducing in any possible way the mental activities of humans and, more particularly, the so called rational or cognitive activities.

The idea of imitating, simulating or emulating human cognitive activities is very old, and can be dated back to the medieval logicians, and particularly to the logic of Ramon Lull and afterwards to the combinatorial logic of Pierre de la Ramée. The works of these authors were based on Aristotle's logic, which basically claimed that human thought (the strictly cognitive thought, at least) develops by logical procedures, like deduction and induction, which could always be reformulated in a rigorous and even formal way. Certainly Aristotelian logic did not include the idea of a reproduction of human cognitive activities, but it contained the idea that such activities are formal procedures, which could have been adapted to any type of cognitive content; in general, this idea was adopted by those who first, in the modern age, formulated the concept of reproducibility of the cognitive activities of man: they considered these activities as the result of formal processes.

Such a conception was also adopted by Ramón Lull, who was certainly the first in the history of philosophy and of logic to pass from considering the cognitive activities as logical-formal activities to believing in the possibility to imitate such activities by mean of techniques or more or less mechanical devices. Lull's logical machines, in fact, were a kind of primitive computers formed by various disks of papers laid one on top of the other, capable to rotate and so to generate different combinations of the symbols which were printed on them. Such *logical machines* allowed to formulate different types of syllogisms concerning any kind of symbolically representable content.

He used these logical syllogistic machines to demonstrate the existence of God, and more particularly the existence and the primacy of the Christian God; his logical machines are the first, even if primitive, material form of the idea of imitating some activities of the human mind; they attracted much interest, and were resumed by other logicians in the following centuries. Such machines were able to formulate mechanically the syllogisms, and their application to theology might have been acceptable within Christian theology as a means to offer a logical proof of the existence of God; unfortunately, however, it proved unacceptable to Muslim theologians, who after a short period of interest decided that it was probably better to abandon these logical machines, and Ramón Lull was stoned to death.

But the idea of the mechanization of the human mind, and more particularly of its logic, was not abandoned, and it survived in various stages till modern logic, the formulation of axiomatic systems, and the construction of intelligent machines. This course, up to the era of the first computers like the Eniac, manufactured among others by Arthur Burks, had many stages, and it is useful to remind some of them. The first was the combinatorial logic of Pierre de la Ramée, and afterwards Leibniz's logical studies and the famous machine devised by Pascal to perform mechanically some elementary arithmetical operations. If mathematical thought was typical for mankind, and if it was mechanically reproducible, then at least some part of human thought was also reproducible by a machine: in other words, a machine could be able to perform some human mental operations.

This concept of the reproducibility of human thought was further used in devising logical or mathematical machines like those of Babbage; since then the possibility of reproducing human thought has made progresses, up to the manufacture of machines capable to perform not only mathematical operations, but also logical operations, in particular deductive operations similar to those of humans: the underlying idea has always been that of considering the mental processes as rational and logical processes grounded on rules and applicable to different contents.

Nevertheless, we must remember that, together with formal logic, in the Middle Age also other kinds of logics were formulated, where the contents of the propositions involved in logical processes were also relevant: just think, for instance, of Boethius' logical conceptions.

Within the formal view the first researchers on Artificial Intelligence (AI) held that it was possible to reproduce, or at least to simulate or emulate, human thought by a machine. Hence, their purpose was not to understand, by the use of machines, the functioning of the human brain, but to use machines to perform the same operations (even if not all) that the human mind is capable to perform. So, they drove to the man-automaton equivalence.

Once again, this conception was based on the idea that human thought was the result of elementary computations which were capable to generate very complex processes. This conception was little by little abandoned within AI, in favor of a more limited objective, the construction of machines capable to realize computations useful to perform specific tasks: therefore, little by little the term 'intelligent' was abandoned, and attention was devoted to the construction of expert systems useful to perform different tasks, like: visual recognition of objects; auditory recognition of sounds; recognition of texts; management and control of very complicated processes, like those of airplanes, of space ships, or space probes; the performance of activities like those of robots, drones, or similar apparatuses.

However, it is useful to remind that computationalism, which took the computations performed by machines as a model of the human mind, brought to a reductionist philosophy of mind; in its strongest form this conception was abandoned as a consequence of results obtained in the neurosciences, according to which the human mind is much more complex than mere computational processes based on few rules applied to basic elements. This point, as we shall see, was stressed by Agazzi since the sixties of the last century.

The first theorists of AI reasoned from the premise that the mind operates always and only on the basis of logical or formal rules; hence, for them simulating the human mind meant building technological apparatuses capable to perform some of its operations, and this entailed that "machines can think". Agazzi dwells upon this statement clearly distinguishing two aspects of the problem.

3 Evandro Agazzi's Approach

Agazzi believes that the claim that an artifact is capable to think like humans think is not in itself wrong, if considered as a guideline idea from which to move in trying to formulate simulations or emulations of human thought. But this thesis, or if you like, the affirmative answer to the question ‘can the machines think?’, cannot be considered as a scientifically validated hypothesis, because this would require supporting it by adequate evidence, not only as to the results, but also as to the procedures needed to achieve such results. But until now no such evidence has been brought up in a conclusive and rigorous way, because even if machines can obtain results similar to those of humans, this does not mean, as Agazzi points out, that minds and machines have the same nature. On the contrary, the neurosciences have revealed that not only inside the black box of the human mind there are components and elementary constituents substantially different from those of machines, but there are also procedures which in most cases are not reducible to merely computational processes, understood in their standard conception: a finite set of formal rules applied to elementary components and relative to any kind of mental information, be it cognitive or not.

Agazzi underlines that often the suggestion of the equation man-automaton and the related statement ‘machines can think’ is due to the use of an anthropomorphic language to indicate operations performed by machines: talking of a machine like an electronic computer we use propositions like ‘it learns’, ‘it remembers’, ‘it makes choices’ etc., but the operations of this automaton are very different from those of a mind; nevertheless, the use of these terms belonging to our ordinary language induces one to imagine that the machine performs precisely the operations which are performed by a mind. According to Agazzi’s this linguistic practice is apparently justified by the analogy with the talk concerning *transitive* operations of machines: *transitive* operations are defined on the ground of the results that any agent (be it a man or a machine) can achieve starting from an initial state of affairs, just as when it is said that a machine sews, another mows and another washes. But when we use expressions like ‘answering’, ‘learning’, ‘remembering’, ‘feeling’, ‘seeing’, etc., we are not talking of *transitive* operations, but of *immanent* operations: immanent operations are not defined through their “external” results but through the modifications they induce “within” the agent; we shall revert further on such distinction when we shall examine the thesis defended by Agazzi in his essay “Operazionalità e intenzionalità: l’anello mancante dell’intelligenza artificiale”.

In connection with the behaviorist perspective, which focuses on the analogy between behaviors/results of a machine and those of the mind, Agazzi states as follows:

> The majority of reductivist positions inspired by artificial intelligence is based on the fact that the actual computers are able to perform a certain number of functions that man can perform thanks to the use of the intelligence, and on the persuasion that even those functions that have not been imitated up to now will be imitated with the progresses of technology. This is the position inspired by the famous apologue of Turing known as ‘the imitation game’, consisting in hypothesizing a machine able to give to a human interlocutor answers that don’t allow her to understand whether such answers come from a machine or from another human interlocutor to which the same questions have been asked. The game is a clear representation of the stimulus-answer scheme of behaviorist psychology, and is subject to the critical remarks received by the latter concerning its adequacy as a methodological attitude to investigate cognitive activities. To these criticisms

> must be added (in case the Turing method is adopted in a reductionist point of view) the mention of a further methodological incorrectness, which consists in holding that the identity of one or more functions justifies the claim of identity of nature between the entities displaying such functions. Such incorrectness can be noticed just on the pure logical level: identity of nature in fact is a condition *sufficient* but not *necessary* for the identity of function, therefore it is not possible to infer the first from the second. Very obvious examples prove this truth in a clear way: both an airplane and a bird can fly, but they have a very different nature; both an old mechanical computing machine and an electronic computer can add two numbers, but they have nearly nothing in common as far as their nature and structure is concerned. The very developments of artificial intelligence are a probatory confirmation of this fact: the progresses of machines in 'emulating' some intelligent operations of man have been obtained by giving up any 'simulation' of the cognitive activities of man: i.e., by using our advancements in electronic engineering more than those in cognitive psychology (Agazzi 1991, p. 5).

Besides this functionalistic or behaviorist perspective Agazzi takes into consideration another, that he names *structuralist*, characterizing it in the following way:

> In order to overcome the difficulties of the behaviorist (or so to say 'functionalistic') formulation the imperative seems to be that of checking what really is inside the 'black box'. But even this proposal is not exempt from reductivist risks: indeed it depends on what one is looking for inside the 'black box', more particularly utilizing the tools of the artificial intelligence. Now it happens that, very often, those who intend to imitate not just the functions, but directly the structure of the human thinking activity, identify such activity with the human brain, of whose functioning the thought would be just a product. If one succeeded in building an 'artificial brain', that would therefore be able to think. At the basis of this perspective there is however an unfounded assumption: the identification of thought with an immaterial product of the brain (a product therefore not merely physiological, like the many products which accompany the metabolism of the cerebral activity). Even in this case a logical mistake can be noticed: exchanging a necessary condition for one which is *also* sufficient: it is fully plausible that there is no thought without brain (at least as long as we deal with human thought), but this is not to say that having a brain is sufficient to have a thought. Even the most spiritualistic philosophies, in fact, have never denied that the brain is an essential *condition* for the exercise of the thought, but this neither means that it is the *cause*, nor that it coincides with thought. But there is something more: strictly speaking the construction of an artificial brain will never consist in the exact reproduction of a natural brain, and all the progresses of the bionics can succeed to produce is an artifact which *works* in a way very similar to the natural brain. But then we really fall in the earlier situation: even the 'structuralist' path in the end is a 'functionalist' path aimed to a lower objective, i.e., to 'simulating' the neurophysiological functions, rather than the cognitive ones (ibid., pp. 5–6).

However, there are some relevant aspects of the functionalist thesis which are focused on by Agazzi in the following way:

> the Western philosophical tradition not seldom has held the immateriality of thought (hence also the spirituality and the immortality of soul) because humans are able to perform activities which do not imply any contribution of materiality (typically, knowledge of the universal), and from this it was possible to infer the existence in humans of an immaterial principle necessary to perform such activities (ibid., p.7).

This conception is grounded on the metaphysical principle *operari sequitur esse*. From this principle Agazzi draws a question crucial to the topic of the difference between humans and machines: "if it is incorrect to infer from the identity of the functions of two entities the identity of their nature, is it not equally incorrect to infer, from the fact that a certain activity leaves out matter, that also the entity capable to perform such an activity can exist without matter?" (ibid.). And he answers this crucial question as follows:

> Notwithstanding the appearances the questions are not on the same level. '*Operari sequitur esse*' means in fact that any being must possess in its nature all the conditions necessary and sufficient to perform its functions, so that if for a certain function a particular condition (in this case that of materiality) is not necessary (besides not being sufficient), the related necessary condition will be of different nature (immaterial in the specific case). On the other hand, if it is granted that the first condition is not necessary, and it is evident that the function takes place, it is possible to infer that the second condition, beside being necessary, is also sufficient. Naturally the application of this principle brings to the acknowledgment of two 'distinct' conditions for the exercise of a certain function: that they are also ontologically 'separated' is a much more complicated problem which is of no interest here (although rather important for the statement of the separation of the soul from the body) (Agazzi 1991, pp. 7–8).

With reference to the simulative aspect of the functionalist thesis in the discussion on the artificial intelligence Agazzi points out that

> if really the simulation of *all* the most intellectually elevated human activities (i.e., those activities which usually are considered not 'depending from matter') would perfectly succeed without residuals by means of material electronic calculators, one couldn't any more claim that the materiality is *not sufficient* to perform such functions; so, it would no longer be necessary to find conditions different from matter for the realization of those functions in humans. Certainly it would still be true that the identity of functions does not logically imply identity of nature, but in order to affirm the difference of nature one should produce *other* arguments, different from those classically considered the more conclusive (ibid., p. 8).

Agazzi, therefore, has pointed out that the two theses (behaviorist/functionalist and structuralist) fail to show that machines think in a way similar to the mind, not only with reference to behaviors or functions, but also to the nature of machines and of the mind.

Agazzi' arguments can also be referred to the strictly logical operations performed by a machine. In this case his analysis appears to be particularly convincing and articulated. There are, Agazzi points out, two essential aspects: on the one side the involvement of the notion of 'truth', and on the other side "the request to take into account all the *possible* interpretations of our assertions" (Agazzi 1967a, p. 19). Neither aspect is present in the same way in humans and in machines. This is how Agazzi argues: "We substantially say that machines succeed in imitating our deductive reasoning since we succeed in reproducing in them that whole of formal schemes and rules of logically thinking that we have considered as typical of our deductive way of *operating*" (ibid., p. 224). But Agazzi points out that

> it must not be forgotten that these formal systems of rules are only an artificial surrogate of what we mean by the terms 'demonstration' and 'logical consequence'. The link of logical consequence has an intuitive nature, and it is possible to say that a proposition P is the logical consequence of a set propositions S when it is not possible to suppose in any case that the S are true and that P is false; in other words, as we say more precisely, when *any* 'interpretation' which verifies the S also verifies P (ibid.).

To avoid the difficulties entailed by the reference to the infinite totality of the interpretations contained in the intuitive notion of logical consequence, this has been replaced by that of *deduction*, that is, "of an *operating* on the ground of rules in which the concept of truth does not appear but which is practically checkable step by step" (ibid., p. 19). For this reason, Agazzi remarks, it is plausible to admit that, as it is possible to manufacture washing or writing machines, it is also possible to manufacture machines to perform material operations on symbols and therefore also logical operations like the deductive ones. But also in this case we only have a surrogate of what the mind does, since also in the case of deduction we always have to do with a living being, with the structure and the contents of his mind, and this makes the nature of these operations different from those of an automaton.

To point out the difference between the human way of operating and that of a machine Agazzi proposes a clear example related to vision. Let's think of the difference between the perceptual image of an object and its image on a photographic plate: in the latter case we don't use the term 'sees', but the expressions 'it fixes' or 'it records' the image. For which reason is this terminological difference made? It consists in the fact that in the case of the visual perception of humans, unlike the recording of an image by a camera, there is something *more* which is indicated by Agazzi by the term *intentionality*; by such term he alludes to the fact that in the cognitive or simply perceptive activity of the living being, and particularly of humans, there occurs a kind of participation or of identification of the subject with the objects "which even remaining themselves, in some way become part of the subject" (ibid., p. 16). In other words, in general, reference is made to the subjectively lived character of perception.

In fact, this argument by Agazzi, in the current state of the neurophysiological researches, allows to claim that the perception of an object is not only the result of the ocular apparatus and various cortical and not cortical areas of the brain used to the reception and to the elaboration of the photons, such as the geniculated nucleus brain or the primary and secondary visual cortex, but also of the involvement of the cognitive and associative cortical areas located in the temporal and frontal regions of the brain, which add information to that received by the stimulus, thus producing the perception.

That something *more* differentiating humans from cameras is constituted by intentionality and consciousness, and this means that in order to have a visual perception we must add to the information coming from the stimuli some information already present in the mind, which allows both to cognitively recognize the object, and to assign the image different meanings, like, for instance, 'it is beautiful', 'I like it, 'it reminds me of a past experience', 'it makes me feel a certain mood'. For this reason we say that an object is *seen* by us (in this wide and conscious sense), while it is simply 'recorded' by a camera.

In even wider terms, according to Agazzi's conception, it is not the eyes, but the whole subject which 'sees', and this is naturally and substantially different not only from what happens in a camera, but also in an artificial apparatus capable of recognizing objects, as in the case of *pattern recognition*; in this sense, in machines there is no subject operating and endowed with intentionality and consciousness. So, the seeing of an object by a human is not a process similar to the recording by an artifact; in this case even the results are different, since the perception of an object involves information possessed by the percipient subject, and such information is not present in the processes of recognition of an object by a machine.

This topic, which is central in Agazzi's analysis of the artificial intelligence, has been more deeply tackled in the essay *Operazionalità e intenzionalità: l'anello mancante dell'intelligenza artificiale* (1991), of which I shall report some passages which allow a more complete understanding of Agazzi's conception of artificial intelligence. First of all, Agazzi introduces the already mentioned distinction between transitive and immanent operations. With reference to the former Agazzi underlines that

> The majority and maybe the totality of man's transitive activities can be expressed by indicating a specific *operation* which has been generated. We have said 'has been generated' and not 'in which consists', because the operation, as we want to define it now, is characterized through two static moments and is independent of the modality of transition from the one to the other. In general, in fact, it is possible to define an operation *O* as *any process whatever* which, applied to an initial state *I*, leads to a final state *F* in which the transition from *I* to *F* doesn't make any difference. For instance 'sewing' is an operation defined by the passage from an initial state in which the two pieces of tissue are separated to a final state in which the pieces are strictly connected with a thread. Just because the operations concern states of the world, they result univocally defined with reference to such states, therefore they remain *the same* whether they are performed by humans or by animals or by machines. On the other hand, in this case they *say nothing* either about the process of transition from *I* to *F*, or about the agents which have performed the process. This is the fundamental reason for which humans, even if having continuously realized machines capable to perform operations identical with those of their transitive activities, never considered this fact as a threat to their human identity, nor they feared to be assimilated to machines, even when these were more efficient than humans in performing certain operations (Agazzi 1991, pp. 8–9).

According to Agazzi, we run into a very different situation when "the machines have started to do some activity which is considered immanent (like those connected to the exercise of thought and of intelligence)" (ibid., p. 9). However, Agazzi points out that in this case it is necessary to carefully consider whether "what is imitated by the machine is really the immanent aspect of such activities" (ibid.). In which way can we answer this question? According to Agazzi we must answer it in the negative, because "even for some of their immanent activities humans have been able to devise some *operations*, and it is only the latter that machines can imitate (or more exactly, can carry out)" (ibid.). An immanent activity is always a change, but it does not concern the state of the external world, but that of the subject which performs such activity. In some cases however, continues Agazzi,

> the subject may reflect on his internal states and associate them with material objects *representing* them as signs, then establishing material *operations* on the signs, so that when from the internal state I he passes to the internal state F, the operation O applied to the material sign which represents I leads to the material sign which represents F. (ibid.).

This condition occurs since humans and their minds operate on symbols, for which they can always produce "the translation into material signs of their internal states and the reconstruction of internal states starting from material signs, in the extremely varied range of the 'languages' in which they express themselves" (ibid., p. 9–10). However this complex activity of the human mind according to Agazzi may be understood only by stressing that it is always guided and controlled by intentionality, which "allows to pass from the internal state 'intentionated' to the material sign which *represents* it, and it is again intentionality which allows to *interpret* some material signs as representing certain internal states" (Agazzi ibid., p. 6).

At this stage of the argumentation we must face a crucial question: is it possible that these mental activities are carried out without the constant control of intentionality? In other words, for Agazzi, we should ask "whether the chain of internal transitions which bring me from a state I to a state F passing through a number of intermediate states (also of intentional nature) can be *represented* by a material *operation* (not intentional as such) taking from the symbol of I to the symbol of F" (ibid., p.10). The answer to this question (which is in general negative) leads to a non-functionalist position and to the rejection of the equivalence between humans and robots. In this sense Agazzi points out that

> the answer to this question is positive only in few cases, that is, in those in which the transition from I to F complies with some well specifiable requirements and in which, in addition, it would be possible to devise some symbolic representations of the operational manipulations on the symbols which, when they are re-interpreted, comply with these requirements. The more relevant example in this sense is represented by formal logic, in which I and F can be respectively considered as the premise and the conclusion of a reasoning in which the requirement of the transition is that the truth of the premise passes to the conclusion. I and F are contents of thought … but they can be appropriately translated in linguistic expressions, that is, in material signs which express them, and formal logic institutes operations which from the expression translating I lead to the expression translating F while complying with the condition of the persistence of the truth. Equally well known is the example of mathematical operations in which one materially operates on symbols according to rules which have been instituted on the ground of an 'intentional' reflection, but which are also applied in a purely material way. From what has been said it is clear that those operations which, following the double application of intentionality, can be performed to *represent* immanent activities, have a transitive nature (they manipulate the external world) and, *qua operations*, can be performed in any way whatsoever by any agent as long as they correctly take from I to F. It is therefore completely obvious that they can also be performed by machines, perhaps even more efficiently than by humans, and it is also clear that their carrying out does not require *intelligence*, just because this, thanks to the intentionality that characterizes it, stays upstream and downstream the carrying out, and it consists in devising the correct manipulations on the material signs, in the appropriate symbolization of I by means of material signs accessible to those manipulations, and in the interpretation of the result of the operation as a certain F (ibid., pp.10–11).

This shows that only a limited and specific number of mental activities, owing to their formal nature, can be simulated by an automaton; therefore Agazzi puts a strong constraint on the concept of simulation of human mental activities by an artifact. There are no difficulties of principle for transitive activities, but there are limitations as to the possibility of current electronic computers to represent immanent activities. In fact such computers "limit themselves to perform material operations which can be used by humans to replace some part or applications of human immanent activities only thanks to intentionality" (ibid., p. 11). A further unavoidable question arises at this point:

> even admitting that intentionality is the characteristic which differentiates the operations of humans from those of a robot, can we consider it as a qualitative difference justifying the application of the principle *operari sequitur esse*? In other words, can we attribute it to a dimension which goes beyond materiality? (ibid.).

From this follows the crucial question for the entire argumentation of Agazzi about artificial intelligence: "is it possible to realize intentionality within a robot?" (ibid.). Agazzi replies negatively, for the following reason: "intentionality is not a physical state, is not the simple 'presence' of something, but a special *way* of such presence" (ibid.). To clarify this remark Agazzi considers the image of a house; "the image of a house is present *physically* in a film and in the retina of an eye, but it is present *intentionally* to the perceptive faculty of a man or of an animal" (ibid.). In this sense, therefore, he points out that "intentionality needs physical conditions to reveal itself, but it does not coincides with them" (ibid.), and this holds both for perception and for the other activities which are considered intelligent. From these remarks Agazzi concludes that it is possible to realize

> for instance by means of sensors and analogic computational procedures governed by specific negative feed-back, according to the ideas of McKay, a 'simulation' of perceptive activities more advanced than the very unsatisfactory attempts realized with digital methods; but it will always be a 'physics' of the perception, even if very interesting. A physics which will help us to understand better *through* which processes intentional cognition adapts itself to the external world and absorbs it in itself, but will not be able to physically explain away intentionality" (ibid.).

For this reason once again we are inside a behaviorist/functionalist point of view of second level "in which … the behavior becomes internal and concerns the evolution of the information and of the programs" (ibid., p. 12), therefore, it still belongs to the physical level and does not belong to the cognitive perspective proper.

The analysis becomes more complex if from perception we pass to the strictly cognitive activities characteristic of thought, where we also find "the capability to intentionate the abstract, the possible, the true and the false, the ought to be, the necessary, the mandatory, the prohibited and all the ranges of the *intentiones secundae* which are at the ground of creativity and of free will" (ibid.). Is it possible that these *intentiones secundae* are realized in an automaton? To this question Agazzi replies in an articulated affirmative and negative way:

> in a certain measure yes: i.e., in the measure in which through acts of *self-consciousness* we *objectify* to ourselves some of those abstract entities, we analyze them and we operate a formalization that afterward becomes symbolizable and translatable into programs for the machine. But the intentionality through which we carry out such objectification and such translation remains behind the shoulders of the program, it does not appear in it, it does not enter into the machine, remaining as unanalyzable and inexpressible than the

intentionality which is at the basis of the perceptive act and which confuses itself with the content of perception (ibid.).

In this regard Agazzi claims that it is this intentionality which allows "the demonstration of all the semantic meta-theorems of mathematical logic (completeness, incompleteness, categoricity and non-categoricity, internal limitations of the formalisms such as Gödel's theorem, etc.)" (ibid.). Hence, in this respect, the answer is negative; indeed, Agazzi holds that:

> it is not difficult to show that reflections and demonstrations of this kind are not replicable in an automaton, since they presuppose the availability of a *semantic meta-language* which the automaton lacks just because is based on intentionality, for without a *starting* intentionality it is not possible to construct the programs which enable a robot to compute algorithms … and without an *arrival* intentionality it is not possible *to evaluate* the obtained results (ibid.).

The arguments grounded on intentionality, however, do not refer only to the immanent activities of humans, since intentionality is actually present in

> *any* human activity, making it specifically human. Even in the transitive activities it is always humans who do something by *using* material instruments which can initially be their two hands, then simple tools like a spade, and finally very complicated machines. It is always humans who *intentionate* the result they meant to obtain, who *plan* the operations to obtain it, who plan and manufacture any machines needed to carry out such operations, who finally *evaluate* whether the result is consistent with what they intended to obtain. The same happens in the case of mental activities: humans *intentionate* what they want to achieve (for instance, the sum of two quantities), they *formalize* it conceptually and *symbolize* it materially (for example, by taking pebbles as proxies for each element of the two quantities, and by the contrivance of an algorithm consisting in throwing the pebbles, one after the other, into one and the same container); finally, once the *program* has been completed, they *interpret* the result (by counting the pebbles which are in the container). If instead of using pebbles they use numerals written on a sheet of paper, or the states of magnetization of electronic components of a computer, and if instead of accumulating the pebbles they use algorithms to manipulate the numerals according to fixed rules or computer programs, nothing relevant changes in this proceeding. Similarly, nothing changes if, in order to solve a much more complicated problem, they must materially memorize some intermediate results, or, during the execution of new algorithms, they must use appropriate sub-algorithms at the appropriate moments, so that at the completion of the process they can read the solution of their problem by *interpreting* the terminal configuration of the material signs. It is always humans which, intentionally using the machines, carry out their operations (ibid., pp. 12–13).

From these arguments follows Agazzi's fundamental thesis according to which

> it is never possible to speak of any human activity without appealing to intentionality, while there is never need of intentionality to describe the operations of a machine, no matter how complicated and improved it is. Not even the computers which perform the fascinating technological programs of the A.I. are an exception to this rule (ibid., p. 13).

Concerning this condition, and the great diversity between man and the machine, Agazzi uses an expression taken from biological evolution: "the *missing link* … between natural and artificial intelligence … undoubtedly seems to be intentionality" (ibid.). Therefore, this complex argumentation of Agazzi not only confutes, as we have seen, all forms of physicalistic, computational and functionalistic reductionism, but it "shows the inadequacy also of those forms of emergentism which consider the most elevated levels of reality only as complexification of those of the lower levels, losing the real nature of the new and of the different" (ibid.).

4 Conclusion

In conclusion, it is useful to point out that Agazzi does not assume an a priori point of view against artificial intelligence, but he limits its feasibility to the cases of transitive or even immanent operations based on formal rules; nevertheless, also in this case formal operations in humans and in machines are substantially different, since they are embodied into largely different systems. Mechanical systems differ from the human mind, in which beside intentionality and consciousness, we find very complex processes, where even simple formal operations involve non logical, or non strictly cognitive aspects. In fact, such operations are expressions of a living being, of a complex cerebral system and of an even more complex mind, in which operations, including formal ones, are always interwoven with meanings and guided by specific goals. Actually, it is just the meaningfulness and teleologicity of mental processes which differentiate them from the processes of an artificial system which does neither elaborate nor generate meanings, nor intentionally proposes goals to itself; the distinctive characters of the operating of a mind are just the intentional goals and the processes which create meanings and interpret the stimuli coming from the world; this does not mean that is impossible to build always more complicated machines which allow to obtain results similar to those of human cognitive activities.

Such machines could also be useful to understand some aspect of the human mind, but one cannot claim that they are intelligent like the human mind, nor that they are a reproduction of the human mind; in fact, in order to be such they should possess a biological mind like that of humans, be inside of a body like that of humans, and finally, have a life like that of humans, with all its existential, psychological, affective, emotional, relational and significant aspects: these last, the significant ones, are characteristic of subjects which possess a mind and intentionality.

However, finally, what is really interesting is not so much emulating the human mind, but building machines able to perform various very complicated tasks that the human mind cannot perform. In this way one can construct artificial minds different from the biological ones, and forms of intelligence different from the biological ones: they will pursue goals that, in Agazzi's terminology, are certainly intentional goals of humans, and this is not transferring human intentionality to the machines, or placing in them an intentionality autonomous from that of humans. Whichever is the plausibility of such scenarios, there is also one more possibility: that the robots can autonomously develop, self-generate, self-reproduce and form their own artificial intentionality. But the analysis of the latter scenario is beyond the analysis of Agazzi's positions, although perhaps he would not exclude it.

References

Agazzi, Evandro. 1967a. Alcune osservazioni sul problema dell'intelligenza artificiale, *Rivista di Filosofia Neoscolastica,* 59: 1-34.

Agazzi, Evandro. 1967b. Simulazione del pensiero e intelligenza artificiale. In *L'uomo e la macchina* (Atti del XXI Congresso Nazionale di Filosofia). Torino: Edizioni di Filosofia, II: 45-48 and III: 109-110.

Agazzi, Evandro. 1978. Alcune osservazioni sul problema dell'intelligenza artificiale. In *Cibernetica e teoria dell'informazione,* ed. Paolo Aldo Rossi, 199-244. Brescia: Editrice La Scuola (reprint of Agazzi 1967°).

Agazzi, Evandro. 1981. Intentionality and Artificial Intelligence. *Epistemologia* IV, Special Issue ("The Mind-Body Problem"): 195-227.

Agazzi, Evandro. 1987. Intenzionalità e intelligenza artificiale. In *Informatica e pastorale*, ed. Pietro Prini, 87-105. Brescia: Morcelliana (translation of Agazzi 1981).

Agazzi, Evandro. 1991. Operazionalità e intenzionalità: l'anello mancante dell'intelligenza artificiale, in *Intelligenza naturale e intelligenza artificiale*, ed. Salvino Biolo, 1-13. Genova: Marietti.

Agazzi, Evandro. 2010. Operazionalità e intenzionalità: l'anello mancante dell'intelligenza artificiale. In *Natura umana, natura artificiale,* ed. Maria Cristina Amoretti, 63-77. Milano: Angeli (reprint of Agazzi, 1991).

Philosophy of Physics and Foundations of Quantum Mechanics

Gino Tarozzi

Abstract I discuss Agazzi's contribution to the philosophy of physics, a discipline introduced in Italy by his groundbreaking treatise of 1969. At a variance with the neopositivistic philosophy, dominant at the time, which appeared concentrated almost exclusively on a formal analysis of scientific languages, he showed that the philosophy of physics should discuss the logical foundations and the epistemological implications of physical theories, addressing also the issues of philosophy of nature, such as the reality and the structure of physical objects, the subject/object relationship, and the role of causality principle. Here I focus on the ontological question of the wave-particle duality, considered by the neopositivistic perspective of the standard interpretation as a metaphysical pseudoproblem. Agazzi, on the contrary, identified it as the new and fruitful experimental evidence from which quantum mechanics originated, as the theory that unified at the elementary level the classical concepts of matter and radiation. I argue that in this way he gave an essential contribution also to the debate on the foundations of quantum mechanics. His ideas, and in particular his appeal to the introduction of new and completely non classical concepts, inspired and influenced certain nonstandard realistic interpretations.

1 Introduction

Among the different areas of philosophy of science, and more generally of philosophical and scientific enquiry, on which Evandro Agazzi has extensively worked, often with innovative and original results, is no doubt philosophy of physics, a discipline introduced in Italy by his groundbreaking treatise (Agazzi 1969). That work soon established itself in the philosophical debate for the depth and

G. Tarozzi (✉)
Department of Basic Sciences and Foundations,
University of Urbino "Carlo Bo", Urbino, Italy
e-mail: gino.tarozzi@uniurb.it

M. Alai et al. (eds.), *Science Between Truth and Ethical Responsibility*,
DOI 10.1007/978-3-319-16369-7_8

robustness of the analysis carried out and for the perspective from which Agazzi interpreted the role and function of philosophy of science, which he believes cannot be limited to a formal analysis of scientific languages, as claimed by at that time dominant neopositivistic philosophy.

This conviction derived to him also from his studies on the history of physics in the late nineteenth and early twentieth century, culminating in the publication of the critical edition of Maxwell's *Treatise of electricity and magnetism* (Maxwell 1973). These studies had shown that the mentioned neopositivistic claim had been largely dismissed by the most advanced developments of modern physics. In fact, a merely formal approach to physical theories left completely unsolved the most problematic issues posed by the theories of new physics, for instance the problem of unobservable entities, like ether, absolute space and time in relativitivistic theories, and the problems of simultaneously unobservable entities, as position and momentum, of the uncontrollable disturbance of the measurement processes, and of the dual nature, both wave-like and particle-like, of micro-objects in quantum mechanics.

For the previous reasons Agazzi believes that the philosophy of physics, and more generally of science, should focus its attention on the study of the foundations of scientific theories, addressing also the issues of philosophy of nature, such as the reality and the structure of physical objects, the subject/object relationship in the measurement theory, and the role of the principle of causality. These issues were considered metaphysical, in the sense of meaningless in the light of the neo-empiricist philosophy. The latter was in fact modeled on the operationalistic methodology developed by the theories of physics of the early twentieth century, without a genuine critical confrontation with them and with their open problems.

As a matter of fact, the identification of the meaning of a concept with the procedures for its measurement, which led Einstein to the elimination of the non-measurable concepts of absolute space and time, was subsequently systematized by Bridgman through the operational definition of concepts in a new conception of science, which became a sort of benchmark for neo-positivist philosophy, which aimed to defend the same anti-metaphysical instance in philosophy.

To achieve this goal, it was necessary to find a linguistic analogue of the operationistic definition, i.e. a criterion through which meaningless propositions could be eliminated, like operationism had banished non-measurable concepts from physics, and later also from other sciences (think for instance of behaviourism in psychology, which required the abandonment of non-overt phenomena as consciousness, feeling and emotion). Neopositivists found a linguistic correlate of operationism with their criterion of verification, or verifiability, according to which the meaning of a statement is given by the method for its verification. Therefore, when this is not possible and, moreover, we are not dealing with an analytic and tautological sentence, we have a meaningless pseudo-proposition, and this was the case of philosophical principles.

2 Beyond Bohr's Complementarity: A Realist View on the Wave-Particle Duality

The most significant example of this itinerary in Agazzi's reflection is undoubtedly constituted by the question of the wave-particle duality, considered by the neopositivistic perspective of the orthodox interpretation as a metaphysical pseudoproblem devoid of meaning. At a variance with such a view, Agazzi on the grounds of his wide knowledge of the history of physics, considers the wave particle duality as the new and fruitful experimental evidence from which quantum mechanics originated, as a theory that unified at the elementary level the classical concepts of matter and radiation and their different descriptions given in classical physics respectively by Newtonian mechanics and Maxwell's electromagnetic theory. Moreover Agazzi stresses the ambiguous and unsatisfactory response to that question given by Bohr's principle of complementarity, which according to him could be viewed from two completely different points of view.

From the first viewpoint complementarity can be narrowly interpreted as synonymous with uncertainty in the sense that the wave properties and the particle properties of a micro-object are to be considered incompatible, as position and momentum, or as energy and time, according to the Heisenberg principle. This interpretation is due to the most definitely antirealistic variant of complementarity, shared by the theorists of the German school in Göttingen like Heisenberg and Jordan, and it establishes a form of unsurpassed incompatibility between the wave description and the particle description of atomic phenomena, an incompatibility that Pauli extended even to the logical and mathematical level.

From the second viewpoint complementarity is more properly considered as a synonym of the wave-particle duality, i.e. in terms of a real dilemma between one or the other of these different representations of physical reality, because on the one hand the use of both concepts seems to be required by an adequate explanation of the dual behavior of micro-objects, and on the other hand such descriptions appear experimentally incompatible within any one physical situation.

According to Agazzi this second interpretation, corresponding to the original formulation of Bohr's principle of complementarity, is however at the origin of the most serious epistemological and ontological contradictions, that would result from the assumption of a different nature of the same object in different situations.

This is one of the problems that led Agazzi to develop his concept of scientific realism, which constitutes the core of his philosophical reflection. The point is to establish the cognitive value of scientific theories, when it appears in question, or at least when it is doubtful whether an objective value can be attributed to our theories. In this regard Agazzi highlighted the existence of three different meanings that can be attributed to the term 'objectivity', i.e. "objectivity as *intersubjectivity*, as *invariance*, and as *correspondence*" (Agazzi 1969: 364). Through an effective and very detailed analysis (Agazzi 1969: 339–357, 1979) he showed that these three senses can be identified.

The conditions that make possible that intersubjectivity, invariance and correspondence (to objects) may coincide from a conceptual and epistemological point of view are essentially, at least for what concerns Agazzi's contribution to the philosophy of physics, three:

1. the operationistic foundation of scientific concepts and the fact that, although based on operations, these concepts cannot be reduced to a purely operationistic dimension;
2. the finding that the meaning of scientific terms is essentially contextual;
3. the fact that scientific objects, though made of properties established in an objective manner through operations, are not simple aggregates of those properties, but well defined structures of relationships between these properties.

As we shall see, these three points are strongly interconnected, especially for what concerns the scientific concepts expressed by the so-called theoretical terms, i.e. concepts of non directly observable entities, like that of the wave function of quantum mechanics. Let's look briefly at each of these three points of Agazzi' perspective.

1. Scientific theories are built on the basis of theoretical terms, but their purpose is to provide explanations for the facts of immediate experience, describable by empirical (or observational) terms. This raises the problem of how theoretical terms could keep a link with the empirical reality (Agazzi 1969: 138–139). A theoretical concept like 'electron' is a "theoretical construct around which we group many properties operationally definable" (ibid: 146). And it is precisely this operational aspect that allows theoretical terms to maintain a contact with experience, and so to have a physical meaning (Agazzi 1997: 49–65). However, these theoretical terms cannot be reduced directly to operational terms denoting sets of operations:

> (…) We would not even dream of saying that theoretical concepts can be reduced to operational concepts: who pretended this, would like those who thought of reducing the house to the bricks that make it up (Agazzi 1969: 147–148).

The various combinations of empirical (operational) terms produce constructs, the theoretical terms, which are no longer themselves directly operational. This point is relevant as it provides the philosophical basis for the attribution of a physical reality to the wave function, although it is not directly "observable" (or measurable), i.e., even if it cannot be defined directly in operational terms. Still, any theoretical entity must be associated with some detectable properties.

2. It follows that the meaning of theoretical terms is always a contextual meaning. As specified by Agazzi, this

> is not to say (…) that the physical meaning [of theoretical terms] comes from observational terms thanks to a context (…), but that it precisely comes from the context in which observational terms are present, but not alone, because the context actually consists even of all the mathematical and logical connections that link the various concepts, observational and not observational (Agazzi 1969: 148).

The context within which the theoretical terms assume a definite meaning is nothing but the theory in which they appear, and which they contribute to form. Only a theory as a whole can be empirically interpreted and then be put in relation with possible observations. This point is of the utmost importance for a realistic interpretation of the quantum mechanical wave function. According to Agazzi, most of the problems in the interpretation of quantum mechanics derive from the attempt to apply concepts of classical derivation to the objects of quantum theory. Because of this contextualistic nature of theoretical terms, the solution cannot be sought through some unusual combination of classic, corpuscular and wave-like concepts: "Not only we can, but we must say that *it is not the same particle*, *it is not the same wave* of which we speak in classical mechanics, because the contexts are different" (ibid: 271).

From this arises the need to explore really new concepts to overcome the difficulties associated with a realistic interpretation of quantum mechanics, concepts that are new "not only, as already happens, for the simple fact that they derive from the combination in a new way of classical concepts, but also for the fact that they replace all or part of these classical 'components' with something really new" (ibid: 285).

3. In relation to the third point mentioned above, we have already recalled that the scientific objects denoted (very often) by theoretical terms, are relational structures of operationally definable properties, but they cannot be completely reduced to such properties. This idea, which as we have seen is fundamental in Agazzi's thought, is closely linked to the contextual nature of theoretical terms:

> The object is always a structure, a structure of relationships, most of which can be the result of operations, but whose co-existence is not justified by any operation, although they should be objectively verifiable (ibid: 374).

Now, trying to reconstruct this structure is really the main task of scientific theories, for "the structure is not what lies underneath the experimental determinations and the characteristics objectifiable, but it is what is made of them: it is precisely the object." (ibid).

On the other hand, it is precisely this structure that makes the world what it is; and it is because of this structure that our theories, as attempts to reconstruct the structure, may be wrong, to the extent that the structure they describe is not that of world, or of the universe of objects that constitute the domain of the theory. This conception of the structural nature of theoretical entities will be crucial to understand what kind of reality can be attributed to the wave function.

We have already mentioned the considerable difficulties connected with the interpretation of this theoretical term and we also saw that Agazzi very pointedly identified the roots of the problem in an inadequate emancipation of the orthodox interpretation of quantum mechanics from classical concepts, deriving from a too narrow operationistic conception that completely identified the meaning of physical concepts with measuring operations performed on them.

Agazzi's rejection of a rigidly operationistic and phenomenalistic view of physical theories is indeed one of the main aspects of his philosophy of physics, based on a refusal of the distinction, dear to positivism and considered by Agazzi as completely artificial, between observational and theoretical concepts. Thus he proposes his original solution, according to which "a physical concept does not denote *a single* operation (or *a single* set of operations), but an *equivalence class* of operations (or sets of operations), which *originates* from an operation, but cannot be identified with it" (Agazzi 1969, p. 128). This solution, improved and formalized in Agazzi some years later (1976), will then be largely developed in a systematic logical approach to the foundations of physical theories (Dalla Chiara and Toraldo di Francia 1979).

A very persuasive example that it is not the use of the instrument to determine in a strict sense the meaning of the physical concept investigated by means of it, is provided from electromagnetic phenomena that can be, and often are, measured through mechanical instruments. Nevertheless this fact does not prevent, as stressed by Agazzi, the recognition of peculiarities of these phenomena and the introduction of new concepts such as charge, current and induction, all of which are clearly non-mechanical, although they are detectable on the basis of their mechanical effects recorded by mechanical instruments.

The explanation lies in his thesis of the contextualist nature of the meaning of physical concepts, according to which, as we saw earlier, a single physical concept is subject to different characterizations that depend on the different levels or contexts in which it appears. In this way, a concept such as "material particle" is seen in the context of classical physics as an object that has both a well defined position and well defined momentum; instead in the context of quantum physics, because of the uncertainty relations, it loses the simultaneous possession of such properties, but it takes on new properties, like spin.

According to his contextualistic perspective, perhaps even the conceptual difficulties of the complementary interpretation of the wave-particle duality could then be solved by assuming that the classical concepts, considered at a formal level, appear as the elements of a semantic combination in which the original contradiction disappears, because it is not formally linked to the concepts themselves, but only to their classic denotation.

3 A New Non-classical Concept for a Testable Realist Interpretation of the Wave Function

Agazzi believed, however, that this view cannot provide a definitive solution to the ontological problem of the nature of micro-objects. He pointed out, with extraordinary intuition, the need to introduce in microphysics concepts that are new not only because they represent the result of a new combination of classical concepts, but also because they are able to replace their classic components with something new. As he emphasized again some years later:

> Only by inventing some new concepts, that is, new in this fundamental sense, we could possibly overcome the present uneasy state of affairs, which is not related to the regret of losing the old concepts, but to the lack of new concepts capable of adequately replacing them (Agazzi 1988).

It seems really surprising that in the same year in which Agazzi emphasized the need of a new philosophical concept to solve the problem of wave particle duality, Franco Selleri proposed a realistic interpretation of the wave function of quantum mechanics based on the introduction of such a new concept. This was the concept of empty wave, later replaced by that of quantum wave, which can be considered as a sort of synthesis with respect to the three different concepts of duality between waves and particles that had been proposed by the main founders of quantum theory.

That notion was reminiscent in the first place of Einstein's point of view: the founder of relativity, despite having reintroduced in physics a corpuscular theory of radiation by his famous hypothesis of light quanta, believed that interference and diffraction phenomena were not explicable on the basis of a purely corpuscular theory, but required also a wave to accompany and guide the quanta in their motion. But the fact that all the energy was concentrated in the quantum, and that the wave associated with it was consequently devoid of this fundamental property, led Einstein to introduce for such a wave the term '*Gespensterfelder*' (ghost field).

When de Broglie, with his wave theory of matter, then extended the duality from radiation to matter, in an attempt to overcome the contradiction due to the "existence" of an entities without the properties that characterize any other physical object, like his pilot waves, he found no other way to ensure their reality, than attributing to them an extremely small portion of energy, almost entirely localized in the corpuscles (de Broglie 1927). But since no one has been able so far to reveal this very small amount of energy, the typical objection to de Broglie's waves is that they are metaphysical rather than physical.

A third conception bearing significant similarities to the preceding one, but also unable to achieve an adequate empowerment towards classical notions, was introduced by Bohr, Kramers and Slater in their attempt to reformulate a purely undulatory theory of radiation in opposition to the corpuscular hypothesis of light quanta (Bohr et al. 1924). This was the concept of *virtual wave*, to which these authors attributed the fundamental characteristic of carrying neither energy nor momentum and of producing only "stimulated transitions" in the atoms the wave interacted with. Atomic transitions would thus allegedly occur in open violation of the laws of conservation of such physical quantities, since any given atom could pass from one energy level to another, without any energy exchange with the electromagnetic field. The concept of virtual wave, however, was soon abandoned as a result of the experiments by Bothe and Geiger (1924) and by Compton and Simon (1925), who provided a decisive confirmation of Einstein's hypothesis of the corpuscular nature of radiation.

None of the above authors therefore, while elaborating concepts quite similar to that of quantum wave, succeeded in formulating a really new concept: neither Einstein, who having contributed more than any other to the definition of the concept of energy, and turned it into the central notion of physics modern, found contradictory to assert the existence of objects without this fundamental property; nor de Broglie, who after establishing the wave theory of matter, could not conceive waves without energy and momentum, and proposed to ascribe them an uncontrollable amount of the previous properties; nor Bohr, Kramers and Slater, whose virtual waves had been introduced as an alternative to the corpuscular hypothesis of light quanta, and who after the failure of their purely undulatory theory of radiation prudently replaced it with Born's probabilistic and strictly corpuscular interpretation of the wave function of 1926, before Bohr's dualistic solution of the complementarity principle.

Selleri's paper of 1969 already contained the basic elements of the decisive conceptual turning point in the interpretations of the wave particle duality, of which Agazzi had clearly highlighted the need in the same year.

Starting from Einstein's and de Broglie's realistic conception that waves and particles exist objectively, and from the fact that experiments show beyond any reasonable doubt that all the energy, momentum, angular momentum and charge are closely associated with the particle, Selleri asked what could be an entity existing without being associated with any directly observable property. He considered unsatisfactory de Broglie's response that all physical properties would be primarily associated with the particle, but that a small fraction of them, so small as to have escaped all possible observations, is associated with the wave. He then proposed that "even if devoid of any physical quantity associated with it", and therefore not directly observable, "the wave function can still give rise to observable physical phenomena" (Selleri 1971: 398). He pointed out that we do not measure energies, momenta, or similar physical quantities only, but also probabilities, as in the case, for example, of the average life of an unstable physical system. According to Selleri, the wave function could therefore "acquire reality independently of the associated particles, if it could give rise to changes in the of transition probabilities of the system it interacts with" (ibid.).

On the basis of this new idea of a non-classical wave, he proposed the first version of his experiment for detecting the physical properties of quantum waves. To this end he considered a piece of matter composed of unstable entities, such as nuclei, atoms or excited molecules, traversed by a flux of neutrinos. The experiment then consisted in measuring the average life of such entities under these conditions, and comparing them with the average life of the same entities in the absence of any particle flux. If any difference was observed, the only logical explanation, according to Selleri, was that such difference was due to the action of the wave function, since neutrinos are particle that interact very weakly and only few of them, in the best case, can interact with the piece of matter (ibid.).

Some years later Selleri improved his original idea by the experiment shown in Fig. 1 (Selleri 1982): instead of a flux of neutrinos we have photons emitted by a Laser, and we have no longer a piece of matter composed of unstable entities, but a laser gain tube LGT. Moreover we have two detectors DT and DR and a semireflecting mirror SM. The latter behaves in the same way as the double slit: the particle is propagated in one direction only, depending on whether it has been transmitted or reflected by SM, whereas the wave, according to Selleri's hypothesis, is both transmitted and reflected.

Selleri proposed to focus on the cases in which DR, located along the reflected beam, detects a photon: this means that in the transmitted beam only the quantum

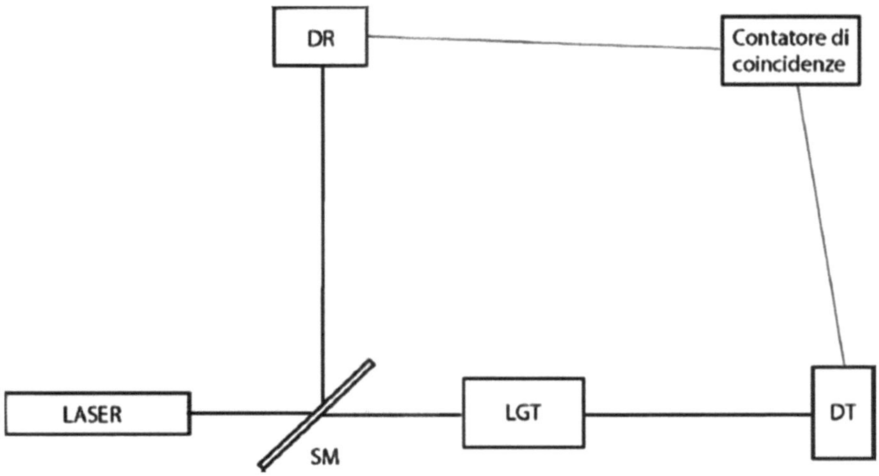

Fig. 1 Selleri's original experiment on quantum waves

wave is present. According to Selleri's hypothesis, however, the wave can reveal its presence by generating the stimulated emission of photons; these in turn can be detected by DT) after passing through the laser gain tube LGT, whose molecules are at an excited level corresponding to the wavelength of the incident wave. In this way the coincidences between the detection of DT and DR would reveal the propagation of a quantum wave, transmitted by SM. The space-time propagation of these entities could be studied by verifying whether the DT–DR coincidences disappear when an obstacle is placed in the transmitted beam in front of LGT. According to Selleri, a positive result of this experiment would have shown that "something having neither energy nor momentum but that can produce transition of probabilities propagates in space and time." (ibid.)

Louis de Broglie, welcomed and endorsed Selleri's idea as an important attempt

> to obtain a more satisfactory interpretation of wave mechanics than that presently adopted, confirming the idea that guided me when in 1923-24 I proposed the basic conceptions of wave mechanics (de Broglie 1969).

However framed this new idea within his old classical conception of the pilot wave: "The experiment you propose to prove the existence of this wave Ψ will be of great interest to prove the existence of this very weak (très faible) wave, which carries the particles" (ibid.).

Selleri's realist interpretation was received with great favor a decade later by another great opponent of the Copenhagen interpretation, the philosopher Karl Popper, who joined to it unconditionally, abandoning its original statistical and closely particle-like interpretation:

> Franco Selleri has suggested, continuing the work of de Broglie, that waves without particles may exist. The consequences (of this proposal) would seem to be revolutionary … They would establish in place of the "complementary" character of particles and waves (wavicles) the interaction of two kinds of real objects: waves and particles (Popper 1985).

But neither de Broglie nor Popper, while expressing the greatest interest and appreciation for the hypothesis of quantum waves, grasped the essential novelty of this concept, which instead was perfectly understood by Agazzi, since it was an instance of the *radically new* concepts of which no one before him had so lucidly pointed out the need in order to resolve the contradiction arising from the wave particle duality.

In his contribution to a volume of Italian studies on the foundations and philosophy of physics (Agazzi 1988), Agazzi noted that this conceptually new hypothesis of quantum waves was important for its refusal of the symmetrical nature of duality, and different from the classical approach of de Broglie:

> The essential novelty of this concept is represented by the acceptance of de Broglie's realist interpretation of the wave-particle duality, but not of the symmetrical nature of this dualism. In Selleri's approach both particles and waves are simultaneously real, but the latter can be characterized only by its relations with the particles, i.e. by the observables properties of producing interference and stimulated emission. Such a possibility implies an ontological priority of particles over waves, which therefore belong to a weaker level of physical reality, containing objects which are sensible carriers of exclusively relational predicates (Agazzi 1988: 73).

Several experiments have been proposed for this new realistic interpretation of the wave function, whose interest, as pointed out by Agazzi, is twofold: on the one hand they allow to test this realist interpretation against the Copenhagen one, experimentally discriminating between two different philosophical interpretation of a given physical theory, an opportunity without a precedent in the history of science. On the other hand, these experiment could also provide the opportunity to control the well known axiom of the reduction of the wave function. Concerning this last point Agazzi noticed that by using the properties of quantum waves it seemed possible to ascertain the paths followed by a photon within an interferomentric device, revealing at the same time the interference pattern in the distribution of their recordings, a possibility utterly excluded by the reduction postulate.

Thus it seemed possible to establish an important connection between the wave-particle duality and the other fundamental problem of quantum measurement. Unfortunately, however, none of the experiments carried out so far has revealed the assumed properties of quantum waves, nor refuted the postulate of reduction.

4 Realism of Properties, Realism of Entities and Their Role in Microphysics

Already during my university studies between 1973 and 1977 I came in contact with the work of Evandro Agazzi, in particular with his work in the philosophy of physics (Agazzi 1969). I remained deeply impressed by his refusal of the identification of philosophy of science with epistemology, and by his consequent belief that the former cannot be limited to matters concerning the form and language of scientific theories, according to the neo-empiricist perspective, but it should also address issues related to their contents, namely problems of philosophy of nature. He held that philosophy of physics must be identified with a survey on the fundamentals, in the sense both of an enquiry in the epistemological foundations of physical theories and an analysis of their philosophical implications, and this soon became the perspective guiding my research on the foundations of quantum mechanics.

In addition to this fundamental methodological lesson on the need to conduct the research in philosophy of science as a study of the foundations of scientific theories, Agazzi influenced my philosophical perspective in an even more direct way, by his conception of scientific realism, and its specific application to the fundamental concept of theoretical quantum mechanics, that of the wave function, which we have discussed so far.

Unlike the neo positivists Agazzi vindicated a substantial autonomy of the philosophical inquiry with respect to scientific research, and as we saw from his discussion of the principle of complementarity he had an approach to scientific theories very different from neopositivism. However, to this philosophy he recognized the merit of not having upheld the cognitive value and the intersubjectivity conception of science:

> Neopostivist epistemology, despite having been deeply influenced by Mach's thought, has come to accept more or less explicitly a realist view of science. We do not care to discuss here how consistently this could happen: it is sufficient to note that such an outcome was imposed by the cultural program of the entire movement, which was characterized by a view of science as the only authentic source of knowledge (Agazzi 1985: 173).

Besides,

> the obsession with which neo-empiricism tried to impose absolute fidelity to experience, and the reducibility to it of the very theoretical components of science, can also be seen as an effort to ensure a solid connection with reality (ibid.).

Moreover we know that the main theses of traditional philosophy, including those of realism, had been refuted by the logical empiricists as meaningless, as generally corresponding to propositions of existential content that are not empirical and for which there is no method for determining their truth value.

Instead research on the EPR paradox showed the possibility of supporting a completely different point of view, by showing a clear form of logical incompatibility (which through Bell's theorem could be turned into an experimental discrepancy) between the quantum description given by some particular state vectors, the so called entangled states, and a very reasonable principle of reality, which identified scientific objectivity with predictability with certainty, considered as a sufficient condition for physical reality (Einstein et al. 1935). It was thus shown uncontroversially that such a realistic principle could meet those requirements of verifiability that the neo positivists believed to be completely inapplicable to philosophical propositions. Thus it became clear that the acceptance of confirmability as a criterion of meaning (but not of scientificity, since scientificity is subject to the stronger requirement of Popper's falsifiability) allows to reformulate some of the main metaphysical theses in terms of philosophical principles endowed with factual meaning. In a nutshell, according to my point of view, scientific propositions must be falsifiable, whereas philosophical ones can only be disconfirmable.

It has also be shown that there are other formulations of realism endowed with meaning. The first was discussed one year after EPR by Carnap, who analyzed a realistic hypothesis proposed by Lewis in terms of the proposition: "If all minds disappeared from the universe, the stars would still go in their courses" (Carnap 1936–1937). Moreover he highlighted that this statement satisfied the most stringent requirements of factual significance, since it is controllable, albeit incompletely.

Other non-metaphysical variants of the reality principle include various probabilistic generalizations of the EPR criterion: for instance, while the original EPR criterion required predictability with certainty (a strong idealization with respect to actual physical situations), I suggested to replace it with predictability with a high degree of inductive probability (Tarozzi 1979); later on, together with Selleri, I modified it by considering the a priori probabilities themselves as real properties (Selleri and Tarozzi 1983).

The common feature of these different realistic principles (EPR, probabilistic EPR, Carnap) is the attribution of reality not to the object but to its predictable properties. This agrees with the logical empiricist refutation—anticipated by Kant's critique of existence as a predicate—of the identification of reality with a (further) property of a physical object, an error that persisted for a long time in the debate on the EPR paradox.

Nonetheless, the shift of reality from the object to its predictable properties allows to preserve the notion of independence from the observer (and from his mind or consciousness), which is at the basis of metaphysical realism. The latter, in fact, as defined by Hume, is the doctrine that reality is what would exist, though we and every sensible creature were absent or annihilated. There is a perfect continuity between metaphysical and empirical realism, and the main difference is that the latter, considering the predictability through our best theories as a guarantee for reality, appears to be based on science, and in our case on physics, whereas, according to the former it is science that is to be based on realism.

It was Agazzi's analysis of the relationship between scientific objectivity and reality, in particular his claim that the latter includes the former (i.e., that being objective takes more than just being real) to be seminal for my investigations, since it enabled me to understand the EPR principle of physical reality in the new light, as I have explained earlier.

He however rightly points out a kind of opposition between this realism of properties or attributes and his realism of objects or entities, and since many years ago and up to the present (Agazzi 2014) he advised me to supplement the reality of the properties, which seems to him rather dim, with that of the object. His exhortation was one of the reasons that led me to investigate, after the EPR problem, also the possibility of an alternative realistic interpretation of the wave function, and to design experiments to test it.

In any case, I feel that empirical realism of the properties and scientific realism of the objects are both fundamental and indispensable issues to any scientific theory; and my dissatisfaction with quantum mechanics stems from the fact that this theory seemed rejects the attribution of physical reality both to its predictable properties and to its basic concepts.

But a recent ideal experiment, which might be easily converted into a real experiment, seems to show that this double anti realistic claim of the standard interpretation is no longer sustainable, and that either Agazzi's realism of theoretical entities, and or empirical realism of (predictable) properties correspond to an essential condition in the interpretation of quantum mechanics.

The experiment aims to assess the possibility that quantum waves produce correlations at distance of the EPR type, identifying in this way a new perspective that would establish a deep and hitherto unsuspected relationship between the two previously discussed ways of interpreting realistically quantum mechanics.

In fact, consider a pair of photons produced by a non-linear crystal, which propagate in the device illustrated in Fig. 2. Any photon can be detected by the two "near" detectors (D1 and D2), which are placed after a shorter path, or by the two "far" detectors (D3 and D4), placed after a longer path. If we do not take into account all the cases in which both photons are detected by D1 or D2, the physical situation will be described by the state vector

$$|\psi\rangle = \frac{1}{\sqrt{2}}[|1\rangle|4\rangle + |2\rangle|3\rangle]$$

that presents some formal analogies with an entangled state, but is actually an ordinary superposition state.

According to the previous description, if detector D1 clicks, we can predict with certainty that D4 will click, and, if D2 clicks, we can predict that D3 will click. In this case the observed correlations can be considered as a consequence of a wave-like behavior.

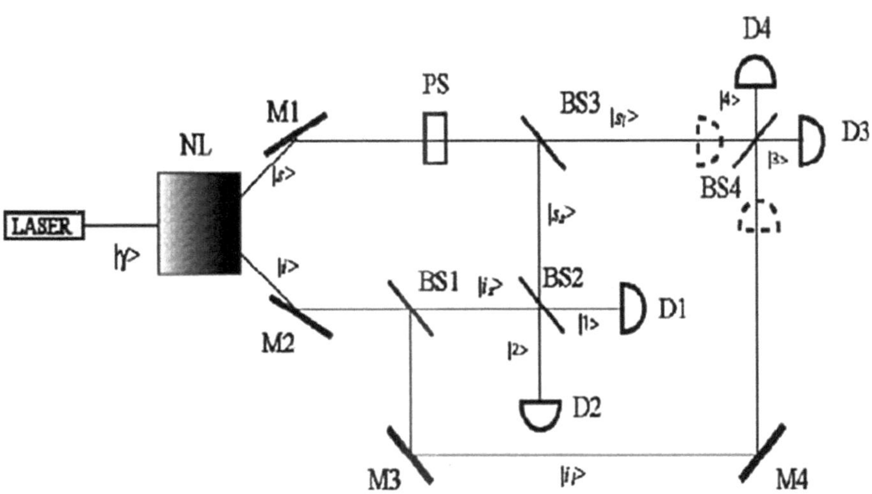

Fig. 2 Another experiment discriminating between the realistic interpretation of the wave function and the reality of the predictable properties

It is interesting, now, to see what happens if we displace detectors D3 and D4 to a position before BS4 (broken lines), once a photon has already been detected by D1 or D2. We have then a delayed-choice experiment (Wheeler 1978), but with an important difference: in our case, an event has already occurred (D1 or D2 has already clicked) before the choice. In this case, we can obtain information about which photon has been detected by D1 or D2 and which photon has been detected by D3 or D4. Now, although we can know which photon has been detected by which detector and therefore the paths they follow, we cannot predict whether detector D3 or detector D4 will reveal the photon after either detector D1 or detector D2 has clicked.

We also observe that, on account of the first interference (by BS2) and of the superposition of the two components of the i-photon and of the superposition of the two components of the s-photons, the latter situation (when detectors D3 and D4 are placed before BS4) is not the classical situation that would arise if both BS2 and BS4 were removed. In this case, if D1 clicks, we know with certainty that the i-photon has been detected and that the s-photon (if not detected by D2) will reach D3. On the other hand, if D2 clicks, we know with certainty that the s-photon has been detected and that the i-photon (if not detected by D1) will reach D4. Our proposed experiment differs from others designed to test the complementarity principle, because in those experiments, in general, many runs are needed in order to obtain an interference (wave-like) pattern at the detectors. In our experiment, on the contrary, the effect of the wave-like pattern is shown in single runs, hence for individual systems.

Now, if we are able to predict something different and new (i.e., whether D3 or D4 will click) when we have wave-like behaviour relative to the predictions allowed by the particle-like behaviour, we see no reason for not attributing an ontological reality to the wave. Still, it is clear that they cannot have the same *kind* or *degree* of reality as particles, which are well localized and possess directly measurable properties. On the contrary, measuring directly waves or quantum states is intrinsically impossible: the existence of these objects can only be inferred.

5 Conclusions

This weakness represents also the strength of our point of view, which allows us to highlight the possibility that an entity could exist without possessing intrinsic properties. For, as we have seen, the wave-like properties of the two photons depend strongly on the experimental context. This means that the decisive reason for which it is not possible to directly detect the quantum waves is that they would belong to a level of reality inherently relational, as it was clearly underlined both by Selleri and Agazzi.

For the orthodox interpretation, which completely denies the physical reality of the wave function, this relational character would be peculiar to atomic particles.

To a certain extent this is true, given that in an experiment of complementary type, such as that considered above, what we reveal depends on our experimental arrangement. However, the very act of detection is by definition the detection of a particle or the recording of an event (and this result can also be stored and communicated), and this explains the asymmetry between ontological recordable events and relational wave-like entities, the assumption that Agazzi, as we saw in Sect. 3, considered as the truly new element in the conception of the quantum wave.

Another strong reason to ascribe physical reality to quantum waves is that between wave-like and particle-wave behavior, there is a *continuum* of possible cases, as it has been shown by the existence of the so called smooth complementarity, i.e. the possibility of a smooth variation between wave-like and particle-like behaviour and consequently of infinite intermediate possibilities between the two extreme alternatives (Mittelstaedt et al. 1987). Obviously, this runs against Bohr's idea that complementarity is a sharp relation in which we have either the wave or the particle.

Our proposed experiment seems to show that perhaps one could distinguish experimentally between EPR's realism of properties and the realism of theoretical entities: the presence of correlations between remote detections of photons would highlight the physical reality of the quantum wave, violating the realism of EPR and confirming the predictions of quantum mechanics, as happened with the experimental controls of Bell's theorem, while the absence of correlations would disprove quantum mechanics in favor of EPR's empirical realism.

The former result, which is certainly the more probable (although we cannot rule out the second one without before running our experiment) would be a direct experimental confirmation of Agazzi's realism of entities, and of the need to find a counterpart in physical reality for fundamental theoretical concepts.

A new feature of our experiment is also that it does no longer discriminate between a realistic and an antirealistic (Copenhagen like) interpretation, but between two different realistic interpretations of quantum mechanics. In my opinion this represents a decisive confirmation of the necessity of a realist interpretation of scientific theories, which Agazzi has always considered an epistemological assumption indispensable to any serious philosophical inquiry.

References

Agazzi E. 1969. *Temi e problemi di filosofia della fisica.* Milano: Manfredi. 2nd ed. Roma: Abete 1974.

Agazzi E. 1976. The Concept of Empirical Data. Proposal of an Intensional Semantics of Empirical Theories". In *Formal Methods in the Methodology of Empirical Sciences,* ed. Marian Przelecki et al. Reidel: Dordrecht.

Agazzi E. 1979. Proposta per una nuova caratterizzazione dell'oggettività scientifica. In *Problemi di epistemologia contemporanea*, ed. Evandro Agazzi. Lanciano: Itinerari, 31-61.

Agazzi E. 1985. La questione del realismo scientifico. In *Scienza e filosofia. Saggi in onore di Ludovico Geymonat*, ed. Corrado Mangione. Garzanti: Milano.

Agazzi E. 1988. Waves, Particles and Complementarity. In *The Nature of Quantum Paradoxes. Italian Studies in the Foundations and Philosophy of Modern Physics*, ed. Gino Tarozzi, Alwin van der Merwe. Dordrecht: Kluwer.

Agazzi E. 1997. On the Criteria for Establishing the Ontological Status of Different Entities. In *Realism and Quantum Physics*. Ed. Evandro Agazzi. Amsterdam-Atlanta: Rodopi.

Agazzi E. 2014. Realism of Properties and Realism of Entities. In *Gino Tarozzi Philosopher of Physics*, ed. Vincenzo Fano: 22-32. Milano: Angeli.

Bohr N., Kramers H. A., Slater J. C. 1924. Ueber die Quantentheorie der Strahlung. *Z. Phys*. 24: 69–87.

Bothe W., Geiger H. 1924. Ein Weg zur experimentellen Nachprüfung der Theorie von Bohr, Kramers and Slater. Z. Phys. 26: 44.

Carnap R.1936-1937. Testability and Meaning. *Philosophy of science*, III: 419-471, IV: 1-40.

Compton A. H., Simon A. W. 1925, Directed Quanta of Scattered x–rays. *Phys. Rev*. 26: 289-299.

Dalla Chiara M.L., Toraldo di Francia G. 1979. A Formal Analysis of Physical Theories. In *Problems in the Foundations of Physics*, ed. Giuliano Toraldo di Francia. North Holland: Amsterdam.

de Broglie. L. 1927. La structure de la matiére et du royonnement et la mécanique ondulatoire. *Comptes rendus de l'Academie des Sciences* 184.

de Broglie L. 1969. *Letter to Franco Selleri* (dated April 11).

Einstein A., Podolsky B., Rosen N. 1935. Can Quantum-Mechanical Description of Reality Be Considered Complete? *Phys. Rev*. 47.

Maxwell J.C. 1973. *Trattato di elettricità e magnetismo*, ed. E. Agazzi. Torino: Utet.

Mittelstaedt P., Prieur A., Schieder R. 1987. Unsharp Particle-wave Duality in a Photon Split-beam Experiment. *Found. Phys*. 17: 891–903.

Popper K. 1985. Realism in Quantum Mechanics and a New Version of the EPR Experiment. In Tarozzi G., van der Merwe A. (eds.) 1985: 3-25.

Selleri F. 1969. On the Wave Function of Quantum Mechanics. *Lett. Nuovo Cimento*.

Selleri F. 1971, "Realism and the wave function of quantum mechanics", in Foundations of Quantum Mechanics. Proceedings S.I.F., Course IL, ed. by d'Espagnat, Academic Press, New York.

Selleri F. 1982. On the Direct Observability of Quantum Waves. *Found. Phys*. 12: 1087-1112.

Selleri F., Tarozzi G. 1983. A Probabilistic Generalization of the Concept of Physical Reality. Speculations in Science and Philosophy 6: 55-64.

Tarozzi G. 1979. The Conceptual Development of the E.P.R. Argument. *Mem. Acc. Naz. Sci. Lett. Arti Modena* XXI: 353-373.

Wheeler J. A. 1978. The "Past" and the "Delayed-choice" Double-slit Experiment. In *Mathematical Foundations of Quantum Theory*, ed. A. R. Marlow: 9-48. New York: Academic Press.

From Physics to Sociology

Giuliano Di Bernardo

Abstract Evandro Agazzi's original proposal of characterizing science through only two requirements, objectivity and rigour, amounts to advocating a concept of science that aims at being general but at the same time admitting a distinction between science and non-science and, in addition, capable of convincingly applying to different sciences. This result he has attained by elaborating an "analogical" concept of science, in the sense that the basic requirements of objectivity and rigour are characterized and satisfied not according to a unique model, but in articulated specific ways from science to science. Therefore, reductionism is the opposite of scientificity, contrary to what has been maintained by several scholars. The social sciences are the domain in which Agazzi has concretely put to test this claim: they do not satisfy many features of the paradigmatic "exact sciences" but, instead of saying that they are not sciences, or that they are sciences but totally at variance with the exact sciences, he has discussed how they have a specific way of being sciences. What Agazzi has discussed in general terms, is analyzed in some details in the present contribution, where physics, biology and sociology are considered in their common elements and in the specificity of their single features, that entail epistemological as well as ontological differences.

1 Premise

Evandro Agazzi's contribution to the epistemology of the social sciences is explicitly contained only in a couple of papers, that are, however, particularly significant. Actually they are in a way a direct application of the most specific traits of his philosophy of science and exhibit a convincing exemplification of his fundamental thesis of the "analogical" nature of the concept of science itself, for which the only

G. Di Bernardo (✉)
Faculty of Sociology, University of Trento, Trento, Italy
e-mail: giuliano.dibernardo15@gmail.com

M. Alai et al. (eds.), *Science Between Truth and Ethical Responsibility*,
DOI 10.1007/978-3-319-16369-7_9

defining characteristics are objectivity and rigor. In his book on the philosophy of physics (1969) he had presented all the essential lines of his theory of scientific objectivity that remained constant during the whole of his speculation, including the rejection of any "ontological" reductionism (that would restrict science only to the study of certain subject matters) and "methodological" reductionism (that would recognize as scientific only a discourse strictly using the methods of the so-called exact sciences). The challenge was that of showing how disciplines different from the exact natural sciences could be objective and rigorous, and Agazzi took over this task first regarding psychology, in a famous paper still often referred to today (Agazzi 1976).

Two years later he addressed the same issue regarding the "*sciences humaines*" (focusing in particular on the social sciences) in the opening lecture of a conference of the International Academy of Philosophy of Science that took place at the University of Trento, and that appeared in its proceeding the following year (Agazzi 1979). In this seminal paper he first outlines his doctrine of scientific objectivity consisting in the fact that any science considers reality only from a specific restricted "point of view", expressed through specific predicates that refer only to a selected set of attributes of reality, that are accessible by means of appropriate operations. This is the ground for rejecting as anti-scientific any reductionism. In the second part of the paper Agazzi analyzes a list of features that are present in the social sciences and are often taken as insuperable objections against the possibility of considering these disciplines as really scientific. He argues that such features correctly reflect certain fundamental attributes of human actions, so that they are constitutive of the specific "point of view" of the social sciences and, therefore, they are the ground for their objectivity.

An equally important paper was presented by Agazzi about 30 years later, again at a conference of the International Academy of Philosophy of Science at a department of the University of Trento in Rovereto, and whose proceedings appeared in 2010 (Agazzi 2010). On that occasion he addressed the issue of replacing the rough pretension of attaining a unity between the natural and the social sciences through reductionism, by a much more elaborated interdisciplinary approach based on general systems theory and the treatment of complexity (relying on his investigations on these two topics that he had performed in the meanwhile). In this approach natural and social sciences appear "interrelated" as dealing each with different subsystems of a global complex system.

It is not by chance that these two outstanding contributions were made public at conferences held at the University of Trento since they reflect the long collaboration between Agazzi and myself that has lasted during four decades and, in particular, has led me (as professor of philosophy of science at the Faculty of Sociology of that University) to organize the two above mentioned meetings of the International Academy of Philosophy of Science (of which Agazzi was the President and to which I was also admitted). Those, however, were only two salient events of a much broader and deeper personal and scientific cooperation: Agazzi had at the University of Genoa, during the 1970s, a group of disciples working in logic and epistemology, while I had a similar group working with me at Trento, and we used to meet alternatively in the one or the other institution at regular time intervals for joint discussions and exchange of results. This interchange was very fruitful and, for example, marked my personal intellectual evolution from a stage in which I was especially concentrated on the logical-linguistic aspect of the social sciences (Di Bernardo 1972), to a broader epistemological approach, and finally to the elaboration of a personal view in which the natural and the social sciences are ordered according to a line of continuity and differentiation that recognizes their links as well as their irreducible specificity. I gladly recognize in this personal evolution the background of Agazzi's fundamental ideas and approaches, a significant testimony of which was the fact that the first work published in the well-known book series "Epistemologia" edited by Agazzi with the published Franco Angeli was my book (Di Bernardo 1979).

For the above sketched reasons I think that a perhaps more interesting way of underlining Agazzi's contribution to the philosophy of the social sciences would be (rather than a summary of his papers) the presentation of an example of how his views have inspired the creative work of another scholar (in this case of myself).

2 From Physics to Biology

The aim of this paper is to investigate the foundations of the human and social sciences. Two interpretations of social reality, those of positivism and hermeneutics (born as a reaction to positivism), confronted each other. However, in recent times certain natural sciences—the neurosciences—have claimed the right to investigate consciousness (primary and of higher order), intentionality, the self (individual and collective), and free will. They have thus occupied domains that traditionally pertained to philosophy and had been assumed as the foundations of the social sciences. This incursion by the natural sciences into the social sciences has had consequences in the epistemological domain as well.

I assume that physics is the prototype of the natural sciences and that sociology is the prototype of the social sciences. I shall seek to show not only their shared bases but also and especially their specificities. In doing so, I shall consider biology to be a science intermediate between physics and sociology, in that it possesses features that can be related to both the former and the latter. The transition from physics to biology will proceed upwards: at every step the specific nature of individual sciences will emerge. As a consequence, any type of reductionism will be avoided. Particular importance will be given to the concept of 'reality' in physics, biology and sociology. It will thus be seen how the ontology of the social (social being) can be introduced into the ontology of the external world (of physics and biology). In this regard, I shall show that the reality of physics and biology is independent of the observer (it is ontologically objective) whilst the reality of sociology is dependent on the observer (it is epistemologically objective and normative). I shall examine the certain and uncontroversial foundations of physics and biology, and on these foundations I shall base sociology as a science, of which I shall provide a preliminary epistemological analysis.

Alternatively, sociology can be founded independently of physics and biology, but this will not be the route followed here.

I begin by describing the essential features of physics at the origins of modern science. The scientific revolution of the sixteenth and seventeenth centuries, whose protagonists were Galileo, Descartes and Newton, today represents the beginning of what we call ‘science’. At that time, science coincided with mechanics and astronomy. Galileo, in particular, was convinced that mechanics was the supreme science, the foundation and origin of all the sciences. Since mathematics performs an essential role in mechanics, not surprisingly it was a decisive and essential component of Galileo’s conception of science. Famous in this regard is the definition that Galileo gave of ‘nature’ in *The Assayer*: “The book of nature cannot be understood unless one first understands the language and recognises the characters by which it is written. It is written in a mathematical language, and its characters are triangles, circles, and other geometric figures, without which means it is humanly impossible to understand a single word of it; without them it is like wandering hopelessly through a dark labyrinth”.

Galileo’s mechanics is a science formed by laws expressible in the language of mathematics. Mathematics is therefore its necessary and sufficient condition.

Physics (mechanics and astronomy) becomes the archetype, the model of science in general. Every discipline that aspires to becoming a science must, like physics, have natural laws, and these must be mathematizable.

The fundamental concepts of physics are the following: observation, experimentation, laws, theories formed of laws, mathematization, closed world, determinism, causality, reductionism.

After physics, however, other disciplines were born, such as cosmology, geology, psychology, linguistics, philology and history. The first problem that arose in their regard was establishing whether they were sciences in the same way as physics was a science. Some philosophers, mainly of German culture, broadened the concept of ‘science’ to encompass the social and historical sciences as well. Thus a distinction was made between natural sciences and human sciences, and the task was to draw a demarcation line between the former and the latter.

Opposed to this distinction was logical positivism, which maintained that only the model of science elaborated by Galileo and Newton could be the basis for a discipline which aspired to becoming a science. The positivists believed that the social sciences were still in their infancy and that they could develop by adopting the models used by the most advanced sciences, like mathematical physics. This entailed that the social sciences must have general laws, nomological models of explanation and prediction, and axiomatic theories. It was precisely the transfer of the hypothetical-deductive method from the natural sciences to the social sciences that gave rise to difficulties which severely strained the positivist theory and fuelled criticisms against it.

In identifying the relationship between physics and sociology, both the positivists and their critics ignored biology, as if that science were an embarrassment to both of them. Yet biology—at least as its nature and method have been recently formulated—can shed a great deal of light on the concept of ‘science’ from physics to sociology.

Today biology is a science which enjoys equal dignity with physics. The theory of evolution, genetics, and molecular biology have definitively dispelled doubts concerning its scientificity. However, before achieving its current status, biology had to overcome numerous difficulties.

Since antiquity, philosophers had sought to define life and the characteristics of living beings, and they had put forward the most disparate solutions. Descartes, for example, proposed that the problem of life could be solved by cancelling it: a living organism, he maintained, is nothing other than a machine. Philosophers with backgrounds in mathematics, logic or physics supported Descartes and sought to erase the difference between animate and inanimate nature.

The majority of naturalists, however, were reluctant to accept this position, and in order to vindicate the autonomy of living beings they concocted the concept of 'vital force': just as the planets and the stars were controlled by the invisible force which Newton called the *force of gravity*, so the motions of living organisms were controlled by an invisible force called the *vital force*. Those who believed in the existence of this force were termed 'vitalists'.

Vitalism immediately became popular, and it represented a qualified reaction to Cartesian mechanism. Among its numerous proponents were H. Bergson (1859–1941) and H. Driesch (1867–1941), who sought, authoritatively but in vain, to demonstrate the existence of a vital force. Lately it has been genetics and molecular biology which have definitively confuted that hypothesis.

Teleology was another obstacle that biology had to overcome before achieving the same scientific status as physics. Vitalism disappeared from biology when it was clearly understood that the experiments intended to demonstrate its existence in reality had failed to do so. But eliminating teleology proved more difficult, mainly because the term 'teleological' was applied to diverse natural phenomena. Thus there arose the need to examine the biological and philosophical literature and find a way to classify the term's different meanings.

E. Mayr demonstrated that four of the five phenomena traditionally considered to be teleological could be entirely explained by science, whilst the fifth phenomenon, cosmic teleology, did not exist.

The elimination of vitalism and finalism from biology was a first important step towards its foundation as a science with the same dignity as physics.

A second and equally important step was the demonstration that it was impossible to apply certain fundamental principles of physics to biology. Physicalists and positivists like Carnap, Hempel, Popper and Kuhn continued to argue that disciplines aspiring to be sciences could be reduced to physics. And biology, even if they neglected it, was no exception. In the 1970s and 1980s authoritative philosophers like Hull, Ruse and Sober based the philosophy of biology on physics. But their training was logical-mathematical rather than biological. So it became clear that the philosophy of biology could no longer be founded on logic and mathematics, but on concepts unique to biology (the biological specificity). This led to the definition of 'biology' as an autonomous science.

After these centuries-long philosophical vicissitudes, biology now divides into two distinct parts: *mechanistic* biology (genetics and molecular biology) and *evolutionary* biology (theory of evolution). The former deals with the physiology of living organisms, in particular the cellular processes (including those of the genome) which can be explained in terms of chemistry and physics. The latter instead has to do with aspects of the living world which concern historical time and evolution. These cannot be explained by the laws of physics or chemistry, but require a specific methodology founded on the *historical narrative* and on hypothetical scenarios. The biological specificity not reducible to physics is given by evolutionary biology.

Having defined the twofold nature of biology, now it is to be established what principles and concepts of physics are applicable to it. From what has already been said it is evident that biology is partly similar to physics and partly different from it.

If biology, with its mechanistic and evolutionist parts, is a science, then it is necessary to revise and enlarge the concept of 'science' adopted by Galileo, Newton and the positivists, so that it includes the characteristics typical of evolutionary biology.

Unlike physics, biology does not have a mathematical basis. This means that there exist sciences which do not satisfy the requirement of mathematization imposed by Galileo, Newton and the positivists.

Every science is constituted by theories. And theories in their turn are constituted by laws or by concepts. Whilst the theories of physics are constituted by laws, those of biology are constituted by concepts. The most important concepts of biology are those of 'evolution', 'biopopulation' and 'natural selection'.

The difference between physics and biology is evident if we compare the nature of living beings with that of inanimate ones. Because of their complexity, biological systems are endowed with the capacities of reproduction, metabolism, replication, regulation, adaptability, growth and hierarchical organization. Nothing similar exists in the inanimate world of physics.

The concept of 'biopopulation' is perhaps the one which best characterizes the difference between the inanimate and animate worlds. The former is constituted by classes whose members are identical, so that apparent variations among them are random and therefore irrelevant. Conversely, in the living world represented by a biopopulation, every individual is unique and unrepeatable. Variation is not irrelevant but instead crucial for evolution.

From the twofold nature of biology derives a twofold causality: the first causality is constituted by the natural laws that hold for physical and inanimate phenomena; the second consists in the *genetic programs* which characterize solely the living world. There is not a single living phenomenon or process that is not controlled by a genetic program contained in the genome. Nothing similar exists in the inanimate world.

A process absolutely unknown in the inanimate world is the *natural selection* propounded by Darwin to confute the concept of 'design' put forward by the natural theologians, and according to whom it is thanks to God's design that organisms are perfectly adapted to each other and to the environment in which they live. Natural selection, unlike the deterministic laws of physics, was the result of interaction among numerous factors, the principal among them being randomness. Because evolutionary biology—or simply biology, since the specificity of biology resides in its evolutionary part—is not reducible to physics, it cannot use the latter's methodology. Biology's methodology must instead take account of the uniqueness of the phenomena that it studies, like the extinction of the dinosaurs or the origin of the human species. In explaining such phenomena, it cannot resort to laws, nor can it conduct experiments. The extinction of the dinosaurs was a unique occurrence which cannot be derived from a general law nor be subjected to experimentation. The method used to explain it is that of *historical narrative*, which constructs a scenario whose explanatory capacity is verified on the basis of the existing evidence.

It is thus obvious why reductionism, though essential for physics, cannot be applied in biology. Biological systems are constituted by parts structured into levels which interact with each other. The interactions take place among genes, between genes and tissues, between cells and other components of the organism, between an organism and the inanimate environment in which it lives, and among different organisms. According to physicalism, the higher levels should be reducible to the lower ones, so that their properties can be determined and the system as a whole explained. Applying reductionism to biological systems would deprive the individual levels of their specificity: everything would assume the meaning of the lowest level, namely physics.

The attempt to create a philosophy of biology based on physics was bound to fail. It was therefore necessary to leave the narrow ambit of physicalism to assert the autonomy of biology as a science enjoying equal dignity with physics. The twofold nature of biology has entailed enlarging the concept of science as understood by Galileo, Newton and the positivists.

If we were to draw a boundary between the natural sciences and the social sciences, we would find that this boundary traverses biology in its middle, connecting its mechanistic part (genetics and molecular biology) to physics, and its evolutionary part to sociology.

3 From Biology to Sociology

These reflections on the foundations of physics and biology help to set up the bases for an epistemology of sociology. The foundation of sociology can now be viewed as an extension of physics and biology. I shall describe this process step by step.

With Galileo and Newton, physics (mechanics and astronomy) became the model of science in general. Every discipline that claimed to be a science had to exhibit the same characteristics as physics: the existence of laws, and their translatability into mathematical statements.

Biology, with its twofold nature, created more than a few difficulties for the positivist proponents of this view of science—so much so, indeed, that they preferred to ignore it. Today, nobody would dispute the scientificity of biology: neither of mechanistic biology (genetics and molecular biology), nor of evolutionary biology (theory of evolution). However, this has required an enlargement of the concept of 'science'. The model of science developed by Galileo would not have been able to comprise the evolutionary part of biology, which is a science despite the fact that it does not fulfil all the requirements of physics—in particular having laws and being mathematizable.

The development of the concept of 'science' that starts from physics and traverses biology must continue to sociology as well. Just as biology was born from an extension of physics, so sociology must be an extension of both physics and biology. Thus to the twofold nature of biology corresponds the threefold nature of sociology. Just as biology has a specificity irreducible to physics, so sociology has a specificity irreducible to either biology or physics. A philosophy of sociology must be founded on that specificity. It must proceed from the bottom (from physics) upwards (to sociology). Hence the procedure in reverse, from the top down, is invalid because it would justify forms of reductionism like Wilson's proposal to reduce sociology to biology.

As I have compared the characteristics of biology with those of physics, so I shall now compare the characteristics of sociology with those of evolutionary biology. Such comparison reveals similarities in epistemology and methodology (the method of historical narrative). However, sociology differs profoundly from biology when it is examined in terms of the concept of 'reality'. Does social reality exhibit the same characteristics as biological reality? If the answer is 'no', in what does the difference consist? The answers to these questions will evince the specificity of sociology.

When we speak of biological reality, we refer to living organisms, concretely existing and observable. They exist objectively in the same way as the objects making up physical reality (mountains, trees, rivers, stars, etc.) exist. They are horses, fishes, reptiles, people, etc. They are constituted by matter, and we can perceive them with our senses. From this point of view, the objects of biology are like the objects of physics. The difference between the two is that, whilst biological reality is animate, that of physics is inanimate.

Does social reality display the same characteristics as the realities of biology and physics? Is it too perceivable through our senses? Is it objective and pre-existent to humans? Answering these questions requires an analysis of the characteristics of social reality.

The point of view of the positivists on social reality is clear and precise: since sociology is a science like physics, the objects that make up its reality display the same features as do physical objects (they objectively exist independently of humans). It is precisely this objective existence of social reality which makes the identification of its laws and their mathematization possible. This world is known passively by the subject through his/her senses: the weaker the influence exerted by the subject, the more rigorous becomes the knowledge acquired by means of controllable instruments. If the meaning that the subject confers on the world is not based on experience (and therefore on verifiability), not only is it nonsensical, but it is an obstacle to scientific knowledge.

The philosophers who sought to give the social sciences a positivist basis (scientific in the meaning specified above), for instance A. Comte and H. Spencer, embraced the above epistemological assumption in its entirety. Hence they sought to give social reality a foundation utterly similar to that of physics. Difficulties soon arose, however. The first and perhaps most important of them concerned the distinction between natural facts and human facts. Do human facts (spiritual, cultural, mental, historical, etc.) have characteristics different from natural ones, or are they ultimately reducible to the latter? Positivists argued, with all the means at their disposal, for the latter thesis, because it enabled them to avoid undesirable consequences conflicting with the general principles of positivism: the unity of reality, methodological monism, the empirical criterion of meaningfulness, etc.

It is here that resides the positivist foundation given to the social sciences by E. Durkheim, which profoundly influenced one of the most important traditions of contemporary sociology. Durkheim's main assumption was that, ontologically, social facts are 'things' and therefore similar to natural facts. As a consequence, social reality possesses an objectivity which can be investigated using the methods of physics. Durkheim was convinced that the concrete processes of society could be uncovered in light of this concept of 'objectivity', and as social scientists carried out this task they had to describe social facts and their reciprocal relationships as if they were extraneous to them: that is, they had to eliminate everything that might inhere in their subjectivity. Hence the science that studied society was independent from that society. This independence was the fundamental premise for identifying the social laws. And it was these laws that made individuals, groups and institutions meaningful.

Contrary to what the positivists thought, however, social reality is a human creation. It exists as long as the people who have created it believe in it; it stops existing when they no longer believe it.

In my book *Le regole dell'azione sociale* (1983), I showed—especially in the seventh chapter entitled "La fondazione della realtà sociale"—how social reality is built by humans by means of constitutive rules. Some years later, in 1995, J. Searle published a work of great importance, *The Construction of Social Reality*, where he envisaged the use of constitutive rules for the creation of social reality. Compared with the treatment made in my 1983 book, Searle's investigation is broader, deeper and more exhaustive. I agree with the fundamental theses that he has proposed and developed in his works, and I shall relate them to my personal contributions to the epistemological foundation of sociology.

The construction of social reality, according to Searle, starts from the distinction between natural facts and social facts. In order to illustrate how social reality is constructed, I shall cite an example provided by Searle. He writes:

> Consider a simple scene like the following. I go into a café in Paris and sit in a chair at a table. The waiter comes and I utter a fragment of a French sentence. I say, "*un demi, Munich, à pression, s'il vous plaît*". The waiter brings the beer and I drink it. I leave some money on the table and leave. An innocent scene, but its metaphysical complexity is truly staggering, and its complexity would have taken Kant's breath away if he had ever bothered to think about such things. Notice that we cannot capture the features of the description I have just given in the language of physics and chemistry. There is no physical-chemistry description adequate to define "restaurant", "waiter", "sentence of French", "money" or even "chair" and "table", even though all restaurants, waiters, sentences in

> French, money and chairs and tables are physical phenomena. Notice also that the scene as described has a huge, invisible ontology: the waiter did not actually own the beer he gave me, but he is employed by the restaurant which owned it. The restaurant is required to post a list of the prices of all the *boissons*, and even if I never see such a list, I am required to pay only the listed price. The owner of the restaurant is licensed by the French government to operate it. As such, he is subject to a thousand rules and regulations I know nothing about. I am entitled to be there in the first place only because I am a citizen of the United States, the bearer of a valid passport, and I have entered France legally.
>
> Notice, furthermore, that though my description was intended to be as neutral as possible, the vocabulary automatically introduces normative criteria of assessment. Waiters can be competent or incompetent, honest or dishonest, rude or polite. Beer can be sour, flat, tasty, too warm, or simply delicious. Restaurants can be elegant, ugly, refined, vulgar, or out of fashion, and so on with the chairs and tables, the money, and the French phrases.
>
> If, after leaving the restaurant, I then go to listen to a lecture or attend a party, the size of the metaphysical burden I am carrying only increases; and one sometimes wonders how anyone can bear it (Searle 1995: 9–10).

This example is one of the innumerable cases that we experience every day and which overall constitute our social lives.

The first important consideration in this regard is that social reality has a two-fold ontology: a *visible*, observable one constituted by the waiter, the beer, the table, the money, and an *invisible* one constituted by the meaning of the money, the rules on operating the restaurant, judgments about the beer, the waiter, the place, etc.

The second important consideration, which follows from the first one, is that every ontology of social reality must be based on both its visible and invisible part. The visible part is similar to the ontology of physics, whilst the invisible part, which is not reducible to physics, is that specific to sociology. The problem which then arises is how to incorporate the specific ontology of social reality into the general ontology.

Schematically, we may state that the ontology of the reality external to humans is based on two theories: the *atomic* theory of matter and the *evolutionary* theory of biology, which respectively explain inanimate and animate matter. From this it follows that reality is constituted by physical particles organized into systems like mountains, planets, rivers, and humans. Certain living systems evolve according to natural selection. Some living systems have developed a brain, and the brain has developed consciousness, as in humans and in the higher animals. Consciousness is expressed through intentionality, or the ability to represent to oneself objects and states of the external world. The question that now arises is this: how is it possible to insert social reality as described here into this ontology?

The third important consideration, which ensues from the first two, is that in the world there are both characteristics independent of us, and others that depend on us. Mountains, stars and rivers exist independently of the representation that we can have of them. However, there also exist objects in the world which depend upon us. Consider, for example, an object which is constructed partly from wood and partly from metal. These characteristics are intrinsic to the object and they do not depend on me. But if I describe this object as a knife, the characteristic of the knife is not constituted by atomic particles, as its wood and metal are. The object 'knife' exists in dependence on the subjects who have invented it and use it. Considering the knife as a union of wood and metal does not add any material object to those that already exist, but it adds epistemically objective characteristics which depend on the users of the knife. We may also say that the knife expresses a subjective ontology.

One constructs social reality from this ontology by specifying the notions of 'collective self' and 'constitutive rule'. The *self* (individual and collective) derives from the *me*. It is therefore important to define the concept of 'me'. However, this task would require entering a labyrinth of philosophical analyses, substantially different and conflicting (from Hume's scepticism to Husserl's transcendental foundation), and from which it would be difficult to emerge with a clear and precise notion of 'me'. I shall therefore abandon philosophy to see what the neurosciences tell us in this regard.

According to G.M. Edelman, the neural changes manifest at the origin of language are the same as those from which higher-order consciousness emerges. This enables a self to be constructed from social and affective relationships. The emergence of higher-order consciousness made possible by language finds necessary support in social relationships. If people did not communicate with each other, there would be no development of language and therefore of intentionality and the self. Hence it follows that the *me*, the *self*, the collective *self*, and *intentionality* are at the basis of the development of higher-order consciousness and regulate social relationships. If we consider real-life experiences like the performance of a concert, a game of chess, a religious ceremony or a university lecture, we see the collective self in operation.

The collective self (also in its expression as collective intentionality) represents social facts. However, there exist some social facts which exhibit specific characteristics that require, for the representation, the use of constitutive rules.

We owe the notion of constitutive rules to J. Rawls, who, in his 1955 essay *Two Concepts of Rules*, drew a distinction between regulative and constitutive rules. Regulative rules are those which discipline activities that exist independently of the rules: for example, the ban on smoking in public places or the obligation to obey the highway code. In such cases, the public places and the highway exist prior to the rules that regulate them: the rules control forms of behaviour that exist before them. However, not all rules are regulative. There are some that do not regulate but constitute: they create what is regulated. These are constitutive rules. A classic example is the game of chess. In order to play chess, it is necessary to know not only the regulative rules that concern the strategy with which to checkmate the opponent but also the constitutive rules by which the chess pieces (king, queen, knight, bishop, etc.) have been created. We will say, for instance, that the "bishop" is that piece which, in the game of chess, moves diagonally. This means that any object (a piece of woods, stone, glass) that moves diagonally in the game of chess is a "bishop". Vice versa, if I place a real bishop, with sceptre and mitre, on the chessboard, but he does not move diagonally, that bishop is not a bishop. It is precisely the constitutive rule that creates the object "bishop" in the game of the chess. The same holds for all the other pieces, their moves, etc. The set of all the constitutive rules creates something that did not exist before and is denominated the "game of chess". It is clear that, although the constitutive rules are necessary, they are not sufficient to play chess: it is not enough to move the bishop diagonally to play. In order to play chess we also need the regulative rules that state the strategy of the game, which is to checkmate the opponent. The set of the constitutive and regulative rules defines the game of chess. Classic examples of constitutive rules are those that concern baptism and Masonic initiation. A person is not born a Christian but becomes one with baptism, which confers upon that person a dimension (Christian) that s/he did not possess before. In this case, the rule constitutes a Christian at the moment when the priest utters the sentence: "I create you Christian". The same happens in Freemasonry. One becomes a freemason at the end of the initiation ceremony when the Venerable Master of the Lodge utters the sentence: "I constitute you, I create you freemason". From that moment on, the neophyte acquires a dimension (Masonic) which he did not possess before and will characterize him for the rest of his life.

Just as constitutive rules create the game of chess, so they create the social facts that have been denominated 'institutional'. Institutional facts can only exist within a system of constitutive rules. If institutional facts are precisely those facts that allow the birth and development of societies, then the importance of constitutive rules is understandable. Typical examples of institutional facts are governments and all state institutions, marriage, and money.

The logical form of constitutive rules is as follows: "X equals Y in context C". Thus, if X is an object (made of wood, iron, glass, etc.) and Y is a bishop, we will say that object X is a bishop in the context (in the game) of chess. For applications of constitutive rules to society, see my above-cited book *Le regole dell'azione sociale*.

In conclusion to this brief inquiry into the foundations of sociology, I now summarize its main points.

1. The construction of sociology starts from physics and proceeds upwards. Hence it enlarges the concept of 'science' without losing the specificities of the individual sciences. Vice versa, if one follows the reverse procedure, of reductionism from sociology to physics, one loses, at every reduction, the specificities of the individual sciences. All attempts to reduce sociology to biology, including the recent one by E. O. Wilson are therefore to be rejected.
2. The consequence is that sociology must be founded on its threefold nature: physical, biological, and its specific invisible dimension created by constitutive rules. Since the invisible dimension can be characterized as normative, it brings into discussion the relationship between 'is' and 'ought to be', in which the ought-to-be should be understood as normative. In this case, however, it is necessary to revise the relationship between 'is' and 'ought-to-be', since the formulations given to it in philosophy are inadequate. I refer in particular to the analyses produced by analytic philosophy and to the inconclusiveness of their results. Apart from the critical rethinking of this relationship by authoritative scholars like Putnam, if it is considered outside ethics, to which it has been confined, but related to the way in which social reality is understood here,

then the reality in question, that social reality constructed by constitutive rules, assumes a completely new and different meaning. Between a normative (ought-to-be) fact and a social and an institutional one (is), there is not the 'logical leap' that Hume declared and repeated in a thousand ways, but rather a direct relationship of constitution and regulation. Consider the case cited by Searle of drinking a glass of beer in a cafe.

3. The previous results require a revision and extension of the ontology founded on physics, chemistry and biology. The closed world characterizing that ontology should be opened up in a manner such that it also encompasses social reality in its invisible specificity.

References

Agazzi, E. 1969. *Temi e problemi di filosofia della fisica*. Milano: Manfredi, 2nd ed. Roma: Abete, 1974.

Agazzi, E. 1976. Criteri epistemologici fondamentali delle discipline psicologiche. In *Problemi epistemologici della psicologia*, ed. G. Siri, 3-35. Milano: Vita e Pensiero.

Agazzi, E. 1979. Problèmes épistémologiques des sciences humaines, *Epistemologia*, 2 Special Issue "Spécificité des sciences de l'homme en tant que sciences": 39-66.

Agazzi, E. 1992. *Il bene il male e la scienza. Le dimensioni etiche dell'impresa scientifico-tecnologica,* Milano: Rusconi.

Agazzi, E. 2010. A Systems-theoretic Approach to the Interrelation of Natural and Human Sciences. In *Relations Between Human Sciences and Natural Sciences/Relations entre sciences humaines et sciences naturelles*, eds. Evandro Agazzi and Giuliano Di Bernardo, Fascicoli speciali di Epistemologia: 25-40. Genova: Tilgher.

Di Bernardo, G. 1972. *Introduzione alla logica dei sistemi normativi*, Bologna: il Mulino.

Di Bernardo, G. 1979. *L'indagine del mondo sociale*, Milano: Angeli.

Di Bernardo, G. 1983. *Le regole dell'azione sociale,* Milano: Il Saggiatore.

Di Bernardo, G. 2010 *La conoscenza umana. Dalla fisica alla sociologia alla religione*. Venezia: Marsilio.

Edelman, G.M. 2004. *Wider Than the Sky. The Phenomenal Gift of Consciousness*. New Haven and London: Yale University Press.

Mayr, E. 2004. *What Makes Biology Unique? Considerations on the Autonomy of a Scientific Discipline*. Cambridge: Cambridge University Press.

Putnam, H. 2002. *The Collapse of the Fact/Value Dichotomy and Other Essays*. Cambridge, MA: Harvard University Press.

Searle, J. R. 1995. *The Construction of Social Reality*. London, etc.: The Penguin Press.

Between Education and Pedagogy

Giuseppe Bertagna

Abstract Agazzi has been very active in education as teacher, as president of educational institutions, as promoter of educational initiatives and as editor-in-chief of the important pedagogical journal «Nuova Secondaria». He has published some books and many papers on different educational and pedagogical issues, but never a comprehensive work on his pedagogy. This, however, can be easily reconstructed from the study of a great deal of his publications. The present paper outlines certain fundamental parts of this pedagogy, showing how they harmonize with his more general epistemological views regarding the nature of scientific knowledge. It also presents a conception of pedagogy as a science that, although not explicitly advanced by him, is in keeping with Agazzi's epistemology.

Evandro Agazzi, as everyone, has met education and dealt with its various problems during all his life. He has first *received* education as a son (the first of three brothers). His father Aldo, teacher at primary school, then at a secondary school for school teachers, and finally full professor of Pedagogy and dean at the Catholic University of Milan, was the editor-in-chief of the journals «Scuola e didattica» and «Scuola materna» published by Editrice La Scuola of Brescia. He has been a protagonist of the school reforms in the Italian Republic, from the Gonella Commission (1947–1950) till 1991. Evandro's mother Emma, teacher at primary school and then at secondary school, graduated as educational psychologist, being a recognized expert in religious education.

He has met education in various forms as a student: primary school, secondary and high school in Bergamo, before enrolling at the Faculty of Philosophy of the Catholic University in Milan. After graduation he continued his formation as a student of Physics at the State University of Milan, then as a postgraduate in Oxford (1960), for an advanced research in Philosophy of Science, and finally in Münster at the Institute for Mathematical Logic and Foundational Research in Mathematics.

G. Bertagna (✉)
Department of Human and Social Sciences, University of Bergamo, Bergamo, Italy
e-mail: giuseppe.bertagna@unibg.it

M. Alai et al. (eds.), *Science Between Truth and Ethical Responsibility*,
DOI 10.1007/978-3-319-16369-7_10

He has *exercised* education to earn his living from 1957 to 1964 as teacher in high school and, for some time, also as headmaster of a high school.

He has finally met and exercised education for his own choice, and not only as a profession, from 1965 to 2006 as professor of young university students in Genoa, Pisa, Milan (Catholic University), Fribourg; as a visiting professor or lecturer in dozens of universities and research centers in all continents, as well as a member of the editorial board of philosophical journals and various national and international associations of Philosophy teachers.

However, Agazzi has never formally cultivated Pedagogy as discipline. At home, he had received a definition of Pedagogy from his father: Pedagogy as «theory and practice of education». According to his father, not everyone—though receiving and doing education—intends to reflect professionally on its conditions, problems and nature. Moreover, not all of those who take up this profession, because of weakness of character or of thought, manage to formulate their own systematic thesis, defensible in public debate, by critically formulating the grounds of the educational problems they have studied, observed or stumbled upon. There are even less people who consider how, when, why, whether and with which limits their theory of education can be translated, without degenerations, into a personal, interpersonal and social action able to improve the educational status of everybody.

As is well known, those who succeeded in this enterprise would realize the steps of the heuristic chain, typical of any «science»: (1) starting from experience, in this case educational experience; (2) reflecting critically on it starting from hypotheses that explain and justify it; (3) proposing a systematic theory of educational practice; (4) comparing one's own pedagogical theory with others, in order to attain a unique theory that could optimize the various available proposals; (5) acting intentionally in order to improve the existing educational experience in the light of the criteria taken from the accepted theory(ies); (6) verifying and evaluating the results obtained on quantitative and qualitative bases; (7) eventually, reformulating the theory(ies), also through a serious confrontation with those who have opted for different ones.

In 1983, Editrice La Scuola, hoping in a forthcoming reform of the Italian secondary school (still essentially based on the Gentile reform of 1923), decided to publish «*Nuova secondaria*», the first monthly journal devoted to the discussion of educational, didactical and cultural problems of this key school level. The Editrice asked Evandro Agazzi, at the time professor of Philosophy of science (then, as is known, he passed to the chair of Theoretical Philosophy in 1997), to be the editor-in-chief of this journal. He accepted and he exercised this function for 28 years, till 2011, when he was appointed director emeritus, writing 280 editorials and several other articles dealing with important issues regarding the education of young people as well as theoretical debates on Pedagogy.

Despite this task, that led him at the center of Italian Pedagogy, and even if he has always acutely considered the steps (1), (2), (5) and in part, (6) of the heuristic chain mentioned above, Agazzi never intended (perhaps because of mere lack of time or of other academic priorities) to review organically the other three steps, in particular (3) and (4), that would have been perfectly up to his potential. Yet, even if he has never formulated an explicit «systematic theory of education», he had one such background theory, though made explicit only through clues. Otherwise, it would be very difficult to understand that pedagogically unitary mark that he has left on 28 years of «*Nuova secondaria*», as even the most inattentive and occasional reader easily perceives. The indirect reference to this background theory, moreover, can be found in various of his works of cultural or philosophical nature.

Working on "underlying meanings" is always a difficult enterprise, since the interpreter risks to overlap, even unwittingly, his own ideas to those of the interpreted theorist. In my case, however, the risk is a bit restrained as, in addition to Evandro's works, I can make reference to a customary dialogue with him, since 1983,[1] on pedagogical principles and issues.

With these warnings, I think, however, it is possible to identify at least the following pillars that could constitute the framework of a pedagogical theory formulated by Agazzi in order to, on one hand, enlighten educational practice and justify it and, on the other hand, learn from practice itself, then acting critically on theory.

1 Genealogy of the Subject-Person

Every human being, as both «hard» and «soft» current human sciences tell us, is always "Embodied" (inherent, incorporated, diluted in its constitutive matter), "Embedded" (relational, or rather social, dependent on the human relations that he/she experiences) and "Extended" (extended to the world, co-existing with it). Therefore, not surprisingly, biogenesis, family, setting, culture and society influence us, the quality and quantity of our education, as well as the way in which each of us structures his/her relations with the world and with him/herself.

It follows that "a subject constitutes him/herself in subordination, and this subordination represents the ongoing condition of possibility" of his/her potential being and his/her being as he/she is (Butler 1997 [my translation]).

No pedagogy can underestimate these genealogical dynamics of the subject becoming person. They reveal an opaque and passive dimension of educational processes that cannot be eliminated and, in addition, that is also fundamental for the following emergence of people's transparent, free and active dimension. Natural attitudes and genetic predispositions need, indeed, an environment of life and relations at least allowing its emergence, if not always supporting it.

This discourse, as is documented by his educational path (Agazzi 2008a, b), matters for Evandro's personality as well as for everybody. And it engages Pedagogy to explore in a more and more accurate and preliminary way these genealogical dynamics of the self-person. Evandro has often underlined the importance of this pedagogical principle in the understanding of education, in order to set it in a realistic and concrete way. Starting, self-reflectively, from his own education.

[1] I have been chief of the editorial staff of «*Nuova secondaria*» from the beginning, and I had the honor to be appointed by Evandro at the Editrice La Scuola first as associate editor, and then as his successor as editor-in-chief of the journal.

2 Conditions of Education

Up to this point, anyway, it's still difficult to talk about a real «education». Concepts like «training», «imitation», «shaping», «communication», «care», «teaching» (also in the Latin etymological meaning of "leaving marks"), «learning», «sociability», «ecological coevolution», «development» and so on.

Of course, all of them are important concepts in Pedagogy, but they may be also elaborated by disciplines that don't study the human education, but only the animal behaviour. Consequently, they can be considered necessary conditions, but not enough to qualify education as «human». In fact, education doesn't exist without a «subject-person» who educates, educates him/herself and is educated while educating someone else or him/herself.

For this reason, Agazzi doesn't rediscuss long-standing issues that affected the Pedagogy of the Twentieth Century (starting from his father), such as: "is education, for everybody, a self-education of the subject-person or is it always a hetero-education, in the sense of supposing the intervention of an external educator? Is self-education the result of the education received from another subject-person, so that it would exist only after the end of the educational process, or is it a dimension that intersects, for nature, the education received from another subject-person all life long? In other words: is there an age in life, not only on a psychological or juridical level, but also on an ontological one, in which it is possible to affirm, pedagogically, that one is not yet a subject-person, and an age in which at last one can become a subject-person thanks to the received education?"

Agazzi, instead, moves his attention towards the exploration of the features necessary to talk about the subject-person in general and the subject of education in particular, where the genitive is both objective and subjective. In that way, he rejects the idea that the subject-person could be reduced to his/her brain and to the peculiarities of his/her relationships with other subjects and the environment. There wouldn't be education if it was the brain that, self-governing his interaction with the natural and social environment, decided the behaviours and choices of each of us. In education (and in self-education), the subject-person cannot be considered just the product of his/her brain, i.e. a pleonasm over a neuronal order. This rule could be valid, at most, to make the subject-person a «subject of training, of imitation etc.», but not the subject of «an education pedagogically legitimated». It doesn't mean to deny the discoveries of contemporary neurosciences. It is just avoiding to uncritically support a mere reversal of the Cartesian setting, in which the *res cogitans* would not be something different from the organic functionality of the *res extensa*. Agazzi considers this way of thinking a regression "to a philosophical anthropology of pre-Socratic kind", which leads man to be nothing more than a part of nature (Agazzi 1995a, b).

On the contrary, human intentionality cannot be reduced to a unique naturalistic-materialistic explanation. In fact, it mediates and overcomes experience. The human subject captures, in his/her intellectual intuition, the empirical individual only in the framework of an abstract concept, in other terms of a "universal" that transcends it. S/he is not able to understand the multiplicity of experience except through a unity that can't be found within experience itself, and that has nothing sensible, even if it has links to what is sensible (Agazzi 1981). The semantic, analytic logos goes together the apophantic, syntethic one, as often stressed by Agazzi. Indeed, it is its necessary completion (Agazzi 1964, 1975, 1995a, b) and indicates an exclusively human specificity. The perspective of the "whole" to make reasons of the "part" and of the "synthetic" to talk sensibly of the "analytic" cannot be eliminated, even in science in the modern meaning. The ontological distance that separates humans from other living beings and from their own more advanced technological productions, by which they attempt to extend and strengthen their own nature (up to artificial intelligence), remains without a persuasive explanation, if we stay within a naturalistic perspective.

Two other fundamental characteristics of the subject-person are rooted on this specific human ontological openness, that makes him/her a full-fledged subject of education: freedom and inner identity. Without inner identity, freedom to choose between good and bad, true and false, beautiful and ugly would be without personal responsibility. The subject-person would have neither faults nor merits. At this point, talking about each human person's dignity, justice, truth, or falsehood would become impossible. The very civil cohabitation between subject-persons would be irreparably compromised: that is, compromised in language, relationships, thought (Agazzi 1993). Surely, freedom and inner identity of the subject are never absolute, in the historic and empirical context. They are not self-sufficient and closed. In fact, they also refer to a whole that founds them while goes beyond them.

Only a subject provided with intentionality, that means with logos, freedom/responsibility and inner identity, can be followed by the subjective and objective genitive of education. Otherwise, it remains in the horizon of a human subject who could train and be trained, imitate and be imitated, shape and be shaped, communicate and be communicated, teach and learn, care for and be cared for, and so on. Nevertheless, it could not integrate all these dimensions in a personal unit, taking them on critically, with intentionality, logos, freedom/responsibility and identity. Agazzi has proven and experienced several times in person this transition, as confirmed by some examples, also very simple, that, in different ways, represent paradigmatic experiences of all people who grow up.

Since he was a child, he had a keen interest in the natural sciences: flowers, animals, chemical and physical phenomena. "As a refugee in a remote village in the Bergamo valley, in order to escape to the air raids", his father had put in his hands the entomology and science books by the well-known French naturalist Jean Henri Fabre, who was recommended at that time for his combined scientific rigor, anti-scientism (maybe a little parenthetical) and elegant divulgation. Similarly, on physics, in order to sort the disordered experiences that Evandro was making on his own account, his father had given him to read *800 facili esperienze di fisica*, written for Editrice La Scuola by Mons. Angelo Zammarchi, a very fine science man and a great popularizer. Therefore, once finished secondary school, he intended to enroll himself in Physics. His family, in particular his father, pushed hard for philosophy. When Gustavo Bontadini moved from the University of Pavia to the Catholic University of Milan, Aldo Agazzi didn't hesitate and imposed himself upon his son in a somewhat authoritarian way (by that time, he was lecturer at the University of Padua with a contract provided by Luigi Stefanini). But eventually his university years, thanks to the encounter with that great theoretical philosopher, persuaded Evandro he had made the right choice. A suffered decision was endorsed: it became something he wanted and, more importantly, a course of action he should have adopted.

In his family and in the social and parish environment, attended when he was young, he breathed every day the firm belief in the Thomistic "natural desire to know God", in the reference that transcendence was get in the immanence and in the idea that these would be just the visible signs of the biblical anthropology about man created in the image and likeness of the Creator. In 1947, his mother, a very sensitive person, published a book that collected and reordered, in an itinerary offered to all the families, the methods used for growing these beliefs in Evandro and his brother Albert, their second son (Agazzi Carminati 1947). Moreover, the book was an enviable publishing success, and validated his mother as an expert in religious education of children within Editrice La Scuola. His encounter with the philosophical teachings of Gustavo Bontadini, Mons. Francesco Olgiati and Sofia Vanni Rovighi at the Catholic University gave Agazzi the critical tools to justify, with the concepts of intentionality, logos, freedom/responsibility and identity, what appeared to him by the time as a "fact" in his inner "ought to be" of consciousness, thanks to the education he had received. "Logos does not intervene to ascertain", wrote Agazzi (1981). And the ascertaining is not only a matter of external sensitivity, but also internal and, therefore, moral and religious one. It intervenes, instead, "to give a reason for what is already certain". In this case, to give a reason for certain moral and religious beliefs grasped by contagion from the formative, religious and historical-environmental devices involved in his education since childhood. Nevertheless, if this goal is not achieved, it isn't a paradoxical ending that, as Parmenides says, we are faced with "the truth of the illusoriness of the sensitive consciousness". «Feeling» deeply a belief as true is not enough to declare it as such: it would remain, after all, only a comforting but deceptive certainty. Instead, if this «feeling», rather than being taken for granted, passed the strict scrutiny of *logos*, the logically conquered truth could be put in front of the freedom and identity of the *ego*, in order to be chosen as good or rejected as bad. This is education in the full sense.

Agazzi reproduces this maturation process also with respect to many contents of that *Zivilisation* which was typical at the time of his youth. These contents, i.e. that set of rules, concepts, behaviours, and external and conventional values, would be taken on critically in order to give them a new form, with new reasons. They become fully his *Kultur*, his *Bildung*, a justified personal way of acting to educate himself more and more while he was educating himself and, at the same time, educating others.

Agazzi has often justified all his philosophical and cultural work in the light of this well-known quote by Hegel: "philosophy is its own time learnt with thought. Therefore, it is just as foolish that a philosophy can transcend its present world, as that an individual could leap out of his time (Hegel 1821)".

3 Between Philosophical Whole and Pedagogical Wholeness

One of the most present categories in Agazzi's thought is «whole». A philosophical way, one might say, to mean the God of theology and Jesus Christ's God of Christian religion; or to shrink from partiality, to dread reductionism, to bring back the parts to their entirety. This means also the contrary: to find each part in the whole, in order to avoid the "absolutization", the maximization of something against the most reasonable choice of optimizing the position and role of anything.

Agazzi uses this category at every level of his activities: theoretical, in particular, but also pragmatic, social, existential and even concerning too human academic trafficking. He is never for an *aut aut*. Always attentive to an *et et*, though always affirming unflinchingly his positions. Therefore, he is an explorer of the reasons of composition, rather than of conflict. Also taking on responsibility for his own choices and, if it is the case, for his own mistakes.

It is easy to imagine how this category was important as a directive of his educational action in teaching and, accordingly, how it may also be legitimately referred to as the general principle of a theory of education that, subsequently, it is essential not to betray in practice.

At the pedagogical level, his father and the environment of Editrice La Scuola of Brescia,[2] thanks to the encounter with Christian activism, had developed this category under the name of «wholeness» ("*integralità*", i.e., whole education of the person). However, Agazzi uses this category sparingly. He prefers his own concept. In substance, however, the concordances between the philosophical concept of the whole and the pedagogical one of wholeness are obvious.

The principle of placing every partial educational action or every auto-educative maturation of some single aspect into the entirety of the whole education of the person, inserted into history, is a valid principle even at the pedagogical level. And then, to get in contact this historical whole, which is actually once again a part, with the entire whole.

Let us take, for example, the subject-person of education. It is not pure spirit, God's breath mentioned in the Bible. It is also the subject of a body with a mind and a brain, connected to a world in history, all without solution of continuity, by its own nature (*physis*). Education, therefore, is not «whole» or «entire» if it abstracts the subject-person from his/her environment and from history. Nor if one aims only at a rationality and at a mind able to explain everything. This is a Pascalian theme *par excellence*,[3] but also very dear to Agazzi, who used it mainly to deal with the problem of evil and the defeats of theodicy (Agazzi 1992a, b, 2014).

---

[2] In particular, Marco Agosti and Vittorino Chizzolini. The latter, since 1936 operative shadow of Mons. Angelo Zammarchi at La Scuola, was also his godfather for Confirmation, May 16th, 1946.

[3] "The last step of reason is to recognize that there are infinite things that transcend it. Only a weak reason does not recognize it. And if natural things exceed it, what shall we say of the supernatural ones?" (cfr. B. Pascal, *Pensées*, ed. Brunschvicg, nn. *272 e *267).

In order to practice the «integral» education of the subject-person, it is necessary to use at the same time world, history, environment, spirit, forms of rationality, psyche, corporeality, sensitivity, motility, sociability, manual dexterity, expressiveness, in a harmony that must be regularly sought and pursued in its «right moments», starting from the genealogy of each human person and from the environmental and historical context within which it is stratified. We must hold together, with the balance of the «right time», all the just mentioned dimensions. But every dimension has its own «wholeness» of parts that should not be forgotten or, worse, betrayed.

Rationality, for example, cannot be limited to the identification with the *nous*, it must always involve the *logos*. And the *logos* cannot be only the theoretical one, it must also be extended to the technical-technological and practical-moral one. The same applies to the body, psyche, sensitivity, motor skills, sociability, expressiveness, dexterity, sociability, memory, and, not least, for the established cultural forms that in every time draw the boundaries of the encyclopaedia of knowledge (sciences, philosophy, humanities, techniques, technology, arts, history, religion, myth).

An education that forgets the duty of these continuous ironic intersections and that theorizes the preliminary resection of some of them, would violate the pedagogical category of wholeness and would end up impoverishing the quality and the sense of each one's and everyone's education (this issue doesn't matter the "world" of practice, where you cannot do everything all together in the same time).

This general perspective explains Agazzi's insistence on three themes that he has always presented as educationally strategic for students, in schools and universities, for the professional journal he directed («*Nuova Secondaria*»), and last but not least, for his own scholarly production.

The first theme is interdisciplinarity in research and teaching. Agazzi has never underestimated the identity of object, method and language, typical of every science, and consequently, of the disciplines featuring in school and university curricula. He never tolerated, in this regard, dilettantism and superficiality. There is no way to carry research in the sciences or teaching in school subjects unless both of these are thoroughly and rigorously understood. He himself, in order to write about philosophy of physics studied physics, in order to write about mathematical logic studied mathematical logic, and so for geometry, etc. And he did so in a brilliant way. There cannot be philosophy "of" something unless these direct and deep knowledge of that "something" (Agazzi 1992a, b): otherwise, it would be only an empty talk. With the same determination, however, precisely in order to do well in scientific research and in teaching school subjects, it is essential to hang out with interdisciplinarity (Agazzi 1994). One needs interdisciplinarity, first, as an attitude of mind, a heuristic style of thinking, an ethical and cultural custom, as openness to newness. Secondly, as an epistemological perspective: reaching the most problematic boundaries of each science and cultivated discipline, discovering the historical and theoretical relationships that they have with other sciences and contiguous or distant disciplines, identifying similarities in the objects, methods and languages used to verify the opportunity of conscious contaminations that enable unprecedented views in scientific research or cultural and educational strategies increasingly integrated into the school. But also, thirdly, interdisciplinarity is necessary as orderly organization of cooperative research between scientists and of didactics between teachers of different school subjects. It is not a fanciful temptation of everybody doing everything, by giving origin to a modern, empty know-everything attitude, in which the skills and objects of study of the individual sciences and other teaching disciplines are confusingly interchangeable. It would be paradoxical, for example, asking a teacher of physical education and sports in high school to teach kids maths and literature. However, it is essential that also the teacher of physical education and sports has been enabled—by his initial training and by opportunities for institutional in-service training—at least to understand, for example, limits and integrals, or the narrative structures of a literary text. Otherwise it would be impossible for him simply to understand what his maths and literature colleagues are talking about. So, he would be precluded from any form of willing collaboration. Besides, if he himself does not understand these concepts, he will never even suggest to his students the connections and references that his perspectives and those of his maths and literature colleagues can and must ensure, each one for his/her own part, when they respond to real problems, as such always transdisciplinary; or when they interpret life situations which are naturally irreducible to the partiality of disciplinary perspectives; or, finally, when they carry out some unit tasks in situation, which are complex for definition, or they develop shared plans.

The second theme often discussed by Agazzi regards the humanistic value of science and technology and, reciprocally, the scientific and even technological value of the humanities. This circularity is suggested by the very structure of operationalism and of Agazzi's epistemological objectualism. If it recognizes the existence, according to an old classification, of the mathematical, physical and natural sciences on one side, and of the human sciences (including philosophical, social, historical, literary, psychological, artistic sciences) on the other, it sees in operational objectualism the assurance that both the former and the latter constitute, in different ways, «scientific» and intersubjectively strong knowledge. They are not as weak as they could be hold by the various contemporary forms of phenomenalism, conventionalism, naturalism, positivism, verificationism, falsificationism, subjectivism, idealism, deconstructionist irrationalism, connected to the epistemology of the Twentieth century. So, on the one hand, there is no longer a privileged form of knowledge to which the others should refer as to a paradigm (i.e., the various forms of more or less physicalist or mathematizing scientism, but also, conversely, by the different varieties of more or less historicist-literary anti-scientism). Indeed, every science has its own way to assess and justify its assertions *erga omnes*. On the other hand, it becomes clear that you cannot ask from each science, be it mathematical, physical and natural or «human (of the spirit)», more than what it can give.

It is no coincidence, then, that the third theme that has always characterized Agazzi's concerns and interventions is that of the centrality of a «philosophy of the human being». This cannot arise from any particular science, either natural or, though it might seem more reasonable, human or social (such as psychology, sociology, cultural anthropology, linguistics, history, etc.). Nor it can arise from the sum of all the different scientific elements of knowledge, that are already existent or will exist in the future. In a way or another, these elements always have something to do with experience and transform the «things» of experience, in order to study them, into «objects» endowed with a restricted number of properties. Now, if we could extend scientific exploration to the whole of human experience in the world, two problems would still remain unsettled: (a) sciences of whatever nature do not study «things», but «objects with certain properties», that are related to «things» without exhausting them; (b) the whole of these «objects», which have an empirical reference that allows us to «know» them in a certain and reliable way, still does not solve the problem of the whole, because the whole of the sensible does not include that of the supersensible, which is essential, as mentioned, to mediate cognitively the former of the two problems. Therefore, the philosophy of the human being has two tasks: «the effort to rationally understand the complex "world of life" in order to find a rationally justified solution to the "problem of life"» (Agazzi 2013); and the attempt to «provide a global image of man», where certain and reliable knowledge provided by sciences regarding the world of senses and the subject-person may be «harmonized and receive sense, taking into account at the same time other aspects of human reality», i.e. those related to the intuition of the supersensible (Agazzi 2007). This means making sense of the whole, although aware that we can never possess it entirely; making sense of the efforts that human subjects have always made to solve this problem, in itself aporetic, not so much through scientific knowledge, as through religion, myth, prayer, poetry, art, literature, music; discovering that the sense you can give to the whole of empirical reality is in itself not empirical, but not, for this reason, less «real», «true» and «crucial» for the life of people (1994).

4 Educational Sciences and Pedagogy

From Dewey's and James' pragmatism onward, it is quite common in Italy to replace the word «pedagogy» with the expression «educational sciences» or, in some cases, to use «pedagogy» and «educational sciences» as synonyms.

The latter expression, however, does not have the meaning of science and scientific knowledge that Agazzi's operational objectualism has explained and defended. On the contrary, we can discern in it a not too hidden group of meanings, referring to three epistemological perspectives that we should briefly mention.

The first perspective is logical positivism. According to this epistemological interpretation, philosophy is nothing other than an expansion of the natural sciences. So, also human nature, and consequently the education of subject-person and the subject-person him/herself, can be known only through an explication of these sciences. The result is the so called naturalism. Also philosophy and pedagogy must be wholly explained through processes of naturalization of their objects, methods and statements.

The second perspective is hermeneutics, which divides the world of certain and reliable knowledge into two: on the one side the neopositivist world occupied by the natural sciences, on the other side the hermeneutic world with the human sciences (or sciences of man, society, history, culture, and consequently of the subject-person of education). From this point of view, the sciences studying the subject-person of education (in the subjective and objective meaning of the genitive), cannot be anything else than human sciences, methodologically based on empathy and the narrative tools of individual reconstruction and understanding, and so completely opposed to natural nomological-deductive sciences. In the end, we are facing a specular subversion of neopositivist proposal.

The third perspective refers to the turn occurred in analytic philosophy inspired by Wittgenstein: philosophy is an investigation of the subject-person's intentionality and of the relations of his/her intentionality with human actions that are finally observable. From this third viewpoint, the neopositivist perspective and the hermeneutic one can be considered as complementary, even if different. For many reasons Italian pedagogy has generally neglected the investigation of this third perspective, which on the contrary has had important philosophical developments. By supporting, on an epistemological ground, the first and the second perspectives, it has involuntarily confirmed the exhausted polarization of many pedagogists. In this way, there were, on the one hand, scholars who supported the identification of pedagogy with «educational sciences» studying the subject-person as a «thing» to be explained in a naturalistic way. On the other hand, there were scholars who supported the wrongness of this perspective, because they put attention to a pedagogy considered as one of the most important human sciences, that is suspicious of each nomothetic knowledge and it is aimed only at the understanding, for somebody even ineffable, of the subject person's uniqueness, with results swinging between lyrics, verbosity and partisan closing arguments. Agazzi's operational objectualism, however, mixing analytic philosophy and classical metaphysics, allows for the analytic discourse (Agazzi has always declared himself an analytic philosopher), not in order to save the neopositivist approach as separated from the hermeneutic one, but to demonstrate that they both need to reshape the concept of scientific knowledge in a more adequate way (Agazzi explicitly speaks of a "hermeneutic dimension of science"). This point of view is very important in order to distinguish «pedagogy as science» from the other «educational sciences» (natural or human ones, not matter).

Therefore, one does not deny that they both want to study the subject-person of education as a subjective and objective genitive, starting from the historical experience of education, already lived by subject-persons, in order to explain and/or understand it, according to the different points of view. What really matters is attributing to «pedagogy as science» a specific object, hence a specific discourse, an operational structure and a purpose, which distinguish it from the other «educational sciences». In this sense, if it supposes, for its own constitution, the scientific analysis of education, already experienced by subject-persons, it couldn't be reduced to that analysis produced by the other educational sciences, for a simple reason: its object is the subject-person who, with his/her specific characteristics of intentionality, *logos*, freedom/responsibility, identity, "has not acted yet, but is going to act", "must act", "will have to act", "is called to act". And s/he is called to act on him/herself and on the others, choosing among all the possible actions those which better correspond to two specific aims of pedagogy.

The first aim is to value the above mentioned conditions of education, without reducing it either to the single, though important, aspects of «training», «imitation», «shaping», «communication», «care», «teaching», «learning», «sociability», «ecological coevolution», «development» and similar, or to their sum.

The second aim consists not only in considering the educational actions that have already occurred, with their shapes, reasons, intentions, rules, meanings, values, aims in the contexts in which they happened, but also and particularly in introducing those that «must occur», moment after moment, by now, for «those» subject-persons who educate and educate themselves in «that» new and determined social and environmental context, with their history and shapes, and with the reasons, intentions, rules, meanings, values, aims that they not only «want», but also «should» have.

In order to legitimate pedagogy as science of the «subject-persons of the education that has yet to occur and to be acted», it is useful to reconsider, more extensively, three clarifications.

Human experience and educational experience. The first clarification concerns the meaning of the term «human experience» and, in particular, of the expression «educational experience». Human experience is always "particular", "individual", "single". There is no experience identical with another one, because experience is always successive and it flows away.

These characteristics increase when we add the adjective «educational». In this case, the subject-person is introduced not only as a passive being, touched by the river of the world and of others, but as an active being, acting freely and consciously on the river of the world and of others, changing its flow, capacity, temperature in a more or less meaningful way. If «human experience» can be compared to constantly new water that laps the subject-person plunged in it, «educational experience» carrying with itself the features of intentionality, *logos*, freedom/responsibility and identity, could double the originality and unpredictability of the flow: the subject-person of education adds to the natural flow, for which human beings are like all animals and plants, another flow that s/he has decided, more or less, by him/herself, through and thanks to the already mentioned characteristics. Therefore, when educational sciences choose only some empirical (natural) features of educational experience in order to simplify its double complexity and to study it better, they risk to reduce educational experience to the human natural one. This is not useless, as every subject-person is a socio-environmental combination and individual empirical embodiment; but this effort, left alone, cuts strongly the richness and the complexity of educational experience.

«Perfect» educational experience. The second clarification concerns the meaning of the term «educational experience» in educational sciences and in pedagogy.

Educational sciences, as we have said, «objectualize» education as «something already happened, as it has happened», according to peculiar empirical characteristics changing on the basis of different points of view and intersubjectively repeatable checking operations, in order to explain or understand its reasons. This circumstance is valid for sciences studying education as a *datum* of «nature» (neo-positivist approach), for sciences studying it as a *datum* of «spirit» (hermeneutical approach) and, finally, for sciences studying it on «analytical» and «analytical-objectualistic» basis. In any case, psychology, sociology, anthropology, neurology, physiology, biology, genetics, hygienics, ecology of human development, geography, history, jurisprudence, economy etc., to use Bergson's words, «crystallize» the experience of education that has occurred; they «ossify» it, because their epistemological framework of explanation and understanding making them reduce it to its past or, to use a Latin expression, to its «perfect». However, in this way, educational sciences obtain more or less formalized explications of the causes of their data, that, once formulated and justified, can also have a predictive function on future educational experience. The initial retrospective perspective becomes a prospective one.

In the case of those sciences studying education according to the modern post-Galilean scientific paradigm, this perspective leads to the formulation of real general laws able to explain the dynamic and processes of educational experiences that have not occurred yet, but which will occur. The result is possible thanks to the great work of reduction of the complexity of the objects of study, insomuch as educational experience is often identified with some empirical features of human experience. Explanation and prediction of future educational experiences are also possible, thanks to the reduction of the plural and complex classical principle of causality to the principle of deterministic mechanism, that makes the role of efficient causes absolute, and transforms teleological causality into teleonomic causality (Agazzi 2008a, b).

Things are different for sciences studying education, through the other paradigms briefly mentioned above. By «understanding» the irreducible and complex singularity of education made by subject-persons, they mature in the educator a "consciousness", that gives him sensibility, ability to discriminate, and a critical judgment that are very useful to face, freely and responsibly, the task of acting as subject-person of education and to read in a deeper and more suitable way the new, original education produced and being produced by the future subject-persons.

Also pedagogy, if it wants to be a «science», particularly if it intends to be based on the analytical-objectualistic option, it couldn't avoid the comparison with these themes. Therefore, it has to start from the past, the perfect, the already happened, in educational experience (retrospective view) and be opened, naturally, to ante-spective, pro-spective, pre-dictive views. *Non sunt moltiplicanda entia sine necessitate*, however. It is good for pedagogy as science to collect the results and teachings coming from educational sciences. From this point of view, the more pedagogy becomes scientific, the more it considers contributions of educational sciences working not only on the basis of the Galilean hypothetical-deductive epistemological paradigm, but also on other paradigms. Without fundamentalisms and, at the same time, with no fear of seeming opportunist. It has only to use the results of natural or human sciences in their limits and possibilities in order to use them properly. Nothing more nor less than that.

The risk, here, is that of a pedagogy considered as «science of educational sciences». Some evident pretentiousness and excess. After all, could pedagogy realistically claim the power to produce a synopsis, not even a summary, of all the knowledge developed by «educational sciences»? Even if it were possible, what would we do of a pedagogy as science that is nothing more than an encyclopaedia, the fruit of others' work? Therefore, it is obvious that pedagogy can be creditable as science if it doesn't pretend to the throne of the unification of «sciences» but if, on the contrary, explicates the point of view from which it interrogates «educational sciences», produced by the critical use and selection of their results.

Pedagogy as science of the im-perfect educational experience. From which point of view does pedagogy become, with full rights, a «science», different from the other «educational sciences»? Briefly, we can summarize it as follows: it consists in using the sound conclusions taken from the educational sciences to face, in a theoretical and methodologically adequate way, the educational «im-perfect», in order to make of it a better «perfect» than the «perfects» that have already been identified.

In particular, pedagogy is called to be prepared to: (a) intervene on the education that «is going to happen», «is happening», «will happen», suffered or chosen in its unique (free) way by subject-persons in their relationships; (b) lead the subject-person to become more and more able, in the present, to use intentionality, *logos*, freedom/responsibility and identity in front of the happening of human and educational experience, not yet accomplished.

The specific characteristic of pedagogy consists in being oriented to the future, though starting from what exists in the present and what has happened in the past. Its distinctive feature is, in fact, the future that comes towards the present, in order to transform it into a past educationally better than the others established until now.

However, the future of the «single cases», as the Greeks reminded us (and the subject-person of education, is, as we have seen, a «doubled single case») has two main characteristics that is ill-advised to neglect. The first is ontological unavoidability: it happens anyway, nothing can stop it, not even remotely. It can be said that, from this point of view, though indeterminate, it is pre-determined, necessary. There is no young that, consuming the future, doesn't get old.

The second characteristic is gnoseologic inscrutability or unpredictability: the future cannot be known because it has not happened yet and, therefore, it cannot be experienced and elaborated through our logos. It is possible to explain the past (*perfectum*), but it is impossible to explain an experience that has not occurred yet (im-perfect). Particularly, if we move from human experience to the educational one. Greek tragedies (like the Theban cycle), mythology, the various more or less millennial eschatologies of many different cultures have taught that extremely well.

Those reasons clarify the continuous effort of people to «tame» (make domestic) the «not yet» in the «already». Two extraordinary strategies have been developed to follow both the education that «is happening and will happen» thanks to already educated subject-persons, and the genealogic growth of subject-persons that are more and more able to educate: (a) *technai* or arts (technical-artistic-poietic rationality); (b) *phronesis* or practical rationality (Bertagna 2010).

Technical rationality is the most reliable of human strategies to «tame» future events, both on the ontological and the gnoseological level. It seems that the future depends on it, because it is able to realize, actualize some goals, i.e. possible events and things that have yet to happen and exist. Indeed, thanks to technique, people firstly imagine them in their minds, then they check whether there are concrete possibilities of instantiating them, then they achieve this «incarnation» using suitable methods and tools; in the end, they control the quality of the results and, if they consider them unsatisfactory, they repeat the process and improve it gradually.

However, not everything that technically can be realized (i.e., that is possible) should be also carried out (i.e., that is also good); technical rationality, in order to face in a positive educational manner the «not yet», needs the intervention of practical rationality.

Phronesis examines the future possibilities, more or less close to the present, that technical rationality can conceive, and it chooses for concrete realization only those it considers «good and just» for the education of subject-persons in the given contexts (the logic of *kairos*: just interventions, in the appropriate ways, times and spaces). In this sense, *phronesis*, choosing between what technical rationality can do and what it should do, directs the good will of subject-persons and obliges them (from the Latin *ob-ligo*: to tie) to do something good: if before, for several reasons, they have lost or never reached the orientation to do what is good, now, thanks to practical rationality, they become involved in the process of improving «their being» (what they are) and the «being» (what exists) through the «potential being» (what should exist and can be realized, if one decides to do so).

Therefore, pedagogy is surely a theoretical science because, thanks to educational sciences, it: (a) examines past educational experiences (retro-spection); (b) justifies the present educational experiences (spection); (c) finds general explanations that could be valid, until proven otherwise, also for the future and/or for those theories that explain the existing problems, making who knows them ready to face future problems (ante-spection, pre-diction). However, it is above all a practical-poetic science because, on the basis of theoretical knowledge available to solve the current problems of education, it uses technical rationality (*techne*) and practical rationality (*phronesis*) to direct its interventions (actions) on what happens, is happening, or will happen to the subject-person in the matter of his/her «good» education, in the context in which s/he is growing up and in the situation s/he is experiencing. Agazzi has never devoted an explicit treatment to this idea of pedagogy as a practical-poetic science. However this is fully in keeping with his approach to the ethics of science and technology, and with his reflections on the necessary presence of "wisdom" in technology. In particular, he has analyzed on several occasions the relations between technical rationality and practical rationality, to which he devoted a whole chapter of Agazzi (1992a, b). Therefore the characterization outlined here of pedagogy as a science is consistent with his general philosophy of science.

References

Agazzi Carminati, Emma. 1947. *Due bimbi incontro a Dio*. Brescia: La Scuola.

Agazzi, Evandro 1964 *La logica simbolica*. Brescia: La Scuola.

Agazzi, Evandro 1975. Scienza e metafisica oggi. In *Studi di filosofia in onore di Gustavo Bontadini*, ed. Francesco Carlomagno. Milano: Vita e Pensiero.

Agazzi, Evandro 1981. Considerazioni epistemologiche su scienza e metafisica. In *Teoria e metodo delle scienze*, ed. Carlo Huber. Roma: Università Gregoriana Editrice.

Agazzi, Evandro 1992a. *Il bene, il male e la scienza. Le dimensioni etiche dell'impresa tecnico-scientifica*. Milano: Rusconi.

Agazzi, Evandro 1992b. *Filosofia e filosofia di. Orientamenti culturali per l'insegnamento della filosofia nella scuola secondaria superiore*, ed. Evandro Agazzi. Brescia: La Scuola.

Agazzi, Evandro 1993. L'essere umano come persona. In *Bioetica e persona*, ed. Evandro Agazzi. Milano: Franco Angeli.

Agazzi, Evandro 1994. *Cultura scientifica* e *interdisciplinarità*. Brescia: La Scuola.

Agazzi, Evandro 1995a. *Filosofia della natura. Scienza e cosmologia*. Casale Monferrato: Piemme.

Agazzi, Evandro 1995b. Dimostrare l'esistenza dell'uomo. In *Interpretazioni attuali dell'Uomo: filosofia, scienza, religione*, ed. Evandro Agazzi. Napoli: Guida.

Agazzi, Evandro 2007. Philosophy and Human Understanding (Maimonides Lecture). In *The Proceedings of the Twentieth-first World Congress of Philosophy*, ed. Ioanna Kuçuradi, vol. 13, 381–389. Ankara: Philosophical Society of Turkey.

Agazzi, Evandro 2008a. *Le rivoluzioni scientifiche e il mondo moderno*. Novara: Fondazione Achille e Giulia Boroli.

Agazzi, Evandro 2008b. *Scienza. Intervista di G. Bertagna*. Brescia: La Scuola.

Agazzi, Evandro 2013. Che cosa è dentro e che cosa è fuori dalla scienza. Una riflessione filosofica. In *Filosofi italiani contemporanei,* eds. Giuseppe Riconda, Claudio Ciancio. Milano: Mursia.

Agazzi, Evandro 2014. La ragione e la fede di fronte al dolore e alla sofferenza umana. *Studium* 110: 68-85.

Bertagna, Giuseppe 2010. *Dall'educazione alla pedagogia. Avvio al lessico pedagogico e alla teoria dell'educazione*. Brescia: La Scuola.

Butler, J. 1997. *The psychic life of power. Theories in subjection*. Stanford: Stanford University Press. Italian edition: *La vita psichica del potere. Teorie della soggettazione e dell'assoggettamento* (trans: Bonini, E., Scaramuzzi, C.). Roma: Meltemi 2005.

Hegel, G.W.F. 1821. *Grundlinien der Philosophie des Rechts*. Berlin: Nicolaischen Buchhandlung. Italian edition: Hegel, G.W.F. 1974. *Lineamenti di filosofia del diritto* (trans.: Messineo, F.). Bari: Laterza.

Part III
Logic, Hermeneutics and Philosophy of Language

Philosophy of Mathematics and Logic

Pierluigi Graziani

Abstract In the present work I attempt to describe Evandro Agazzi's research on philosophy of logic and mathematics. In particular, after a general introduction to his works, I focus my analysis on the philosophical implications of Gödel's Incompleteness Theorems. This is because they have always remained a constant point of interest in Agazzi's research. In particular, I wish to offer a tribute to Agazzi's mastership by analysing Gödel's thought on the consequences of his Incompleteness Theorems for the philosophy of mind. That is part of the research developed by Fano and I (Epistemologia 36(2):207–232, 2013) after Agazzi's conference in Cesena (2006) and it is for me a way to honour Evandro Agazzi's constant stimulus to the researches in philosophy of logic and mathematics.

1 Agazzi's Impact on the Philosophy of Mathematics and Logic

The first stage of Agazzi's professional work was almost entirely devoted to logic, philosophy of logic and philosophy of mathematics. Only four years elapsed between Agazzi's graduation in philosophy (1957) and the publication of his first book, *Introduzione ai problemi dell'assiomatica* (*Introduction to the problems of axiomatics*) of 1961, and only four more years later he published *La logica simbolica* (*Symbolic logic*), another very fortunate book that had four successive editions and was also translated into Spanish (Agazzi 1964).

Agazzi himself has explained in some autobiographic interviews (the most extended and accessible of which is Agazzi 2008[1]) the reasons for this beginning of

[1]See also Agazzi and Alai (2012c).

P. Graziani (✉)
Department of Philosophical, Pedagogical, and Quantitative-Economic Sciences, University 'Gabriele D'Annunzio', Chieti-Pescara, Italy
e-mail: pierluigi.graziani@unich.it

M. Alai et al. (eds.), *Science Between Truth and Ethical Responsibility*,
DOI 10.1007/978-3-319-16369-7_11

his philosophical journey: mathematical logic, after receiving pioneering developments by Peano[2] and his school at the beginning of the twentieth century, had remained nearly ignored in Italy during almost 40 years, so that the only books available in Italian until 1960 were the old handbook by Peano's disciple Burali-Forti (2nd edition 1919) and the booklet *Nove lezioni di logica simbolica* (*Nine lectures on symbolic logic*) by Bocheński and Józef (1938). Therefore, when Agazzi, still an undergraduate, began to enter (practically self-taught) the field of mathematical logic, he was obliged to study several of the most reputed books of this discipline, especially in German and English, coming into contact with different approaches and contents.

At the same time, Agazzi was inevitably influenced by the fact that, during the first half of that century, the developments of mathematical logic have been strictly interwoven with the research on the foundations of mathematics. As a consequence of this fact, logic and foundations of mathematics became the core of Agazzi's intellectual interests for a few years. In particular, very soon Agazzi became fascinated by the project of understanding correctly and in depth Gödel's Incompleteness Theorems, and it was the realization of this project that moved Agazzi to write his first book (Agazzi 1961).

The first part of this work is devoted to a historical-philosophical reconstruction of the axiomatic method, considered as the most natural explicit realization of the idea of a "demonstrative knowledge" proposed since the time of Plato and Aristotle. This classical view was deeply modified towards the end of the 19th century owing to the crisis of mathematical intuition that produced the so-called "foundational crisis" of mathematics, and led to the formal conception of this method with the related metatheoretical problems.

The second part of the book is devoted to Gödel's Incompleteness Theorems, with a careful progressive introduction of all the necessary notions and technical prerequisites, that culminates in a simplified but completely rigorous reproduction of the different theorems of the original Gödelian paper of 1931.

An Appendix contains the Italian translation of the original German paper, that actually was the first translation appeared in any other language and had the privilege of adding a footnote communicated by Gödel himself. The clarity of exposition and the pedagogic skill that allowed also readers with no technical background to understand the complex structure of this famous set of theorems was undoubtedly the main ground for the great success of this book.

The conclusion of this book specifically attracted the interest and appreciation by the philosophers: here Agazzi, though underscoring all the merits and advantages of formalization, argues against the purely formal conception of logic and mathematics by critically discussing several metalogical results, like those concerning syntactic and semantic completeness and incompleteness, undecidability and categoricity, and by considering their philosophical implications.

In particular, Agazzi analyzes Gödel's Incompleteness Theorems, that have always remained a constant point of reference for his research. This is confirmed by a number of papers (Agazzi 1966, 1978, 1994, 1997, 2007, 2010, 2011, 2012a, 2012b, 2014) and by the fact that in 2006 he was invited at various international

[2]See Borga's paper in this book.

symposia to celebrate the hundred years of Gödel's birth (for example at University of Giessen and the University of Urbino). The just mentioned titles constitute an evidence of the persistent interest in the philosophy of logic and in particular in Gödel's Incompleteness Theorems that characterizes Agazzi's thinking from his first book for more than half a century.

From a historical point of view, we can say that Agazzi was the philsopher who, with Ludovico Geymonat and Ettore Casari, did most to launch and stimulate the study of logic and philosophy of mathematics in Italy. A complete account of Agazzi's research in these areas should offer details at least on four topics of his groundlaying work: (a) the relationships between form and matter[3]; (b) philosophical reflections on Henkin Theorem[4]; (c) philosophical reflections on Löwenheim-Skolem Theorem[5]; (d) the study of Gödel's Incompleteness Theorems and their implications. Such a complete account would show how the suggestions and cues coming from Agazzi's work in the 60's in the following years have been developed, more ore less independently, by other Italian and foreign scholars. Unfortunately it is not possible for me to develop in this paper the whole project, and I will limit my analysis only to point (d).

Here I will just propose some reflections on Agazzi's study of the consequences of Gödel's theorems. In a review of *I Problemi dell'Assiomatica*, Ludovico Geymonat[6] emphasized the importance of this publication, which filled a gap in the Italian culture. In fact, Agazzi's publications devoted to the problems of the axiomatic, and in particular to Gödel's Incompleteness Theorems, enabled many Italian scholars not only to approach and understand these important results, but also to problematize them theoretically, developing for example the so called *Gödelian arguments*, that is, arguments using Gödel's Incompleteness Theorems to show that minds cannot be explained in purely mechanist terms. In fact, Gödel's Incompleteness Theorems were almost immediately seen as tools for refuting the mechanistic thesis, whether we consider it in an *extensional* way (mind's procedures and results are mechanizable), or in an *intensional* one (human intelligence is a particular machine).

Turing himself understood such implications of these theorems[7]; beside him, in the 1950s Nagel and Newman (1956, 1958),[8] developed argumentations hinged

[3]In Agazzi (1994), a paper on formalism, we can read: "The entire history of philosophy could be rebuilt in light of the different ways in which the complex relationship between form and matter was conceived (note that the distinction between form and 'content' constitutes merely a particular example of this dichotomy)". This is a very interesting idea that in recent time, indipendently from Agazzi, has been deeply analysed in logic: see, for example, the literature on *logical hylomorphism*, MacFarlane (2000); Dutilh Novaes (2011). With regard to (a) see also the paper by Carlo Penco in this book.

[4]Henkin (1949, 1950). In particular see: Agazzi (1978, 2011). See also Manzano (2014).

[5]Löwenheim (1915); Skolem (1920, 1922, 1928, 1929). See Agazzi (1994, 2012b) and Carlo Penco's paper in this book. See also Arsenijević (2012); Bays (2014).

[6]Geymonat (1962).

[7]Turing (1936, 1947, 1950); for an interesting analysis of this work, see: Piccinini (2003); Bruni (2004, Chap. 3).

[8]See Feferman (2009). It is in 1961 the first Italian translation of Nagel and Newman's book.

upon the idea that Godel's Incompleteness Theorems could provide a logical tool to refute the philosophical thesis of mechanism. Despite this tradition, Gödelian anti-mechanists argument is associated with the name of the English philosopher J. R. Lucas. In fact, in 1961, he developed an argumentation aimed at demonstrating, on the basis of Godel's theorems, that it is not possible to represent human intelligence by a Turing machine.[9] Agazzi developed his Gödelian Argument[10] almost simultaneously with Nagel and Newman, and Lucas,[11] but independently from them. Obviously, just as any other scholar, at that time Agazzi couldn't know Gödel's own ideas on this topic, which were published later on[12]).

In 2006, during a conference in Cesena (Italy), Agazzi explained his thoughts on Gödel's Incompleteness Theorems, inviting the audience to continue to reflect on the great results in logic and mathematics, because they are a great gymnasium for our minds and theories, teaching us important lessons about our cognitive capacities and our representation of the world. In that occasion, Fano and I decided to take up this challenge and write a survey on the implications of Gödel's Incompleteness Theorems in the philosophy of mind. We published our papers in 2011 and 2013, the latter in *Epistemologia*, a journal founded by Agazzi in 1978. This little story captures an important side of Agazzi's thought and intellectual life: his capacity to propose important problems and stimulate scholars to their solution.[13]

Following this example, the tribute I wish to offer to Agazzi's mastership will be an analysis of Gödel's thought on the consequences of his Incompleteness Theorems for the philosophy of mind. That is part of the research developed after Agazzi's conference in Cesena and it is for me a way to honour Evandro Agazzi's impulse to the researches in philosophy of logic.

2 Gödel's Argument[14]

In 1951 Gödel held one of the prestigious *Gibbs Lectures* for the American Mathematical Society. The title of his lecture was *Some basic theorems on the foundations of mathematics and their implications*. The theorems in question were precisely those of incompleteness, and the philosophical implications concerned the

[9]See Fano and Graziani (2013) for a reconstruction of this history and analyses of most important Gödelian Arguments.

[10]See Bianca's paper in this book.

[11]See Fano and Graziani (2013).

[12]See Gödel's *Collected Works* (1986-2003), especially vol. III, IV, and V.

[13]We can remember the so-called 'seminarione', a seminar organized by Agazzi at the University of Genoa where were discussed important open problems in logic and philosophy of science. Another evidence are the titles of Agazzi's main books where the term 'problems' occurs in a fundamental way.

[14]Parts of this section are already published in Fano and Graziani (2013).

nature of mathematics and the abilities of the human mind (Gödel 1951).[15] This was one of the few official occasions in which Gödel expounded his opinion on the philosophical implications of his theorems. Without going into details about Gödel's paper, what is interesting here is the first part, where he derives the thesis of essential incompleteness of mathematics from his famous theorems. Such a thesis was sanctioned by the second theorem. Gödel's idea is that if one perceives with absolute certainty that a certain formal system[16] is correct (sound), s/he will also know the consistency of the system, that is, s/he will know the truth of the statement which establishes the consistency of the system itself. But, by Gödel's second theorem, the formal system considered cannot prove its own assertion of consistency, therefore the system does not capture all arithmetical truths, and for this reason "if one makes such a statement he contradicts himself" (1951: 309). But what does all of this mean? Does it mean perhaps that a well defined system of correct (sound) axioms cannot contain all that is strictly mathematical? Gödel believes that such a question has two possible answers:

> It does, if by mathematics proper is understood the system of all true mathematical propositions; it does not, however if someone understands by it the system of all demonstrable mathematical propositions. [...] Evidently no well-defined system of correct axioms can comprise all [of] objective mathematics, since the proposition which states the consistency of the system is true, but not demonstrable in the system. However, as to subjective mathematics it is not precluded that there should exist a finite rule producing all its evident axioms. However, if such a rule exists, we with our human understanding could certainly never know it to be such, that is, we could never know with mathematical certainty that all the propositions it produces are correct; or in other terms, we could perceive to be true only one proposition after the other, for any finite number of them. The assertion, however, that they are all true could at most be known with empirical certainty, on the basis of a sufficient number of instances or by other inductive inferences. If it were so, this would mean that the human mind (in the realm of pure mathematics) is equivalent to a finite machine that, however, is unable to understand completely its own functioning. This inability [of man] to understand himself would then wrongly appear to him as its [(the mind's)] boundlessness or inexhaustibility (Gödel 1951: 309–310).

Therefore, not only does the previous question pose the problem of the inexhaustibility or incompleteness of mathematics considered as the totality of all true mathematical propositions; but it also raises the question as to whether mathematics is in principle inexhaustible for the human mind, that is to say, whether the human mind's demonstrative abilities are extensionally equivalent to a certain formal system, or to the Turing Machine (*TM*) connected to it (the *TM* which enumerates the set of theorems of the corresponding formal system).

The question, then, requires due consideration precisely of the relation between what Gödel calls *objective* and *subjective* mathematics. First let T be the set of mathematical truths expressible within first-order arithmetic, and call this 'objective arithmetic', or following Gödel, 'objective mathematics', that is "the body of

---

[15]A very accurate analysis of this paper is proposed by Feferman (2006), Tieszen (2006), van Atten (2006).

[16]It is understood that, in this paper, the expression "formal system" indicates a formal system which is adequate to derive incompleteness theorems.

those mathematical propositions which hold in an absolute sense, without any further hypothesis".[17] By Tarski's theorem *T* is not definable within the language of arithmetic, hence *T* is not recursively enumerable. Let us then define *K* as the set of arithmetical statements which a human being can know and prove absolutely and with mathematical certainty, that is what one can derive[18] and know to be true. Let us call it 'subjective arithmetic' or, following Gödel, 'subjective mathematics', which "consists of all those theorems whose truth is demonstrable in some well-defined system of axioms all of whose axioms are recognized to be objective truths and whose rules preserve objective truth" (Feferman 2006: 135–136). What is then the relation between *K* and *T*?

Quoting Feferman we could synthesize Gödel's answer by saying: if *K* was equal to *T* "then demonstrations in subjective mathematics [would not be] confined to any one system of axioms and rules, though each piece of mathematics is justified by some such system. If they do not, then there are objective truths that can never be humanly demonstrated, and those constitute absolutely unsolvable problems" (Feferman 2006: 136–137). That is, if the equivalence $K = T$ held, the human mind would not be equivalent to any formal system or *TM* connected to it. In fact, having established characteristics of *T*, for each formal system there would be a provable statement by the human mind, but not within the formal system. Hence, the mechanistic thesis would certainly be false: *T* non-recursive enumerability entails, in fact, the non-existence of any effective deductive system whose theorems are only and all truths of arithmetic.

If, on the contrary, *K* did not coincide with *T*, and thus the human mind were equivalent to a given formal system or to the *TM* related to it, the existence of arithmetical statements humanly undecidable in an absolute sense would follow. In fact, as underlined by Gödel, the second incompleteness theorem does allow this conclusion: the proposition expressing the consistency of *K*, say Con_K, is true but is not provable within the system itself; the negation of Con_K is false and is not provable in *K*. Having established the equivalence between the human mind and a formal system, Con_K is not even provable by the human mind. Finally, since Con_K can be put in the form of a Diophantine problem,[19] it is an absolutely undecidable problem. Such a proposition is, thus, an unknowable truth.

---

[17]Gödel (1951: 305).

[18]As Feferman (2006: 140) emphasizes, Gödel believes that "the human mind, in demonstrating mathematical truths, only makes use of evidently true axioms and evidently truth preserving rules of inference at each stage".

[19]The expression "absolutely unsolvable problems", or Gödel's expression "Diophantine problems which are undecidable" refers to the following fact: Gödel's unprovable proposition which expresses the consistency of a formal system within the same system (with the formal system satisfying the first incompleteness theorem hypothesis) has the form $\forall(x)R(x)$, where *R* is a primitive recursive predicate and each statement of such a form is equivalent (Gödel proved it) to a statement of the form

$$\forall x_1, \ldots, \forall x_n \exists y_1, \ldots, \exists y_m \left[\mathrm{p}(x_1, \ldots, x_n, y_1, \ldots, y_m) = 0 \right]$$

where the variables vary on natural numbers, and "p" is a polynomial with integer coefficients: that is, it has the form of those *problems* faced by the Greek mathematician Diophantus of Alexandria in his book *Arithmetica*. See Gödel (1934, 193?, 1964).

Such questions and arguments lead Gödel to the idea that from the incompleteness results one can at the most derive the following *disjunction*:

> Either [subjective] mathematics is incompletable in this sense, that its evident axioms can never be comprised in a finite rule, that is to say, the human mind (even within the realm of pure mathematics) infinitely surpasses the powers of any finite machine, or else there exist absolutely unsolvable diophantine problems of the type specified (where the case that both terms of the disjunction are true is not excluded, so that there are, strictly speaking, three alternatives) (Gödel 1951: 310).

So, following Tieszen 2006, and considering the translatability between the concept of a well defined formal system and that of a *TM*, we can say that Gödel's Incompleteness Theorems show that it could not be true that:

The human mind is a finite machine (a *TM*) and there are for it no absolutely undecidable Diophantine problems.

The incompleteness theorems show that if we think of the human mind as a TM then there is for each TM some 'absolutely' undecidable Diophantine problem. The denial of the conjunction (i) is, in so many words, Gödel's disjunction. In formulating the negation of (i) Gödel says that the human mind 'infinitely surpasses the powers of any finite machine'. One reason for using such language, I suppose, is that there are denumerably many different Turing machines and for each of them there is some absolutely diphantine problem of the type Gödel mentions. So Gödel's disjunction, understood in this manner, is presumably a mathematically established fact. It is not possible to reject both disjuncts (Tieszen 2006: 230–231).

So the disjunction leaves open the three following possibilities:

I. human intelligence infinitely surpasses the powers of the finite machine (TM), and there are no absolutely unsolvable Diophantine problems (see Gödel 1951: 310).
II. human intelligence infinitely surpasses the powers of the finite machine (*TM*) and there are absolutely unsolvable Diophantine problems. That is, although human intelligence is not a finite machine, nevertheless there are absolutely irresolvable Diophantine problems for it.
III. human intelligence is representable through a finite machine (*TM*) and there are absolutely irresolvable Diophantine problems for it.

Gödel was convinced that (I) held, but he was also aware that his incompleteness theorems did not make the existence of a mechanic procedure equivalent to human mind impossible.

Gödel, however, as I explained, believed that from his theorems it followed that if a similar procedure existed we "with our human understanding could certainly never know it to be such, that is, we could never know with mathematical certainty that all the propositions it produces are correct".[20] But this established Godel's idea that "the human mind, in demonstrating mathematical truths, only makes use of evidently true axioms and evidently truth preserving rules of inference at each

[20]Godel (1951: 309).

stage",[21] and this exactly means that "the human mind (in the realm of pure mathematics) is equivalent to a finite machine that, however, is unable to understand completely its own functioning".[22] This argument, as it can be noticed, reminds those presented by Benacerraf (1967) and Chihara (1972).[23]

3 Conclusion

In 2012 Agazzi, responding during an interview, said: "I have always conceived philosophy as an effort to find answers to the fundamental problems of human existence, situated in its historical and cultural context" (Agazzi and Alai 2012c). This principle also applies to the research in the philosophy of logic and mathematics. Therefore, I would like conclude this paper following this line and putting Gödel's argument in an evocative way. Quoting Paul Benacerraf (1967: 30), I would say: "If I am a Turing machine, then I am barred by my very nature from obeying Socrates' profound philosophic injunction: KNOW THYSELF". This conclusion does not have any great relevance to the science. Nothing prevents one from building computational models, which would simulate ever-increasing parts of our intelligent behaviour. One day, we could even build a Turing machine, which will simulate in every way human intelligent behaviour, but we will not know this with absolute certainty! I believe, then, that the significance of this conclusion is more anthropological, than scientific: it simply reasserts the fundamental incompleteness of human self-knowledge.

This conclusion should be, I think, very interesting for Agazzi who has devoted much of his work on Artificial Intelligence to the analysis of the problem of intentionality.[24]

References

Agazzi, Evandro. 1961. *Introduzione ai problemi dell'assiomatica*. Milano: Vita e Pensiero.

Agazzi, Evandro. 1964. *La logica simbolica*. 5th revised and enlarged edition 1990. Brescia: La Scuola.

Agazzi, Evandro. 1966. Riflessioni su alcuni nuovi orizzonti della logica matematica. *Logica e analisi*, special issue of *Archivio di Filosofia* 47-69.

Agazzi, Evandro. 1978. Non-contradiction et existence en mathématique. *Logique et Analyse* 21: 459-481.

Agazzi, Evandro. 1994. On Formalism. In *Philosophical Problems Today*, ed. G. Fløistad, 1: 75-173. Dordrecht-Boston-London: Kluwer.

---

[21]Feferman (2006: 140).

[22]Godel (1951: 310).

[23]See Fano and Graziani (2013).

[24]See Bianca's paper in this book: this contribute analyzes Agazzi' ideas concerning the role of intentionality in artificial intelligence.

Agazzi, Evandro. 1997. On the criteria for establishing the ontological status of different entities. In *Realism and Quantum Physics*, ed. E. Agazzi, 40-73. Amsterdam- Atlanta: Rodopi.
Agazzi, Evandro. 2007. La rilevanza dell'opera di Gödel nella filosofia della logica e della matematica. In *Herr Warum. La musica di Gödel*, ed. F. Pollini. Cesena: Società Editrice Il Ponte Vecchio.
Agazzi, Evandro. 2008. *Scienza* (intervista di Giuseppe Bertagna), Brescia: La Scuola.
Agazzi, Evandro. 2010. Vlijanie Gödelia na filosofiiu matematici. *Epistemologija & filosofija nauki (Epistemology and Philosophy of Science)* XXV/3: 16-41. Moscow: Institute of Philosophy of the Russian Academy of Sciences.
Agazzi, Evandro. 2011. Consistency, Truth and Ontology. *Studia Logica* 97/1:7-29.
Agazzi, Evandro. 2012a Meaning between sense and reference: Impacts of semiotics on philosophy of science. *Semiotica. Sepcial issue "Semiotics and Logic"* 188-1/4: 29-50.
Agazzi, Evandro. 2012b. *Ragioni e limiti del formalismo, Saggi difilosofia della logica e della matematica*, ed. F. Minazzi. Milano: Franco Angeli.
Agazzi, Evandro, and Alai Mario. 2012c. Conversazione con Evandro Agazzi. *APhEx* 6, Interviste, available also at: http://www.aphex.it/public/file/Content20141117_06.APhEx6,2012IntervisteAgazziAlai.pdf.
Agazzi, Evandro. 2014. *Scientific Objectivity and its Contexts*. Heidelberg, New York, Dordrecht, London: Springer.
Arsenijević, Milos. 2012. The Philosophical Impact of the Löwenheim-Skolem Theorem. In *Between Logic and Reality. Modeling Inference, Action and Understanding*, eds. M. Trobok, N. Miśćević, B. Źarnić 59-82, Heidelberg, New York, Dordrecht, London: Springer.
Bays, Timothy. 2014. Skolem's Paradox. *The Stanford Encyclopedia of Philosophy*. http://plato.stanford.edu/entries/paradox-skolem/.
Benacerraf, Paul. 1967. God, the devil and Gödel. *The Monist* 51: 9-32. Also at: http://www.univ.trieste.it/~etica/2003_1/3_monographica.htm.
Bocheński, Józef M. 1938. *Nove lezioni di logica*. Roma: Angelicum.
Bruni, Riccardo. 2004. *Riflessioni sull'incompletezza. I teoremi di Gödel tra logica e filosofia*. Firenze: Ph.D. Thesis, Università degli Studi di Firenze. Also at http://www.philos.unifi.it/CMpro-v-p-88.html.
Burali-Forti, Cesare. 1919. *Logica Matematica*. Second edition, Milano: Hoepli.
Chihara, Charles S. 1972. On alleged refutations of mechanism using Gödel's incompleteness results. *The Journal of Philosophy*, 69:507-526.
Dutilh Novaes, Catarina. 2011. The Different Ways in which Logic is (said to be) Formal. *History and Philosophy of Logic* 32,4: 303-332.
Fano, Vincenzo and Graziani, Pierluigi. 2011. Gödel and the fundamental incompleteness of human self-knowledge. *Logic and Philosophy of Science* IX(1):263-274.
Fano, Vincenzo and Graziani, Pierluigi. 2013. Mechanical Intelligence and Gödelian Arguments. *Epistemologia* 36, 2: 207-232.
Feferman, Solomon. 2006. Are there absolutely unsolvable problems? Gödel's dichotomy. *Philosophia Mathematica* 14: 134–152.
Feferman, Solomon. 2009. Gödel, Nagel, Minds, and Machines. *Journal of Philosophy*. 106, 4: 201-219.
Geymonat, Ludovico. 1962. Review of Evandro Agazzi 'Introduzione ai Problemi dell'Assiomatica. Soc. Ed. Vita e Pensiero, Milano 1961'. *Bollettino della Unione Matematica Italiana* III, XVII: 408-10.
Gödel, Kurt. 1931. Über formal unentscheidbare Sätze der Principia mathematica und verwandter Systeme. *Monatshefte für Mathematik und Physik* 38: 173-198; also in Gödel K., *Collected Works* I: 144-195. Oxford: Oxford University Press.
Gödel, Kurt. 1934. On undecidable propositions of formal mathematical systems. In *Collected Works* I:346-369.
Gödel, Kurt. 193?. Undecidable Diophantine propositions. In *Collected Works* III:164-175.
Gödel, Kurt. 1951. Some basic theorems on the foundations of mathematics and their implications. In *Collected Works* III: 304-333.

Gödel, Kurt. 1964. Postscriptum to Gödel 1934. In *Collected Works* III: 369-371.
Gödel, Kurt. 1986-2003. *Collected Works* vol. I-V, eds. S. Feferman, et al.. Oxford: Oxford University Press.
Henkin, Leon. 1949. The Completeness of the First-Order Functional Calculus. *Journal of Symbolic Logic* 14: 159-66.
Henkin, Leon. 1950. Completeness in the Theory of Types. *Journal of Symbolic Logic* 15:81-91.
Löwenheim, Leopold. 1915. Über Möglichkeiten in Relativkalkül. *Mathematische Annalen* 76: 447–70.
Lucas, John R. 1961. Minds, machine and Gödel. *Philosophy* 36:112-127. Also at http://users.ox.ac.uk/~jrlucas/mmg.html.
MacFarlane, John. 2000. *What does it mean to say that logic is formal?*. Pittsburgh: PhD Dissertation, University of Pittsburgh.
Manzano, Maria, Sain, Ildikó, and Alonso, Enrique. 2014. *The Life and Work of Leon Henkin. Essays on his contributions.* Studies in Universal Logic. Springer Birkhäuser Basel.
Nagel, Ernest R. and Newman, James R. 1956. Gödel's proof. *Scientific American*; reprinted in Newman (1956), 3:1668-1695.
Nagel, Ernest R. and Newman, James R. 1958. *Gödel's Proof.* New York: New York University Press; revised edition, 2001, edited with a new foreward by Douglas R. Hofstadter.
Newman, James R. 1956. *The World of Mathematics: A small library of the literature of mathematics from A'h-mosé the Scribe to Albert Einstein, presented with commentaries and notes.* New York: Simon and Schuster.
Piccinini, Gualtiero. 2003. Alan Turing and the Mathematical Objection. *Mind and Machines* 13:23-48.
Skolem, Thoralf A. 1920. Logisch-kombinatorische Untersuchungen über die Erfüllbarkeit oder Beweisbarkeit mathematischer Sätze nebst einem Theoreme über dichte Mengen. *Skrifter utgit av Videnskapsselkapet i Kristiania I, Matematiska- Naturvetenskap Kl* 4:1–36.
Skolem,Thoralf A. 1922. Einige Bemerkungen zur axiomatischen Begründung der Mengenlehre. *Wissenschaftliche Vorträge gehalten auf dem 5. Kongress der skandinavischen Mathematiker in Helsingfors* 4:217–232.
Skolem, Thoralf A. 1928. Über die mathematische Logik. *Norsk. Mat. Tidsk.* 10:125–142.
Skolem Thoralf A. 1929. Über einige Grundlagenfragen der Mathematik. *Skrifter uigit av det NorskeVid. Akad. i Oslo I,* 4:73–82.
Tieszen, Richard. 2006. After Gödel: mechanism, reason, and realism in the philosophy of mathematics. *Philosophia Mathematica* 14: 229–254.
Turing, Alan. 1936. On computable numbers, with an application to the Entscheidungsproblem. *Proceedings of the London Mathematical Society* 42:230-265.
Turing A.M. (1947). Lecture to the London Mathematical Society on 20 February 1947, in The collected worksof A.M. Turing, II, Amsterdam, North Holland, 1992, pp. 87-105.
Turing, Alan, 1950. Computing Machinery and Intelligence. *Mind,* 59: 433-460.
Van Atten, Mark. 2006. Two draft letters from Gödel on self-knowledge of reason. *Philosophia Mathematica* 14: 255-261.

The Issue of Alethic Logic

Antonio Livi

Abstract Evandro Agazzi's epistemological consideration of the objectivity in the sciences fully legitimates the rationality of metaphysical inquiry as well as the embedding of science into broader contexts of moral, social and political nature. His main argument is the intrinsic limitation of any object assumed by a particular science, when human knowledge always takes into account reality as a whole. Science cannot exclude the questioning of the whole as such; much more, each specialized field of scientific research suffers a kind of contingency. Such a logical defense of the legitimacy of metaphysics was for me a strong support in building my own theory on the relationship between common sense and metaphysical research, whose goal is the rational mediation of the immediate certainties about the whole, namely the discovery of the causes of the existence of beings and the determination of their true nature. Unifying Agazzi's epistemological justification of metaphysics with Gilson's theory of realism as the very method of metaphysics, I was able to give a critical definition of 'common sense' as the primary truth in approaching reality—a truth that, although pre-scientific, is absolutely undeniable and so makes metaphysics as a science possible.

1 Agazzi's Notion of 'Objectivity' and the Search for Truth in Sciences

When in the year 1990 I published my first essay on the logic of scientific reasoning and its metaphysical presuppositions (see Livi 1990), I quoted in many pages what Evandro Agazzi had written about the principles of logic in general (see Agazzi 1961,1981, 1998) and much more about the realistic background implicitly present and effectively performing in all kind of scientific research (see Agazzi 1974, 1978,

A. Livi (✉)
International Science and Commonsense Association, Rome, Italy
e-mail: antoniolivi.isca@gmail.com

M. Alai et al. (eds.), *Science Between Truth and Ethical Responsibility*,
DOI 10.1007/978-3-319-16369-7_12

1981, 1987, 1988, 1991). The essay I mainly quoted is *La questione del realismo scientifico* (cf. Livi 1990: 70–78), an essay in which the limits of logical formalism are clearly detected (see Agazzi 1985). In the following years I noted that the Italian philosopher was developing this issue in many papers and especially in two important books: the first one in Italian, *Ragioni e limiti del formalismo* (Agazzi 2012), and the second one in English, *Scientific Objectivity and Its Contexts* (Agazzi 2014). In the first of those books Agazzi expounds his point of view about the question of truth in science and clarifies the necessary epistemic relationship between every scientific demonstration and its alethic presuppositions, intrinsically implied by the logical function of the inference. Commenting on Aristotle's doctrine of syllogism, Agazzi notices:

> In order that a syllogism (i.e. a correct argument) becomes a demonstration, certain rigorous requirements must be satisfied by its premises: they must be true, certain, more certain than their conclusions and 'causes' of the conclusions themselves. Leaving aside here these and similar further requirements [...] it is sufficient for us to point out that the domain of logical inference is bound to truth in highly committed sense. [...] This also entails a strict relation with *evidence,* since the truth of the first premises of a 'scientific syllogism' cannot be established by means of an argument and must be obtained, in the last analysis, through an intellectual intuition offered by the *nous* (Agazzi 212: 178)

Those considerations led Agazzi to build, in the latter book, an original and coherent theory of what he calls 'objectivity'. According to him, objectivity should be understood in a weak sense (as inter-subjective agreement among the specialists) or in a strong sense (as having precise concrete referents). In both cases objectivity relies upon the adoption of operational criteria designed within the particular perspective under which any single science considers reality. The 'object' so attained has a proper *ontological status*, dependent on the specific character of the criteria of reference. Agazzi shows how this theory can be applied equally well to the natural sciences and philosophy (metaphysics and ethics). Assuming the most correct methodologies from traditional, analytic and continental philosophy, Agazzi brings them to a fruitful complementary interplay in the philosophy of science, and so he offers a clear methodological framework for interdisciplinary investigation. This justifies a form of *scientific* realism, in basic agreement with what we know as *philosophical* realism. Both are grounded in the *natural* realism of common sense knowledge. According to Agazzi's theory, the awareness of such a "historical determinacy" of science justifies including in the philosophy of science the problems of ethics of science, the relations of science with metaphysics, and the social dimensions of science that overstep the traditional restriction of the philosophy of science to an epistemology of sciences (physics, mathematics, biology).

1.1 How Any Reference of Scientific Statements Presupposes a Realistic Background

Agazzi then moved to the study of foundational problems in the empirical sciences, at the same time elaborating his own original philosophy of science.

Its core is just his theory of scientific objectivity, which is based on the epistemic distinction between common sense things and scientific objects, the latter being structured sets of selected attributes expressing the special point of view from which a given science considers reality. These attributes are expressed through specialized 'predicates' and the 'basic predicates' of an empirical science are equipped with standardized operational criteria of reference that allow for the empirical test of statements. In such a way scientific objectivity has a weak sense, according to which it consists in the inter-subjective agreement among specialists secured by the use of standardized criteria for referring to something. But it has also a strong sense, according to which it consists in the fact of having precise concrete referents, equally attained by means of the same operational criteria. This doctrine has very important consequences. It vindicates the legitimacy of scientific truth, recognizing that it is relative to the actual referents of the scientific theory concerned, and thanks to this fact it advocates a realist conception of science, including the admission of the existence also of theoretical (non observable) entities. In addition, this view presents an analogical concept of science that does not imply the reduction of scientific truth to one single model. The awareness of the partiality of the point of view of any science opens the way to the consideration of broader points of view on reality and on science itself. Thanks to an original approach based on general systems theory, all these dimensions can be harmonized with a substantial respect of the freedom of science. Let's quote what Agazzi himself says presenting his last essay on philosophy of science:

> This position, in turn, has led me to vindicate a fundamental role for *truth* in science (something that had been almost banned from philosophy of science) and to study how truth can be attained, either by direct reference, or by argument, and this offers a foundation for admitting also the truth of non-observationally testable statements. Finally, the referential commitment of truth justifies a (carefully and duly specified) realist view of science. In my perspective, scientific objectivity is not context-dependent in a purely linguistic sense, but in a historical sense (of which the linguistic dependence is only a very particular aspect). The exploration of such a historical contextualization (that does not amount to relativism) opens the way to a due appreciation of all the right points stressed by the sociological interpretation of science, without falling into its excessive conclusions, and at the same time it justifies the consideration of those problems (Agazzi 2014, x).

1.2 How the Theory of 'Objectivity' Leads to the Justification of Metaphysics

Agazzi's epistemological consideration of the objectivity in the sciences fully legitimates also the rationality of metaphysical inquiry (also regarding the metaphysics of science) as well as the embedding of science into broader contexts of moral, social and political nature. For the Italian philosopher metaphysics has the epistemological status of a true science, the science of the whole. His main argument is the intrinsic limitation of any object assumed by a particular science, while trough common sense we can know reality as a whole and make it the object of an inquiry. In an essay published in 1999 Agazzi expounds his argument as follows:

> A science is never concerned with the entire domain of 'reality'; rather, from this it designates its specific domain of 'objects' by resorting to some 'predicates' which can be thought of as representing its 'viewpoint' on reality. [...] It can be maintained, therefore, that every science characterizes its objects or determines its proper 'domain of objects' by means of its specific predicates. It follows that whatever is not characterized by these predicates falls outside the competence of this science while, on the other hand, everything which can be characterized by them falls within its competence. Every such set of specific predicates determines 'the whole' of physics. By adjoining to this the whole of chemistry, the whole of biology, etc., one obtains the whole of natural science. In a kind of limit considerations, by considering the complex of all possible scientific 'wholes' one obtains the 'whole of science', which may be considered as characterized by the totality of all possible empirically definable predicates. For this reason, we could say that the specific domain of science is 'the whole of experience.' This is because the 'objects' of science in general are built up by means of primitive empirical predicates, which fact automatically limits the competence of science to what can be described by such predicates. The 'choice' of each set of primitive predicates is itself contingent. While this determines the whole of a certain science, it cannot prevent other sciences from being both different and equally legitimate 'viewpoints' upon reality. The choice of such viewpoints is in fact a matter of 'decision' and 'interest' [...]. If we apply this remark to science, we must say that adopting a scientific attitude towards reality amounts to taking the decision to place oneself from the viewpoint of the 'whole of experience', as we have already discussed. This decision is certainly fully legitimate. It does not, however, state a necessity, but is contingent; nor can it exclude other decisions and viewpoints from being equally legitimate. In particular, one could be interested in investigating reality from the viewpoint, not of the 'whole of experience', but of the 'whole' without further specification. In this case, he would not be obliged to limit himself to statements which could be traced back to experience. Such a condition is compulsory for science only because the 'whole of experience' constitutes its specific domain of inquiry, but this cannot be the condition for admitting statements which are concerned with the 'whole' without limitation. If now we qualify metaphysics as the effort to investigate reality from the viewpoint of the 'whole', which is different from investigating 'the whole of experience', the verification principle cannot constitute an objection because it is simply a 'demarcation' criterion which circumscribes only the domain of science (i.e., the domain of the 'whole of experience'). What does not fulfill this principle can be said to fall outside science, but not outside all meaningful inquiry (Agazzi 1989: 45).

For Agazzi science cannot rationally stop questionsonthe whole as such; much more, there are moments when the viewpoint of the whole comes into play within the scientific discourse itself. As Agazzi said above, each specialized field of scientific research suffers a kind of contingency. This implies the characteristic of 'refutability' for scientific statements, as Popper suggested. One can never be sure that reality (or even only physical nature) can be described fully by means of those predicates which are selected in order to establish a certain domain of inquiry. Hence, one must always expect to be confronted with aspects of reality which cannot be treated by means of the accepted tools of inquiry. When such cases appear, one is faced with the problem of the 'whole', in relation to which he must measure the inadequacy of his previous viewpoints. So Agazzi can say:

> Speaking more generally, whenever one is concerned with the problems of the 'foundations' of science—and this happens not only in the philosophy of science, but at times also in science itself—one cannot help being involved with the viewpoint of the 'whole'.

> These clarifications make possible a clear evaluation of the philosophical position which reproaches metaphysics for neglecting in its statements the continuous control of experience. In order to be correct, that is, in order not to confuse the 'contingent' choice of the viewpoint of the 'whole of experience' that characterizes science with a 'necessary' requirement for every meaningful discourse, those advocating that position must prove that the 'whole' coincides with the 'whole of experience'. Surely, there is no such proof in the entire history of philosophy, and such a claim must be held to be purely dogmatic. What is more, if such a proof were ever to be proposed it would necessarily be metaphysical, for in order to show that the 'whole' coincides with the 'whole of experience' one cannot help taking 'the viewpoint of the whole' which means adopting a metaphysical attitude. What has been said thus far is fair not only to metaphysics, but to science, because it does not claim that science contains at least some metaphysical elements, as some philosophers today seem to maintain. In fact, when we established that science is obliged to admit mediation of experience, to accept non-empirical elements in its theoretical apparatus and to resort to a synthetic use of reason, one might have felt inclined to consider all that as a claim that these are unavoidable metaphysical components of any scientific knowledge. But this is not true because all these elements always concern the 'whole of experience' (Agazzi 1989: 51–52).

In order to show how this can happen, Agazzi gives an example from physics. A concept like that of an 'electron' is clearly obtained by a *mediation* of the empirical evidence, since it does not refer to something *directly* observable: it is a theoretical construct and, as such, non-empirical. Despite that, this concept should not be classified as *metaphysical* because the predicates through which it is characterized are still the usual predicates adopted to circumscribe the whole of physics, like mass, charge, etc. This shows that it is possible to mediate experience, which means to transcend the field of immediate evidence, without leaving the whole of experience as a thematic domain of inquiry. On the contrary, when a metaphysician says, e.g., that God exists, he does not intend that this entity is definable through the same predicates as those characterizing the usually experienced things, but, quite the contrary, that it belongs to a broader whole with respect to the whole of experience, that is the whole of reality as such, without any restriction.

1.3 Metaphysics and the Relationships Between Man and Nature

After discussing the legitimacy of holding the viewpoint of the 'whole' along with the viewpoint of science and after laying a sound foundation for this, Agazzi proceeds to see whether such a viewpoint, besides being legitimate, is somehow required, and he demonstrates that this is actually the case. His point of departure are the relationships between man and nature. He says:

> Can such a question be envisaged correctly with the help of scientific knowledge only, or does it also call metaphysics into play? Beyond all doubt a metaphysical consideration cannot be dispensed with, because every possible proposal about the correct way of conceiving this relationship follows from an 'interpretation' of man and nature respectively, which cannot be attained by means of science alone. In fact, every scientific consideration

> necessarily unifies man and nature, but this happens simply because, as repeatedly noted above, every science must employ its own uniforming criteria or 'viewpoints' or 'specific predicates.' [...] Every science is done by instituting uniformities and deleting differences; i.e., by introducing at least one viewpoint under which things can be considered as uniform even if they differ under many other viewpoints. If this be the cognitive procedure of science, it can be easily understood that one can scarcely expect to discover differences between man and nature by continuously applying tools of inquiry which render only uniform knowledge of the two. On the other hand, if the two terms of the relation are not conceived as distinct the very problem of their relationship becomes immediately meaningless because identity is the only relation that can hold between two indistinguishable. It follows that only a metaphysical perspective, which enables one to consider man as a 'whole' and nature as another 'whole' can provide the correct approach to our question. Moreover, in order to study this relationship we need a broader viewpoint; we must conceive man and nature from the viewpoint of a 'whole' in which there is place for both. Such a viewpoint cannot be the rather general viewpoint of the 'whole of experience' because, from a purely methodological consideration, we cannot be sure that the adoption of this viewpoint, which despite its breadth is still specialized, would not lead us to neglect differences which cannot be perceived within it. The only methodologically correct position is therefore to adopt the genuinely general viewpoint of the 'whole' without specification, i.e., the authentic metaphysical viewpoint (Agazzi 1989: 55).

Such an epistemological defense of the legitimacy of metaphysics was for me an efficient support in building my own theory on the relationship between common sense and metaphysical research, whose essential goal is the rational mediation (i.e., interpretation) of the immediate certainties about the world, namely the discovery of the causes of the existence of beings and the determination of their true nature. Unifying Agazzi's epistemological justification of metaphysics with Gilson's theory of realism as the very method of metaphysics (cf. Gilson 1990), I developed my own theory of common sense as the primary knowledge of reality as a whole, and so a material (not yet formal) metaphysical knowledge, whose truth, although pre-scientific, is absolutely necessary to make metaphysics possible as a scientific approach to the whole of reality (cf. Livi 2010).

2 Can One Detect in Agazzi an Implicit Theory of Alethic Logic as Unifying the Presuppositions of All Kind of Knowledge?

The emphasis that Agazzi puts on the truth-value (which is the central value of knowledge if it is considered from a logical point of view) made me understand that Agazzi's arguments fitted perfectly with what I maintained from the beginning, i.e. that alethic logic should detect the logical foundation of all kinds of knowledge on the common certainties, warranted by the truth of primary immediate knowledge. This thesis was the hardcore of the system of alethic logic I have proposed under the title of a 'philosophy of common sense' in order to defend the epistemic primacy of common sense among all kinds of ordinary and scientific knowledge (see Livi 2013a, b, c). Starting just from Agazzi's thought on the relationship between science and its logical presuppositions, I have tried to build a holistic theory of truth which were capable to support my system of alethic logic with a coherent justification (see Livi 2005, 2013c). In this path toward my final purpose Agazzi's thought was once again quite useful, especially for finding a formal and proper determination of the notion of "common sense". Actually, I discussed this issue with Evandro Agazzi in many occasions (cf. Agazzi 2007a, b, c, d; Livi 2006, 2007). In the general idea of what common sense is in a system of alethic logic, Agazzi had no difficulty to agree with me. Referring to my proposal of a holistic theory of truth based on common sense certainties, he wrote:

> Antonio Livi, a contemporary thinker who has devoted a deep reflection to the theme of common sense philosophy, starts from the point of view according to which philosophy aims at attaining *truth* in a full and absolute sense, by using rigorous logical methods that do not reduce to the pure requirement of *formal* correctness, but aim at attaining the possession of truth (*alethic* logic). In this task philosophy cannot avoid scrutinizing the purport of pre-philosophical knowledge that presents itself as true, and that constitute its starting point (i.e. the "presupposition" in a genuine sense and not simply in a "hypothetical-deductive" sense). Such a set of presuppositions coincides with commons sense, understood as a core of absolute and final "primary truths" that motivate and steer the philosophical inquiry. The author explicitly affirms that this philosophical approach consists in a realist position, which recognizes a full value to common sense as a complex of "empirical evidence" that is uncovered phenomenologically. Common sense consist in a precise *set of judgments* and not in a vague "ordinary knowledge", and its positive valuation expresses the position of a "philosophy of the world" that is at variance with the modern and immanentist "philosophy of thought" explicitly inaugurated by Descartes. The defense of this realist position and of the related demonstration of the "existence" of common sense is presented along the lines of the impossibility of getting rid of any presupposition (that has been defended many times in the history of thought) and in addition by maintaining that truth is a property of any single judgment in which the adequacy with that reality that it intends to express is recognized, with no need of an additional foundation consisting in a justification of such a judgment within the context of a "total truth" attained by the reflecting thinking. Therefore, the truth of the judgment is *recognized* by thinking but not *posited* by it. This is the reason for which philosophical thinking can and must recognize the contents of common sense, though accepting the task of making them explicit and deepening them, and also of integrating them with other truths that might not be given in the immediateness of an intuition (Agazzi 2007c: 194).

In the same essay Agazzi declared himself to be in perfect agreement with me even when the topic was the relationship between common sense and scientific thought (both physical and metaphysical). So he says:

> Antonio Livi underscores the reasons for which metaphysics, understood as a rigorous and truth-bringing philosophical discipline, cannot avoid referring to common sense. In the first place, the certainties of common sense constitute true judgments that represent the starting point of *all* the problems with which metaphysics is concerned. Indeed metaphysics cannot deny these certainties, but takes them as an inexhaustible source of *problems* since these immediate certainties demand to be understood and explained, that is, they demand that the *reasons* for them be given. Metaphysics has the proposal of searching for these reasons up to the bottom, that is, of uncovering the *ultimate reasons*. The philosophy of common sense, besides demonstrating its existence and ineliminability, brings to light those absolutely true "judgments of existence" that must guide metaphysics in its scientific course. They are: the existence of a world of things in movement, the

existence of the I that knows the world, the existence of other entities similar to me, the difference of the relation among the various I and of these with the objects of the world, the existence of an ordering intelligence that is also the ultimate end of the order of the world (Agazzi 2007c: 198).

2.1 What Precisely 'Common Sense' Means in the Alethic System Proposed by Agazzi

The difference between Agazzi's way of conceiving alethic logic and mine can be detected when Agazzi expounds what precisely he means by the term 'common sense'. Actually, the evaluation of how much scientific knowledge is founded on pre-scientific knowledge depends directly on the definition of the epistemic function of common sense. If common sense is meant as the logical system of the first undeniable *truths* based on direct experience, then the truth-value of any scientific research depends on the truth-value of common sense (this is just what I maintain). On the contrary, if common sense is meant as a vague complex of popular *opinions* or pre-scientific *interpretations* on the facts of experience (this is just what Agazzi maintains), then the results of scientific research have to be evaluated as more certain and more founded with respect to common sense views of the world. In order to clarify this epistemic difference, I will *firstly* expound Agazzi's concept of 'common sense' as the primary logical presupposition of any kind of scientific research. *Secondly*, I will compare it with my own notion of "common sense".

The first step of Agazzi's research on this issue was a critique of the traditional opinion held by modern scientists and philosophers of science. Actually, they maintain that all scientific knowledge, while being sometimes merely hypothetic, is *opposed* to common sense, either through a direct falsification or through an indirect demonstration that it is impossible to think it is true. Agazzi, after clarifying the primary criterion of truth as that of 'evidence' (referred both to sensible data and to rational principles), denies the identification of 'evidence' with 'immediate experience': the former is, according to his language, something purely rational, while the latter is a popular belief, something *obvious*. On the basis of this distinction, he deconstructs my theory of common sense certainties as the primary truths, since common sense seems to have a constitutive lack of rational self-justification:

> The contents of common sense have the characteristic of *shared opinions* that are such in a spontaneous and non-reflexive way. Therefore, they are accepted as *obvious*. The convictions of common sense, however, are characterized by the fact of not "exhibiting" the credentials of their soundness, since they are spontaneous and non-reflexive and the only indirect support for their soundness seems to be represented by the fact that they are very widely "shared", and this more or less amounts to saying, "How is it possible that we all are wrong?" and this is obviously not a very strong argument. It can be reinforced, however, when one underscores that such a very wide acceptance of such beliefs depends on the fact that they do not need a special effort in order to be clarified and founded, that they impose themselves *spontaneously*, that is, that they are *obvious*. Precisely this claim, however, lies at the root of the oppositions to common sense that have arisen on the ground of philosophy and science, which we could summarize in the following remark: obviousness

> and *evidence* are not the same thing and the progress of human knowledge has consisted in a kind of continuous struggle against obviousness for conquering evidence. This last can be defined as the characteristic of truth that "imposes itself" to reason in such a way that it cannot be put in doubt and even less denied (Agazzi 2007a: 12–13).

Agazzi offers some examples which show that the scientific explanation of phenomena does notdeny the common sense sensible evidence:

> Let us consider again a common sense "sensible evidence": the fact the sun appears at a certain point of the horizon at morning, it covers an arch in the ski during the day and disappears under the horizons at a different point at sunset. A way for expressing this complex evidence in a compound judgment is the following: the sun moves in a circular motion around the resting earth during a revolution period of 24 h. This offers the "why" or the "rational explanation" of the mentioned sensible evidence. In this judgment, however, are unconsciously used the concepts of motion and rest as "absolute" properties of the bodies and, in particular, the property of being in a state of absolute rest is attributed to the earth ("geostatic" conception). Nevertheless the same sensible evidence can be rationally explained by saying that the sun finds itself in a state of absolute rest and the earth rotates on itself with a rotation period of 24 h ("heliostatic" conception). The two conceptions turn out to be even equivalent if rest and motion are conceived as properties "relative" to a given system of reference and their ascertainment is related with the fact that the observer be linked with the one or the other reference system. Assuming that the earth constitutes the "natural" reference system is again a common sense conviction that, nevertheless had to be abandoned for cogent *reasons* when "heliocentrism" has superseded "geocentrism". However, that at least empty space (and similarly time without events occurring in it) can constitute "absolute" reference systems is a well rooted conviction of common sense that, as is known, is explicitly admitted in Newtonian physics but had to be abandoned with relativistic physics. Let us note, however, that the sensible evidence of common sense has never been denied in giving the *reasons* for it (Agazzi 2007c: 14–15).

I think that Agazzi's argument is in itself perfectly correct. It contributes, together with many other arguments, to the development of Agazzi's continuous and severe critique of what he calls 'scientism' (see Velazquez 2012). But it does not deal with what I mean when I speak about 'common sense'. I don't refer to the popular beliefs existing in each time about the cosmological order. All beliefs of this kind are always changing according to the different ages and cultural environment in which they are born. On the contrary, the beliefs I refer to belong to every man in every time and in every cultural environment. In relation with such kind of beliefs physical science cannot ignore them, because scientists are themselves thinking subject starting from those evident data in their own scientific research. Actually, no scientist can really think that they are not true. A scientist can only correct some popular *interpretation* of those data if they seem inadequate. The scientific aim is not to deny the truth of common sense beliefs, but to furnish human knowledge with a correct causal explanation of what everybody already knows, i.e. to discover, when it is possible, the cause (material, formal, efficient and final) of each event belonging to the physical world. I can quote, in this regard, what an American scholar correctly said:

> Some might object by pointing out that many conclusions of contemporary science seemingly contradict common sense knowledge, e.g., quantum mechanics, or strange astrophysical phenomena. The response is that technical scientific conclusions offer explanations of certain phenomena beyond the proper context of common sense (either in a

very small context or a very large one), yet they do not deny the irreplaceable value of common sense for its own context. To assert that Copernican astronomy or Einsteinian relativity contradict our common sense knowledge of the world implies a misunderstanding of what common sense knowledge is. In the pre-Copernican understanding of the world, the assertion that "the Sun revolves around the Earth" would not have been considered a principle of common sense as fundamental as, say, "the external world exists", but rather as a quite natural and spontaneous explanation of the phenomena we all observe everyday. That "natural explanation" has been replaced with a truer one, even though the observed phenomena did not change in any way and the core principles of common sense were not contradicted. We still refer today to the Sun as moving from East to West in the sky, even though we know that in fact is the Earth which moves; this does not contradict common sense, as the notion that "there are no objects" or "the sun does not exists" would. This is precisely the way in which proper scientific discovery comes about, in continuity (and not in contrast to) common sense knowledge (Larrey 2007: 46).

2.2 *Has Common Sense an Alethic Primacy in the Epistemic System Proposed by Agazzi?*

I think that a dialogue between particular sciences and metaphysics is certainly necessary, and de facto always beginning anew. However, the threats of reciprocal interference or methodological commingling can be avoided, as can attempts to annex one to the other, deriving from epistemological errors (as in the age of classical philosophical cosmology or the age of modern positivism) if a basis of conceptual agreement is found. This basis can be found precisely by returning to the common epistemological derivation of the certainties and contents of direct experience, thus circumventing any attempt at an impossible, unmediated translation (without the mediation of those basic certainties) of the technical language of metaphysics into the technical language of the other sciences, or vice versa. I maintain that the organic body of certainties of which I speak is in itself *qualitatively* superior to science, including within the notion of "science" both the metaphysical as well as the particular sciences, that is, all reflexive and systematic knowledge, mediated by reasoning and by culture and equipped with its own technical methodology. The superiority of those certainties lays precisely in the quality of the certainty itself. While the certainty of those direct and universal statements is unconditional and absolute, scientific certainty always has limiting characteristics: either it is a subjective certainty linked to privileged conditions of experience or intellectual capacity; or it is a certainty available only to members of a specific cultural community equipped with its own research instruments; or it is a certainty that can be reached by all humankind, only in relation to particular historical events, or a certain historical level of technological development, or a particular historical perspective on human events; or it is a provisional (hypothetical) certainty, susceptible of falsification or at least revision and adjustment, when it is not a bold working hypothesis of an instrumental nature (that is, as an efficacious tool for research, for science itself, or a mere instrument at the service of technical work or other practical ends). So, while common sense certainties are incontrovertible, any science certainty is debatable, or at least relative, perfectible, revisable; the former belong to all people at all times, the latter belong to some people and only in specific moments of personal or collective history. Secondly, the superiority of these certainties over science is also a *primacy in truth*: they come first, both logically and chronologically (because every reflection presupposes some direct knowledge to which to return), and are also elevated above other certainties, in the sense that they cannot be contradicted (that is, proven false), to the point that every scientific thesis that contradicts those certainties is for that very reason vitiated by error (even if that error can only be demonstrated at the level of science).

2.3 The Meaning of 'Common Sense' in Agazzi Pertains More to the Sociological Tradition then to the Epistemic One

Summing up, it can be said that in my holistic system of truth I use the term 'common sense' in its properly alethic meaning, while Agazzi accepts the usual sociological or psychological meaning. As many contemporary essays on this topic, he stands in the tradition of past and current common sense philosophers, like Thomas Reid, George Berkeley, Henry Sidgwick, George E. Moore, James B. Conant, Radu J. Bogdan, and NohaLemos, who defend common sense against rationalism in philosophy and in science. Some other thinkers go beyond their accounts by not only defending common sense but also considering what common sense means (cf. Davidson 1984; Boulter 2007; Piccari 2011). Unfortunately, no one of these scholars realized the possibility and the necessity of passing over the limits of the phenomenological research (social psychology, cultural anthropology, sociology of culture, and so on). A clear example of this could be the research performed by the Swedish Marion Ledwig. Besides giving a historical exegesis of common sense in Thomas Reid and showing parallels in Austin, Searle, Moore, and Wittgenstein, he discovered common sense also in Hume's *An Enquiry Concerning the Principles of Morals* and in Kant's *Critique of Pure Reason.* But the final interpretation of common sense made by this author does not reach at all the level of alethic logic. With his essay he aimed only to make clear how far common sense generalizes, whether proverbs are a form of common sense, and whether common sense can be found in the common knowledge assumption in game theory. Also, he holds that folk psychology should be considered as a common sense psychology (see Ledwig 2007). On the contrary, my own notion of common sense pertains to epistemology, which is the main issue of philosophical logic, as Evandro Agazzi himself acknowledges (see Agazzi 1981). Actually, my philosophy of common sense should be understood as something similar to what Roderick Chisholm called 'the foundations of knowing' (cf. Chisholm 1972), and John Searle calls "default position" (cf. Searle 1998: 10).

If understood in this epistemic meaning, common sense should be *the first step* of a theoretical process which leads to overpass the simply *semantic* holism, i.e. the holism of meaning (see Tarski 1944), in order to take into account the *alethic* holism, i.e. the holism of truth. This is made possible by detecting a set of logical connections between judgments based on the truth as the basic value of judgments (see Davidson 1990a, b). The result is an axiomatic system of epistemic logic based on the acknowledgement of the real dependence of every judgment on the truth of its necessary presuppositions, or logical conditions of possibility for it to be true. Then, this is the general framework of what I conceive as the holism of truth. According to this logical system, any thought of truth—and any assertion which can express it—is linked with all the other thoughts in its very epistemic justification, through the need of finding its own premise and presuppositions. In such a holistic system, my notion of common sense retains a very narrow extension, since it refers *only* to few, well determinate primary certainties which are the common presupposition of both ordinary and scientific knowledge in all their forms and degrees. In others words, "common sense" is just the hard core of the holistic system of truth. I reached such a conclusion taking in account not only Agazzi's philosophy of science but also the basic results of the cognitive science (see Smith 1995a), the ontological research performed by some analytic philosophers (see Kripke 1980), and the most advanced studies on the philosophy of mind (see Searle 2004), and the best results of the phenomenology of consciousness—which makes use both of subjective introspection and the analysis of the inter-subjective communication. I realized that in the consciousness of every thinking subject there are some certainties about the 'real world' –certainties whose epistemic justification is founded on the immediate evidence of existing beings which *necessarily* and *always* are present in everyone's experience. In Searle's philosophy of mind such permanent presence of some existing beings is named 'original or intrinsic intentionality':

> Where the mind is concerned we also need a distinction between original or intrinsic intentionality on the one hand and derived intentionality on the other. For example I have in my head information about how to get to San Jose. I have a set of true beliefs about the way to San Jose. This information and these beliefs in me are examples of original or intrinsic intentionality. The map in front of me also contains information about how to get to San Jose, and it contains symbols and expressions that refer to or are about or represent cities, highways, and the like. But the sense in which the map contains intentionality in the form of information, reference, aboutness, and representations is derived from the original intentionality of the map makers and users. Intrinsically the map is just a sheet of cellulose fibers with ink stains on it. Any intentionality it has is imposed on it by the original intentionality of humans. So there are two distinctions to keep in mind, first between observer-independent and observer-dependent phenomena, and second between original and derived intentionality. They are systematically related: derived intentionality is always observer-dependent (Searle 2004: 7).

2.4 How Agazzi's Alethic Logic Is Limited by the Rationalist Roots of His Epistemology

In my system of alethic logic such certainties constitute the very first link in the chain of presuppositions; so that they can in no way be subject to doubt. This means that their non-truth is absolutely unthinkable: actually, no one can ever really doubt them, and one must understand that any affirmations to the contrary are merely verbal posturing: actually, they respond to some pragmatic logic, and not the expressions of a real certainty, endowed with its own adequate epistemic justification. They constitute the nucleus of 'experience', understood not only as a body of sensible data but as the body of any kind of unmediated knowledge. Much more, such certainties are present to consciousness in every moment of the search for truth as the logical presupposition of all knowledge deriving from reflection and inference, both inductive and deductive. For this same reason, such certainties function as an ultimate criterion of truth to verify any hypothesis successively formulated. They therefore constitute the main *alethic presupposition*, that is, the presupposition necessary for any further knowledge to be thought of as true. In fact, on the basis of these original truths, every thinking subject verifies, time after time, the admissibility of any hypothesis—formulated by himself or proposed by other subjects through one of the ways for communicating thought—that presents itself in the search for other truths over the course of his lifetime. As a result, all scientific knowledge should be structured as a system logically compatible with the primary truth of common sense(see Nagel 1961) , so as to place the instruments of dialectics (reflection, interpretation, inference) effectively at the service of the search for further truths(see Gadamer 1960).

Agazzi cannot agree with me on this issue because he holds that the *logical form* of common sense is only that of some popular believes which are largely shared but cannot be founded or proved, while their *content* is very difficult to be determined by the means of sociology of culture. For this reason Agazzi holds that it is impossible to avoid a hard opposition between common sense and science, or almost between common sense and reflexive, methodical and critical thought. So, all science of nature and much more metaphysics find themselves in basic conflict with common sense. Evandro Agazzi, as many other scholars who were followers of Gustavo Bontadini (see Rivetti Barbò 1990; Carlani 1995), establishes a formal distinction between what is believed because it is *obvious* and what is believed because it is *evident*, while identifying evidence with *incontrovertibility*. On the contrary, I think that many evidences—among which the primary evidences which constitute common sense—do not have the epistemic quality of an abstract rational incontrovertibility, since they show simply things existing and facts happening in the world, and this kind of reality is in itself something contingent, non-necessary. Nevertheless, the certainty we have about this is absolute, so that it is impossible that somebody can think that they are false, as I have said above. Something similar happens when Agazzi, adopts the distinction (introduced by Bontadini) between *certainty* and *truth*. For him, certainty is an *epistemic* property of a proposition: it expresses the way we are subjectively sure that what we have said is true, but this has no practical effect on its effective truth or falsity (cf. Agazzi 1997).

References

Agazzi, Evandro. 1961. *Introduzione ai problemi dell'assiomatica.* Milano: Vita e Pensiero.
Agazzi, Evandro. 1974. *Temi e problemi di filosofia della fisica.* Roma: Abete.
Agazzi, Evandro. 1978. Eine Deutung der wissenschafltichen Objektivität. *Allgemeine Zeitschrift für Philosophie* 3: 20-47.
Agazzi, Evandro (ed.). 1981. *Modern Logic. A Survey.* Dordrecht: Reidel.
Agazzi, Evandro. 1985. *La questione del realismo scientifico.* In *Scienza e filosofia,* ed. CorradoMangione, 170-200. Milano: Garzanti.
Agazzi, Evandro. 1987. *Philosophie, Science, Métaphysique.* Fribourg: Éditions Universitaires.
Agazzi, Evandro. 1988. *L'objectivité scientifique.* In *L'objectivité dans les différentes sciences,* ed. Evandro Agazzi, 13-23. Fribourg: Éditons Universitaires.
Agazzi, Evandro. 1989. *Science And Metaphysics Before Nature.* In *Person and Nature*, ed. George F. McLean, Hugo Meynell, 20-55.Washington, D.C.: University Press of America.
Agazzi, Evandro. 1991. *The Problem of Reductionism in Science.* Boston: Kluwer Academic Publishers.
Agazzi, Evandro. 1997. *On the Criteria for Establishing the Ontological Status of Different Entities.* In *Realism and Quantum Physics*, ed. Evandro Agazzi, 40-73.Rodopi: Amsterdam.
Agazzi, Evandro. 1998. *Normatività logica e ragionamento di senso comune.* In *Normatività logica e ragionamento di senso comune,* ed. Francesca Castellani and Luisa Montecucco, 45-66. Bologna: Il Mulino.
Agazzi, Evandro. 2007a. *Introduzione.* In *Valore e limiti del senso comune,* ed. Evandro Agazzi, 9-20. Milano: Franco Angeli.
Agazzi, Evandro. 2007b. *Il senso comune e l'unità dell'esperienza.* In *Valore e limiti del senso comune,* ed. Evandro Agazzi, 25-38. Milano: Franco Angeli.
Agazzi, Evandro. 2007c. *Senso comune e filosofia.* In *Valore e limiti del senso comune,* ed. Evandro Agazzi, 193-200.Milano: Franco Angeli.
Agazzi, Evandro. 2007d. *Continuità e discontinuità fra scienza e senso comune.* In *Valore e limiti del senso comune,* ed. Evandro Agazzi, 341-356. Milano: Franco Angeli.
Agazzi, Evandro. 2012. *Ragioni e limiti del formalismo. Saggi di filosofia della logica e della matematica*, ed. Fabio Minazzi. Milano: Franco Angeli.
Agazzi, Evandro. 2014. *Scientific Objectivity and Its Contexts.* Berlin: Springer International Publishing.
Boulter, Stephen. 2007. *The Rediscovery of Common Sense Philosophy.* New York: Palgrave Macmillan.
Carlani, Paolo. 1995. Senso comune e filosofia: dall'ovvio all'evidente, tra scetticismo e certezza. *Poietica* 12: 18-48.
Chisholm, Roderick. 1972.*The Foundations of Knowing.* Minneapolis: University of Minnesota Press.
Davidson, Donald. 1984. *Inquires into Truth and Interpretation.* Oxford: Clarendon Press.
Davidson, Donald. 1990a. The Structure and Content of Truth. *The Journal of Philosophy* 12: 279-328.
Davidson, Donald. 1990b. *Truth and Predication.* Cambridge, MA: Harvard University Press.
Gadamer, Hans-Georg. 1960. *Wahrheit und Methode. Grundzüge einer philosophischen Hermeneutik.* Gottingen: J. C. B. Mohr Verlag.
Gilson, Étienne. 1990. *Methodical Realism.* Trans. A. Pegis. Front Royal: Christendom Press.
Kripke, Saul A. 1980. *Naming and Necessity.* Oxford: Blackwell.
Larrey, Philip. 2007. *Thinking Logically: a Historical Critique of Trends in Contemporary Philosophy of Logic in the Analytic Tradition.* Aurora, Col.: The Davies Group Publishers.
Ledwig, Marion. 2007. *Common Sense: Its History, Method, and Applicability.* New York & Oxford: Peter Lang International Academic Publishers.
Livi, Antonio. 1990. *Filosofia del senso comune. Logica della scienza e della fede.* Milano: Ares.
Livi, Antonio.2005. *Senso comune e logica aletica.* Roma: Casa Editrice Leonardo da Vinci.

Livi, Antonio. 2006. Discussione con Evandro Agazzi sulla filosofia del senso comune. *Aquinas* 49: 503-516.
Livi, Antonio. 2007. *Evandro Agazzi e la filosofia del senso comune.* In *Filosofia, scienza e bioetica nel dibattito contemporaneo. Studi internazionali in onore di Evandro Agazzi,* ed. Fabio Minazzi, 225-234. Roma: Istituto poligrafico e Zecca dello Stato.
Livi, Antonio. 2010. *Metafisica e senso comune. Sullo statuto epistemologico della "filosofia prima"*. Roma: Casa Editrice Leonardo da Vinci.
Livi, Antonio. 2013a. *A Philosophy of Common Sense. The Modern Discovery of the Epistemic Foundations of Science and Belief.* Aurora, Col.: The Davies Group Publisher.
Livi, Antonio. 2013b. Why Common Sense, When Assumed As the Primary Truth, Furnishes Alethic Logic System With its Very Epistemic Justification. In *La certezza della verità. Il sistema della logica aletica e il procedimento della giustificazione epistemica,* ed. Antonio Livi, 19-30. Roma: Casa Editrice Leonardo da Vinci.
Livi, Antonio. 2013c. Perché ogni ricerca di un'adeguata giustificazione epistemica presuppone una coerente teoria sistemica della verità. In *La certezza della verità. Il sistema della logica aletica e il procedimento della giustificazione epistemica,* ed.Antonio Livi, 217-230. Roma: Casa Editrice Leonardo da Vinci.
Nagel, Ernest. 1961. *Science and Common Sense.* In Id., *The Structure of the Science: Problems in the Logic of Scientific Explanation*, 547-605. Indianapolis, Indiana: Hackett.
Newell David J. 1980. *Philosophy and Common Sense.*Washington: University Press of America.
Piccari, Paolo. 2011. *Conoscenza ordinaria e senso comune*, with an Introduction by Mariano L. Bianca. Milano: Angeli.
Reffes, A. J. 2008. *The Philosophy of Realism, Common Sense, and Ordinary Experience*. Berlin: Progress Publishers.
Renzi, Fabrizio. 2012. *La logica aletica e la sua funzione critica.*Roma: Casa Editrice Leonardo da Vinci.
Rivetti Barbò,Francesca. 1990. Dubbi, discorsi, verità. Milano: Jaca Book.
Searle, John R. 1998. Mind, Language And Society: Philosophy in the Real World. New York: Basic Books.
Searle, John R. 2004. *Mind: A Brief Introduction.* New York: Oxford University Press.
Smith, Barry. 1995a. Formal Ontology, Common Sense, and Cognitive Science. *International Journal of Human-Computer Studies* 43: 641-667.
Smith, Barry. 1995b. The Structure of Common Sense World. *Acta Philosophica Fennica* 58: 290-317.
Tarski, Alfred. 1944. The Semantic Conception of Truth. *Philosophy and Phenomenological Research* 4: 341-376.
Tarski, Alfred. 1959.*Introduction to Logic and Methodology of Deductive Sciences.* Oxford: Blackwell.
Velázquez, Lourdes. 2012. *Verità e certezza. La crisi dello scientismo e il realismo del senso comune.* Roma: Casa Editrice Leonardo da Vinci.

Interpretation and Hermeneutical Judgment

Jure Zovko

Abstract An original characteristic of Agazzi's philosophy of science is that, beside adopting several outlooks of the analytic approach, he also adopts some fundamental perspectives of hermeneutics, obtaining in such a way a very fruitful and stimulating complementation of those two ways of philosophizing, that otherwise have been considered (and often are still considered) to be at variance. The theory of objectivity advanced by Agazzi is based on the thesis that every science expresses a "point of view" on reality, that every scientific theory expresses a certain Gestalt within the respective domain of objects, that interpretation and explanation are indispensable both in natural and human sciences (and that explanation takes place within a certain interpretation), and on the fact that scientific objectivity is always "historically determined", because it develops within a historical-cultural context: these are all elements that would justify the thesis that Agazzi has actually offered a hermeneutic philosophy of science capable of safeguarding also the requirement of objectivity that cannot be overlooked in interpretations. In particular, this is supported by his explicit devoting a section of his major work to the "hermeneutic dimension of science". These facts are recalled in the present contribution, that goes on presenting the basic features of hermeneutics in general, offering in such a way an explicit complement to what is often only implicit in Agazzi's writings.

J. Zovko (✉)
University of Zagreb, Zagreb, Croatia
e-mail: jzovko@ifzg.hr; jzovko@unizd.hr

J. Zovko
University of Zadar, Zadar, Croatia

M. Alai et al. (eds.), *Science Between Truth and Ethical Responsibility*,
DOI 10.1007/978-3-319-16369-7_13

1 Introduction

Evandro Agazzi's philosophy of science has attained a special reputation for the originality of his theory of objectivity, which organizes in a coherent perspective the referential aspect of science grounded in its operational dimension (that links it with ontology and with the domain of truth), together with the linguistic aspect that makes possible all sorts of metatheoretical investigations (and links it with philosophy of language and epistemology in a strict sense), as well as its consubstantial relations with technology (that are at the same time a strong support of a "realist" conception of it and an irrefutable evidence of its structural links with society). There is, however, another element of originality in Agazzi's philosophy of science that may risk to remain little perceived in the wealth of the historical, logical, semiotic, epistemological and ontological discussions that have marked the several decades of the elaboration of his speculation and are now systematically presented in his monumental life-work, *Scientific Objectivity and its Contexts* (Agazzi 2014). This very significant element of originality is the explicit recognition and discussion of the *hermeneutic dimension* of science. This fundamental dimension was already present from the very beginning of Agazzi's conception of science, when he characterized every scientific discipline, and also every theory within a single discipline, as the development of a specific "point of view" on reality: this very way of speaking (repeated and refined dozens of times along his production), already contains the core of a hermeneutic conception, and the interesting fact is that it is not presented at variance with scientific objectivity, but rather as a precondition of it, simply because the standardized operations allow for the intersubjective agreement about the features of reality accessible from the chosen point of view. This fundamental original view has been then deepened in his development of a "perspectivist" view of science, in which also the notion of a *Gestalt* has been widely used and applied.

This has been possible because Agazzi has correctly seen the priority of *understanding* with regard to *explanation* but, contrary to what have maintained several scholars, he did not assign understanding to the humanities and explanation to the natural sciences, but has strongly maintained that in *whatever* science both requirements must be satisfied. The priority of understanding can be expressed in a concise way by saying, as he does, that "it is not possible to explain something that has not been understood", but, in a much more elaborated and expanded form, he shows how a theory (whose specific task is explanation) is always developed within the framework of a given interpretive *Gestalt*. Even "facts" are always affected in science by an interpretation that, however, cannot jeopardize their autonomy based on the operational criteria. We can dispense with resuming these positions of Agazzi since they are systematically presented in a rigorous way in Sect. 3 of Agazzi (2014), titled explicitly "The hermeneutic dimension of science". We want rather to point out an interesting detail. Already in Agazzi (1985) is presented a very stimulating analogy between scientific knowledge and artistic interpretation, by comparing the work of a scientific discipline investigating certain natural phenomena with the different interpretations of one particular musical score that are offered by different interpreters. This comparison (in which similarities and differences are highlighted) is also of interest for a discussion about the possibility of having objectivity in interpretation. Another particularly relevant point where Agazzi's consideration of the hermeneutic dimension of science becomes very clear is his explicit consideration of the *contexts* of scientific objectivity, something that appears in the very title of his major work and is certainly not usual in the mainstream philosophy of science. These contexts are especially "historical" in a broad sense, that is, they are "cultural" and constitute the precondition of every scientific "objectification". Sections 1 and 2 of Agazzi (2014) are thematically devoted to this historical dimension of science. Whoever is familiar with hermeneutic philosophy knows very well how important is the role attributed to the consideration of the "context" in the performance of a correct interpretation, and this not simply concerning the linguistic context of a text, but (also in the case of a text) the appraisal of the influence of the broader cultural context. This is precisely what Agazzi has constantly stressed concerning science, and has called the "historical determinateness" of science meaning by this not a "deterministic" relation that would make of science a "social product", but the more subtle situation of contingent conditions that constitute the background in which the creativity of the single scientists can give rise to hypotheses, theories, methodological innovations, and so on.

It is this strong embedding of science in the cultural environment (that has the characteristics of a double-way feedback loop), that constitutes the foundation of the "cultural value" and "humanistic value" of science, and should inspire also the teaching of science. This is the idea and the ideal that has inspired Agazzi's long commitment in educational issues, and in particular has represented the *leitmotif* of the dozens of editorials and articles he has written for *Nuova Secondaria*, the monthly journal addressed to high-school teachers that he has edited for 30 years and whose aim was to promote a "cultural" maturation both of the teachers and their students. According to Agazzi, the target of this cultural maturation in the young persons consists in helping them to attain critically accepted "criteria of judgment" and the capacity of performing personal conscious judgments. In the promotion of this capacity consists the work of "formation" (*Bildung*) of school education.

These various constituents of an hermeneutic conception of science are in a way diluted along the whole philosophical reflection of Agazzi, especially because they have been independently and originally elaborated by him mainly at times when his conceptual approaches and methodological tools were particularly linked with those of the analytic philosophy of science, whereas his intellectual and also personal contacts with hermeneutic philosophy came only later. Therefore it seems advisable to offer a kind of "complement" to his treatment by presenting a short systematic survey of the main points of hermeneutic philosophy, giving of it a historical and methodological portrayal certainly more concise than that devoted by Agazzi to scientific objectivity, but equally oriented to show the objective character of interpretation, This is the modest aim of the following considerations regarding hermeneutics.

2 Nature and Scope of the Hermeneutic Reflection

The primary intention of my article is to show that the hermeneutical judgment is integral part of the theory of understanding and interpretation. Dilthey's radical distinction between the natural sciences and the humanities is no longer acceptable. Understanding is not only a feature of the humanities, but also a crucial segment of the natural sciences and engineering. However, hermeneutical reflection does not aim only at an interpretation and explanation of the existing expressive forms of human mind; it also considers the possibilities of some new forms of artistic and cultural creation, and seeks reflective answers to both the challenges of contemporary age and the complex issues pertaining to the modern societies. The primary tasks of hermeneutics also include a complex understanding and judgment of a concrete situation as well as the ways to cope with the issue of application of the universal to the particular. Such universal feature of hermeneutical understanding, modelled after the experience of truth through the work of art, supplies one of the key reasons why, on the close of the last century, hermeneutical philosophy reaches the status of "philosophia prima". In the words of its founding father, Hans-Georg Gadamer, hermeneutical reflection poses the questions that relate to "the whole of human world-experience and the life-practice. Put in Kantian terms, it poses the question of how the understanding is possible".[1]

The issue of responsible interpretation arises from the beginning of hermeneutics in the writings of Matthias Flacius Illyricus (1520–1575). Hermeneutics as a theory of understanding of the works was established in early German Protestantism. Prominent representatives of hermeneutical philosophy, W. Dilthey and H.-G. Gadamer, attribute to Flacius the key role in the founding of hermeneutics as a method of reliable textual understanding and interpretation.[2] Flacius resolutely defended the hermeneutical principle of understanding Biblical texts and wrote a massive guidebook which was to serve as the "key" (*clavis*) to the proper understanding Holy Scriptures. In Flacius' opinion, a text needs to be interpreted autonomously, in accordance with its immanent sense, so that the meaning of particular words, sentences, and parts of the text is discerned through the 'scope' of the text and the totality of its context. We owe to Flacius in this connection two principles of interpretation that are of a crucial importance to the further evolution and constitution of a universal method of interpretation: first, the principle of 'scope' as the basic intention of both the text and the author, which remains the primary purpose and task for any interpretation; and second, the principle of the interconnectedness of the whole and its parts vis-a-vis the textual understanding, which will be later designated as the hermeneutic 'circle'. This implies a complementary application of an inductive-synthetic and a deductive-analytical method of consideration and explanation of a text. Within the synthetic procedure, through

[1] Gadamer (1986): XVII.

[2] Cf. Zovko (2007).

the process of understanding, one needs to integrate the separate parts of the text (*membra*) into a meaningful, coherent whole, a process that Flacius calls *dispositio,* and then, through a heuristic procedure, check the degree to which the consistency of separate parts and passages of the text fits into the wider textual coherence (*quomodo singulae partes se invicem cohaereant*).[3]

The hermeneutical analysis of a text explores the integration of its parts into a meaningful and coherent totality. Convenience (*convenientia*) and consistency of the explications of individual parts of the text, merged together within the coherent complementarity, are confirmed in the light of the contextual intentionality of the text, whereas the principal scope (*scopus principalis*) plays the role of a unifying thread that integrates the meanings of words, sentences, passages, and parts of the text, into a coherent, meaningful totality.

The concept of a responsible form of interpretation was characterised by Georg Friedrich Meier, a prominent representative of German Enlightenment, as principle of "hermeneutic equity" ("hermeneutische Billigkeit") in his classical treatise *Versuch einer allgemeinen Auslegungskunst* (*Attempt at a General Art of Interpretation* 1757), whose primary aim is to decipher the meaning of the text according to the standard of interpretation: "*Hermeneutic equity*" is the inclination of an interpreter to take those meanings as hermenuetically true, which correspond best to the perfections of the author of the signs ("*Die hermeneutische Billigkeit* [*aequitas hermeneutica*] ist die Neigung eines Auslegers, diejenigen Bedeutungen für hermeneutisch wahr zu halten, welche, mit den Vollkommenheiten des Urhebers der Zeichen, am besten übereinstimmen").[4] According to Enlightenment theoreticians of the normative interpretation, the process of heuristic-reconstructive interpretation extends from understanding of the meaning of words on the basis of their usage to the unfolding of the complex hermeneutic truth of meaning. This explanatory principle, according to which the meaning of individual *loci* is mediated through understanding of the intention and composition of the entire work, the meaning of the content of the entire work developed in the coherent congruence of its individual parts, was designated by Dilthey as the pinnacle of hermeneutics.

Many critics of contemporary hermeneutical philosophy claim that the hermeneutical request for sense-discernment is indeterminate and vaguely formulated. This, to a degree, also applies to their relativistic notion of truth as advocated by Heidegger, Gadamer and postmodernists who explain the structure and essence of truth through the concept of "play". We all, seeking to learn and realise something, climb up the language games to the understanding of the world: "we, the understanding ones, are caught into the net of the truth-event and arrive, in fact, always too late, if we want to know what we need to believe".[5] Hermeneutical practice of understanding, through which one needs to arrive at the truths that have to be prevented from falling under the rule of the modern notion of science,

---

[3]Flacius (1968): 100–101.

[4]Meier (1996): § 39.

[5]Gadamer (1986): 465.

actually expresses our belonging to what we understand. Since such hermeneutical reflection dispenses with the assumptions that preside over the standard scientific methodology, its relevance to a reliable textual interpretation and understanding remains extremely questionable.

In the process of understanding and interpreting a written work, the individual parts of the text are connected to a meaningful and coherent unity and made plausible from their context. In order to be able to fathom the intention of the author, the "correspondence theory of truth" proves to be the *conditio sine qua non* or norm of interpretation. The interpretation unfolds as a hypothesis which needs to be confirmed, whereby the criterion of coherence, the criterion of logical correctness or connection, serves as a heuristic tool. It is a question here of a "zetetic" interpretation of texts, for which the research instrumentarium of a general theory of science ("Wissenschaftstheorie") is to be employed, i.e. hypotheses regarding the interpretation are to be tested regarding their consistence and their compatibility with the "facts" concerning the texts. This method of interpretation has enjoyed a thoroughgoing presentation and thoughtful application in the important works of Eric D. Hirsch, Wolfgang Wieland, Hans Krämer, Hans Ineichen, and Nicolas Rescher.[6] In their view, proposals for the understanding and interpretation of texts should be understood according to their position in the formation of hypotheses ("Hypothesenbildung") and carefully tested with respect to their plausibility. The task of textual interpretation consists accordingly in "finding out the most probable hypothesis of interpretation", bringing it into harmony with the text itself, and testing its plausibility within the framework of a circle-like process of examining its complementarity with the understanding of the individual elements of the text and the text as a whole. The meaning intended by the author manifests itself thereby as the standard of the interpretation, or as the an ideal which the interpretations should more or less approach. The approximation claimed here demands a necessarily replicable, critical analysis concerning the question as to whether and to what extent the product yielded by the interpretation deviates from the *interpretandum*.

The greatest danger to interpretation in contemporary human and social sciences is hermeneutical and epistemic relativism, for without normative standards of interpretation no interpretation has any advantage over any other and no explanation is possible at all, a condition which is ultimately insupportable to us human beings because of our natural desire and need to know and understand. American physicist Alan Sokal characterized the tendency of mainstream postmodern Philosophy in his parodist Essay "Transgressing the Boundaries: Towards a Transformative Hermeneutics of Quantum Gravity": "The content and methodology of postmodern science thus provide powerful intellectual support for the progressive political project, understood in its broadest sense: the transgressing of boundaries, the breaking down of barriers, the radical democratization of all aspects of social, economic, political and cultural life".[7]

---

[6]Hirsch (1973), Ineichen (1991), Rescher (1997), Wieland (1999:332 ff).

[7]Cf. *Social Text* (1996, 46/47: 217–252; 229).

The American philosopher Nicholas Rescher, an advocate of the coherence theory of truth, is one of the most decisive opponents to "a relativistic indifferentism" in the theory of interpretation. In a critical debate with Derrida, he claims that the task and purpose of interpretation is not to supply as diverse textual interpretations as possible, but to interpret the text stringently and meaningfully, as demanded by its wider context. Derrida and deconstructionists assume that each text carries within itself limitless possibilities of distinct interpretations that should all be posed as equally adequate and worthy: "As deconstructionism sees the matter, the enterprise of text interpretation accordingly confronts us with an inevitable plethora of coequal alternative possibilities".[8] In this regard Rescher ironically points to some limits to variation, and refers to the example from the Talmudic tradition of forty nine different hierarchies of sense in the interpretation of the Torah, which Walter Benjamin advocated vis-a-vis the interpretation of the work of art in particular.[9] Rescher resolutely claims that the theory of interpretation needs to be normative and guided by strictly determined standards and criteria of evaluation in the matters of truth and falsity, a responsible and irresponsible perspective on the text. The text interpretation is legitimate only if it has been performed in accordance with the meaning of the text and with the criteria of coherence. "It is in fact coherence with the resources of context (in the widest sense of this term) that is at once the appropriate instrument of the text interpretation and the impetus to objectivity in this domain".[10] A responsible hermeneutical interpretation never takes place as an imaginative process of creative playing in the search for diverse possibilities and variations of meaning; but it implies an attempt to track down a correct and adequate immanent meaning of the text, the task the classics of hermeneutics, from Flacius to Dilthey, set for themselves anyway. In that sense, Rescher puts it thus: "But who makes the rules of appropriateness? The answer is that they are not *made by* but *given to* us, not something invented but rather something to be discovered by anyone who examines the range of relevant phenomena with sufficient care".[11] The key motive of each responsible interpretation of a text ought to adhere to the principle of so-called "*hermeneutical optimisation*", which is grounded on coherent contextuality, which means that the more an interpretation is consistent, and the more it coheres with the whole of the context, the more it is entitled to obtain our acceptance.

By the examination and evaluation of different forms and products of human creativity in a variety of historical and cultural epochs, hermeneutic methods and hermeneutic philosophy have shown how there is one single process of application of the universal to particulars and how experience is used in the consideration

[8]Rescher (1997): 198.

[9]Benjamin (1978: 524): "And if I need to couch it in a single expression: I was not able to think and explore but in a, if I may say so, theological sense—in accordance with the Talmudic doctrine on the forty nine levels of each paragraph of the Torah".

[10]Rescher (1997): 204.

[11]Ibid.

of fundamental issues confronting contemporary society. The task of the humanities is not only to protect and promote pluralism in the age of the globalization of information, but also to consider the possibility of intercultural dialogue in the creation of one's own identity. In this sense, the task of hermeneutical methods is to preserve the certainty and reliability of understanding. The object of study in the humanities is not something abstract or foreign to us, but that which we necessarily belong to ourselves: culture and the intellectual tradition which is the fruit of the self-realization of the human spirit. In education we recognize and study the cultural goods passed down to us, and by recognizing and assimilating their inherent value we construct our own personality. By our conscious appropriation of cultural goods we become a part of the tradition which lives on in us. Without the mediation with the present, without actualization and reception, tradition begins to lose its significance. Cultural tradition comprises not only texts and artistic works, but also institutions and social forms. The intellectual wealth of a culture cannot, pictorially speaking, be passed on from generation to generation like a treasure in a chest. Cultural tradition is passed on exclusively in the dialectic remembrance, which, as exemplified in the Platonic dialogues, is developed in the finiteness and contingency of our existence in critical reflection and the cultivation of our individual and collective identity.

The advantage of the hermeneutic method is to be seen in its advocacy of a pluralistic dialogue and of the practice of tolerance in the exchange of a variety of experience with regard to the examination and evaluation of existing ethical and cultural norms, with the aim of preserving and enriching our civilization, evermore threatened by the Moloch of a globalizing abstraction from all concrete cultural contents and contexts. The question of the research method is directly related to the discernment and judgment of the researchers. Cultivation is a process of education (*Bildung*, *paideia*), in which the Understanding appeals to us.

Thinkers of the hermeneutic school of philosophy, taking as their point of departure Hegel's definition of identity as self-consciousness, confirm that identity is not only in an abstract sense a fundamental philosophical concept. By self-consciousness the human individual differentiates himself from animals, because by the capacity for self-conscious reflection the human individual can rise above the experience of particulars to the formulation of general concepts and universal norms. Paradoxically, it is through this process of generalization or universalization that it becomes possible to develop and cultivate one's own personal and cultural identity. By the mere fact of their being born into this world, human beings are not already all that they should or could be; it is necessary to differentiate between the natural conditions of existence as a human being (*conditiones humanae*) and their natural potential. It is only by a complex and life-long education process that the individual re-creates the original characteristics by which he can approach the full expression of his integral personality.

Every individual who on the basis of the natural preconditions of his existence is able to rise to the level of rationality and intuition recreates in the educational process the existing substantiality of the cultural heritage by which his identity to a certain degree is predetermined, and in which he is rooted in a specific, individual manner. The process of growth and education is at once a process of personal maturation in the course of which the individual examines and critically evaluates the preexisting cultural and humanistic heritage, as well as the set of institutionalized norms into which he was born. The wealth of tradition becomes in this process of intelligent (re-)integration and re-appropriation a part of our individuality, while whatever is not subjected to this process remains unassimilated and foreign to us.

John McDowell sees Aristotle's and Hegel's concept of a "second nature" as the basis for a "partial re-enchantment of nature" with regard to a successful integration of nature, knowledge of nature, and ethics in human behavior. This presents a possible alternative to predominant positivist and materialist concepts of nature, which have failed to provide a universally appealing basis for the formation of sound and reliable moral judgment. The integration of nature and knowledge by means of "second nature" depends, however, on education, Bildung.[12] In McDowell's view, education completes and perfects our personality with respect to the world, which is never merely empirical, but also "Lebenswelt". McDowell fails, however, to consider the role of the formation of judgment in this process of understanding of the lifeworld.

McDowell writes that "our nature is largely second nature, and our second nature is the way it is not just because of the potentialities we were born with, but also because of our upbringing, our *Bildung.* Given the notion of second nature, we can say that the way our lives are shaped by reason is natural, even while we deny that the structure of the space of reasons can be integrated into the layout of the realm of law. This is the partial re-enchantment of nature that I spoke of".[13] Wittgenstein's later work contains several essential philosophical concepts, such as "forms of life" (Lebensformen), "world picture" (Weltbild), "system of relationships" (Bezugssystem), but also "manner of thinking" (Denkstil), concepts which contain reference to various aspects of human identity and cultural productivity. Taking into account the implications they involve, these concepts offer a wide variety of possibilities for achieving as objective a knowledge and understanding of the "other" as possible, taken in the broadest sense from an understanding of nature and the natural world to an understanding of other peoples and cultures, through whatever the form of communication.

Formation of judgment is rooted in preexisting cultural and historical contexts, as well as in the intentional formulation of our aims, goals, and ideals. The latter, however, does not occur in a vacuum, but is necessarily informed by education and culture, as well as by reflection on the products of culture, as cultivated by the humanities, by a creative encounter with the natural and intellectual world through technology and the arts, and by participation in and acquaintance with the findings of scientific inquiry.

---

[12]Cf. McDowell (1994: 84).

[13]McDowell (1994: 87).

The successful cultivation of judgment requires the informed and reflective encounter with higher-level manifestations of the human spirit and the study of cultural heritage which forms the specific task of the humanities.[14] The cultivation of judgment is conditioned by physical preconditions, natural structures of motivation and preexisting cultural and social circumstances; it depends essentially on specific forms of encounter with manifestations of higher-level reflection in the arts, culture, humanities and philosophy, as well as on forms of creativity promoted and studied by them. Kant's *Critique of judgment* demonstrates that the power of judgment is central not only to human rationality but to the understanding of the integral functioning of our natural and intellectual powers in the production of human experience, knowledge, understanding and action as a whole. Instead of advocating pluralism of interpretation, hermeneutics as a universal theory of understanding should focus on judgment considering the philosophical relevance of the cultivation of judgment.

References

Agazzi, Evandro. 1985. The historical dimensions of science and its philosophy. *Diogène* 132:60-79.

Agazzi, Evandro. 2014. *Scientific Objectivity and its Contexts*. Cham/Heidelberg/NewYork/Dordrecht/London: Springer.

Benjamin, Walter. 1978. *Briefe*. Herausgegeben und mit Anmerkungen versehen von Gershom Scholem und Theodor W. Adorno. Frankfurt/M: Suhrkamp.

Flacius Illyricus, Matthias. 1968. *De ratione cognoscendi Sacras literas, Über den Erkenntnisgrund der Heiligen Schrift.* Übers. eingl. u. mit Anm. vers. von L. Geldsetzer, Düsseldorf: Stern.

Gadamer, Hans-Georg. 1986. *Wahrheit und Methode. Grundzüge einer philosophischen Hermeneutik*.5. Aufl. (durchgesehen und erw.) Tübingen: Mohr (Siebeck).

Hirsch, Eric D. 1973. *Validity in Interpretation.* New Heven: Yale Univ. Press.

Ineichen, Hans. 1991. *Philosophische Hermeneutik.* Freiburg/Br.

McDowell, John. 1994. *Mind and World. With a New Introduction.* London: Harvard University Press.

Meier, Georg Friedrich. 1996. *Versuch einer allgemeinen Auslegungskunst*. Herausgegeben von Axel Bühler und Luigi Cataldi Madonna. Hamburg: Meiner.

Rescher, Nicholas. 1997. *Objectivity. The Obligations of Impersonal Reason.* Notre Dame: University of Notre Dame Press.

Sokal, Alan. 1996. Transgressing the Boundaries: Towards a Transformative Hermeneutics of Quantum Gravity, *Social Text* 46/47: 217-252.

Wieland, Wolfgang. 1999. *Platon und die Formen des Wissens. 2. Aufl.* Göttingen: Vandenhoeck & Ruprecht.

Zovko, Jure. 2007. Die Bibelinterpretation bei Flacius (1520-1575) und ihre Bedeutung für die moderne Hermeneutik. *Theologische Literaturzeitung* 132: 1169-1180.

Zovko, Jure. 2014. The Human Sciences and Moral Judgment. In *Ethics or Moral Philosophy. Contemporary Philosophy* vol. 11, ed. Guttorm Floistad: 131-143. Heidelberg, New York, London: Springer.

---

[14]Cf. Zovko (2014).

Philosophy of Language and Mind

Carlo Penco

Abstract In this paper I will give some remarks on the intersection between Agazzi's work and some topics in philosophy of language and mind. In Sect. 1 I give an introduction to the approach to meaning in Agazzi's works, and to his recent idea of general semiotics. In describing Agazzi's treatment of meaning and understanding it appears that some of his papers antedate arguments later become fundamental in the philosophy of language and mind, mainly by Putnam and Searle. In Sect. 2 I describe his discussion of the limits of intensional logics, which antedates analogous criticism made by Putnam; in Sect. 3 I focus on the link between the operational aspects of meaning and the idea of three level semantics; eventually in Sect. 4 I present what can be considered a forerunner of Searle's argument of the Chinese room in a different setting.

1 Introduction: Three Level Semantics and General Semiotics

Evandro Agazzi has devoted much attention to language, especially in the context of the treatment of scientific theories. Although he never deals with specific problems in the philosophy of language we can say that he has been in deep harmony with many topics and theories in the field, sometimes forerunning some of the most relevant ideas, from the importance of intensional aspects of meaning to the role of intentionality as a criterion of understanding.

C. Penco (✉)
Department of Philosophy, University of Genoa, Genoa, Italy
e-mail: penco@unige.it

M. Alai et al. (eds.), *Science Between Truth and Ethical Responsibility*,
DOI 10.1007/978-3-319-16369-7_14

The mingling of interests in logic and the philosophy of science has brought Agazzi to discuss a main problem which is at the boundary between philosophy of language and science: how is it possible to identify a scientific theory with a set of true sentences while science is intrinsically connected with an operative or procedural dimension? This question is placed on the background of a discussion of general topics in the philosophy of logic, many papers of which are collected in an anthology *On the reasons and limits of formalisms* (2000), collecting paper from the sixties to 2012. One of the leading themes in this context is a constant reference to Fregean views, connected with some Husserlian themes, which appears since his first books on symbolic logic (1961, 1964). The general background is a deep feeling with Aristotelian and Medieval tradition grounded on the distinction introduced by Aristotle between *semantic logos* and *apophantic logos*. The distinction is a constant in his remarks on semantics and supports the claim that, in developing formalisms, we cannot avoid any relation to a non formal (Frege would have said "contentful") dimension in logics:

> Meanings that are contextually defined in a formal system apply to possible systems of objects (referents) through appropriate interpretations. This solution is valid, but we need to understand *why* it is so: a suggestion is that the solution is valid because hidden in the concept of "interpretation" we may find that eidetic meaning, that it the semantic logos as such, which cannot be erased neither in favour of syntactic context nor of the referential level. (…) Semantic logos and apophantic logos cannot be defined but in reciprocal relation and saving their differences (Agazzi 1989: RLF 116).

Semantic logos is interpreted after the character of "noematic" meaning, or of Fregean senses, bypassing, for the generality of the topic, the specific differences between the concept of sense and the concept of noema (as discussed for instance by Dummett 1993). This peculiar mixture of the Fregean notion of sense and the Husserlian noema is present in all his writings since the beginning. However speaking of "meaning" in philosophy of language is not an easy task given that "meaning" is what require explanation with a theory of language. One of the last books in distinguishing different aspects of meaning is Gillian Russell's book (2008), where she shows different ways in which we may speak of "true in virtue of meaning" and describes different meaning relations we have to take into account on any discussion of meaning. Her interesting essay shows the difficulty of contemporary treatment of meaning, which still relies on traditional contrapositions, that at first seem very similar, but actually illuminate from different viewpoint the complexity of meaning. Agazzi has always insisted on the importance of the history of philosophy, where many distinctions antedate contemporary worries in philosophy of language, and in particular the Aristotelian distinction between *semantic logos* and *apophantic logos*, the Stoic distinction between *lekta* e *axiomata*, the Medieval distinction between *significatio* e *suppositio,* and the Port Royal distinction, connected also with Leibniz, between comprehension and extension (or, in Mill's terminology, between connotation and denotation). Are these distinctions all equivalent? Not really, and each of them has its particular character, although they recall one another. Particular care is given by Agazzi to the Fregean distinction between sense and reference, that cannot properly be reduced to the previous ones, and to what appears to be the most fortunate distinction in contemporary philosophy of language, since Carnap (1947), that is the distinction between intension and extension. Recalling Leibniz's possible world Carnap gives the founding ideas of alethic modal logic, which had do be developed by Kripke, Hintikka and Montague. Agazzi does not want to abandon the development of this line of thought, but realizes its shortcomings and tries to propose a repair, recalling the necessity of keeping distinctions inside the realm of meaning.

From a general point of view, while in Agazzi's early writings the focus is against the reliance of mathematical and logical reasoning on mere formal and syntactic structures, in the later work the focus shifts on the conflict between conceptual role semantics and direct reference theories. In this context he starts with the different meanings given to the term 'semantics':

> If one takes semantics to be a general theory of meaning, the anti-Fregeans recognise that according to Frege a theory of meaning is a theory of *understanding*, while for them it is a theory of *reference* based on the social 'functioning' of natural languages. Therefore, the two semantics have very little in common, and the many efforts made on the part of the anti-Fregeans to meet challenges related to their inability to account for the cognitive significance of certain linguistic expressions appear to have been, in a way, misdirected, since these problems are not of the sort as are relevant to their semantics (Agazzi 2012a: 6).

In contrast either with formalistic theories in philosophy of science or with direct reference theorists, Agazzi insists on the idea of a triadic semantics based on the distinction of sense and reference, leaving always a central role to conceptual or cognitive aspects, that now begins to be recognized as important also in the environment of direct reference theory (see Kaplan 2012). In his later writings (2012, 2014) the discussion is developed in a new framework that could be interpreted as an actualization of Locke's original distinction of sciences in semiotics, physics and ethics. While in many context semiotics has become just a container of a huge amount of different studies and remarks on advertisements, fashion, and everything has some vague connection with what we call "signs", Agazzi 2012a recovers the original Locke's idea of semantics as general logic, or, as Agazzi says, "general theory of meaning", and gives a overview of all the general problems of a logical system, where the attention is focused on the different levels of meaning of different kinds of expressions. We will discuss some detail of this new vision in Sect. 3 below.

2 Limitations of Intensional Logics

After having worked on axiomatization of scientific systems (including the first translation of Gödel's theorem), Agazzi treats intensional logic as a starting point for a better treatment of a good formalization, while at the same time rejecting the "sentential view" of theories, according to which theories are sets of true sentences (the ones deducible from a logico-mathematical axiomatization of a scientific system). How to develop the idea of intensional logic while, at the same time, rejecting the idea of a scientific theory as a set of true propositions? Agazzi's starting point is the view according to which a physical concept does neither refer to an intension nor to an extension, but a set of operation, following the original idea of Percy W. Bridgman's operationalism: "the concept is synonymous with the corresponding set of operations. If the concept is physical, as of length, the operations are actual physical operations, namely, those by which length is measured" (Bridgman 1927, quoted in Agazzi 1969: 126).

However, traditional operationalism, after its first introduction by N.R. Campbell and Bridgman's development, has become too dogmatic and naïf, and Agazzi (1969: 132) tries two fundamental corrections of it: (i) on the on side he rejects the idea that every physical concept is operationally defined; (ii) on the other hand he rejects any operationalist reductionism because "actually physically significant measures are always performed inside a theory and get a meaning of physical measure exactly because they are framed inside theories".

Up to this point the criticism to Bridgman's operationalism is similar to Quine's criticism to the strict verificationism of the first Vienna Circle: in analogy with Quine-Duhem view of holistic evaluation of empirical theories we cannot speak of verification as the meaning of a single proposition and therefore we cannot speak of the meaning of an individual proposition as the set of operations defined for it; only in the context of a physical theory a physical measure has its meaning. However there is possible a third shortcoming of operationalism, that is defined by Agazzi (1969: 133) as (iii) the risk to "mistake a semantic problem for a methodological problem". The point is that, from the viewpoint of a standard intensional semantics, we are satisfied to endow a predicate with an intension and an extension, while the problem of the *means through which we check the extension* is no more of semantic nature and has "no bearing on the problem of meaning". But this is restricting intensional semantics to a mere formalism of functions and extensions as if they were "given". Instead of accepting such an assumption Agazzi claims that the means for determining an extension should be taken into account in a theory of meaning of a scientific theory: actually, the intension of an expression of length should contain different kinds of measure operations (for instance comparing rulers, or using a goniometer, or with trigonometry, and so on); therefore referring to such kinds of operations is "not foreign to the meaning of 'length' but it belongs to it, as part of the its intension". The main core of operationalism is therefore saved in the following way: "if a physical concept contains intensional components that have no access to verification, neither directly nor through more or less complicated (but actually identifiable) links with operatively defined predicates, they have no right to citizenship in physics" (Agazzi 1969: 135–136).

This conclusion is not identical with the empiricist criterion of significance, according to which the meaning of a proposition is its method of verification. The point is that, before accepting or verifying a proposition I need to understand a level of meaning which undergoes (or precedes) any verification. To understand this level of meaning is to understand the kinds of operations that are needed for properly using a concept inside a theory. While keeping a critical attitude to the neopositivistic theory of meaning and assuming a basic holistic view (experimental control concerns a theory in its totality, sometimes with reference to supplementary theories that constitute the "indirect" connection between sentences of the theory and observed facts), Agazzi saves some aspects of operationalism inside a theory of meaning. The most specific work on the application of an operational view of meaning is the paper "The concept of Empirical Data. Proposals for an Intensional Semantics of Empirical Theories", published in 1976. In this essay the author claims that the main problem of the intensional logics is that they are unable to univocally determine the intended model. In fact such logics are a representation of what might be called "inferential competence",[1] that is the ability to find, from meaning postulates, the consequences of the axioms of the theory. However, and this is the key point, they cannot choose among different interpretations, and they can only give what is given inside the language of the theory, where a semantic interpretation cannot give what is intended by a working scientist.

The criticism to intensional logics presented by Agazzi (1976) can be considered as an anticipation of the criticism given by Putnam (1981), with a development of the Quinean indeterminacy of reference. Actually Putnam's analysis is even more clearly revealed in advance in Agazzi (1966) (discussed again in 1978b and 1994) where the author works on the philosophical import of well known results by Löwenheim-Skolem: if a first order theory has a (infinite) model, then it will be true in an indefinite number of isomorphic models; therefore a formal system cannot be characterized by a unique interpretation, but at most it can define the structure of the domain; a formal semantics is a *structural* semantics and cannot give but an interpretation of the *kinds* of elements of the domains (what are objects, what properties or classes and how they interact). Agazzi (1966, 1994) and Putnam (1977) discuss the existence of infinite isomorphic models also given the same truth values assignments: no truth value assignments to a class of sentences is sufficient to define the reference of singular terms and predicates; Agazzi (1976) and Putnam (1981) are concerned not only with truth values assignments (extensions) but also truth conditions (intensions). The general claim therefore holds not only for extensional logics, but also for intensional logics: an intensional logic—also if it specifies the truth values of sentences in all possible worlds, restricting the admissible interpretations through meaning postulates—cannot fix the reference of the expressions of language, and always leaves open the possibility of alternative isomorphic interpretations. Basically "notwithstanding linguistic expedients as enriching meaning postulates to the theoretical sentences, a formal

[1]See also Marconi (1997), who discusses the different aspects of structural semantic competence and sees in inferential competence the aspect linked to meaning postulates and inferential role, although he claims that only the combination of inferential and referential competence makes a reasonable project of the idea of semantic competence.

model can never guarantee the uniqueness of the model". Given that the negative results are used by Putnam to demolish metaphysical realism, Agazzi's reaction (although we cannot properly speak of "reaction" given that he writes before Putnam) can be interpreted as a defence of metaphysical realism. His main argument is enriching formal theories not only with meaning postulates, but also with operational definitions grounded on observable criteria connected with scientific instruments in the physical environment.

3 Operative Definitions and Three Level Semantics

Frege conceived logic as ranging over a universal domain, while the algebraic tradition, starting from De Morgan and Boole and developed in classical Tarskian semantics, abandons the idea of a universal domain, and assumes from the start the possibility of different domains of interpretation of the signs of the formal system. Yet the idea of a universal domain (with the idea of quantifying on all objects) is not completely abandoned in contemporary philosophy and metaphysics, and sometimes the idea that we may speak of "everything" is strongly defended (see for instance Williamson 2003). Agazzi distinguishes a universal domain of discourse, where we may speak of things described in common language and specific domains of a theory, where something belonging to the common discourse may become a specific object of a specific science. A sheet of paper may be considered as a different kind object, depending on the sciences that take it into consideration: if we take into consideration its weight, its chemical composition, its spatial properties, then it can become an object of physics, chemistry or topology (Agazzi 1979: 155 ss.). Every theory has its own objects, its own domain; but there is a universal discourse domain on which we may interpret the individual variables of every language: a specific theory would then concern only the subset of the universe of discourse to which we apply certain operative criteria typical of the theory.

An interpretation of Agazzi's point has been given by Bottani (1997) with an argument on the different uses of ordinary language and scientific language. Among the objects of chemistry we don't find sheets of paper, but chemical properties, belonging to parts analysed with specific procedures; therefore sentences like "this sheet of paper has such a chemical composition" are not sentences belonging to chemistry, but "sentences of ordinary language that represent sentences of Chemistry as referred to objects that belongs to the domain of quantification of ordinary language, but not to the quantificational domain of Chemistry". We have therefore two different objectual domains, a set of sheets of papers (described in everyday language) and an equivalent set of sheets of papers *modulo* chemical compositions. Operative procedures accepted in Chemistry can therefore be applied to sheets of papers from the point of view of quantification in ordinary language.[2] Therefore the problem becomes that of giving a precise definition of how to characterize an operative procedure, that is the meaning of predicates with which something is assumed as an object of a theory. This is the best way to introduce the topic of the *meaning* of operative predicates or of operative definitions.

A predicate is an expression that has a class of objects as extension and a set of possible worlds as intension (or a function that maps a class for every possible world). But here we have the problem posed by Agazzi and Putnam and discussed in the previous paragraph: how can we define the "intended" interpretation? How to distinguish among different interpretations of isomorphic models, among different classes of objects? Intensional logic, although it gives more than extensional logic, cannot give a specification of predicates, because it can give only a discrimination concerning the relations among predicates; and different interpretations may keep the same structural relations (as clearly presented by Quine with his permutations). If we follow this line of though we are bound to accept the well known consequences on the incommensurability problem: we cannot determine which of two theories better fits the reality (the intended model) through a description of the meanings (as intensions) of their predicates.

Agazzi's proposal is to insert in the definition of the meaning of a predicate also the operative procedures linked to the use of the predicate in a scientific and experimental context. Operative procedures are therefore intended as belonging to the meaning and not as a mere methodological aspect of the theory. Following this idea only two theories with the same operative procedures speak of the same objects; while we do not have any possibility of real comparison between two theories if we simply assume that their predicates have the same intentions, we can compare theories whose intensions are associated with operative procedures, connected with specific scientific instruments. Besides, the commensurability of theories is granted by the fact that they share at least *some* operative predicates, given that it is really rare that there were no operative predicate shared by two theories on the same domain; in fact basic predicates can always be reduced to similar inter-theoretical operative procedures.

From a logical point of view Agazzi (1976, 1979) does not assume a priori a domain of individuals, against the standard logical assumptions according to which

[2] It is not easy to say whether this is a correct interpretation of Agazzi's argument. However this interpretation has the merit of avoiding the problem of the dogma of the conceptual as discussed by Donald Davidson. With this solution we have two kinds of ontologies, the ontology of ordinary language (with universal domain) and the ontology of specific sciences, like chemistry, that are subparts of the universal domain. Let us call them "external" and "internal" ontologies. With this solution we do not have any more and we do not need the notion of "empirical content" as "something neutral and common that lies beyond all schemata"; we have on the contrary a dichotomy where empirical content is given by the external ontology given by ordinary language and conceptual schemes are given by the expressive resources of the different internal ontologies. The connection between a conceptual schema and an empirical content is given by the operative procedures. On this line of thought see Bottani (1997: 246–249).

(i) the individuals of the domain are "given" and (ii), while the relations are not always decidable, the domain is decidable (given any individual we may know if it belongs to the domain or not). But the undecidability of relations is one of the main sources of undecidability, and the source of semantic ambiguity. Agazzi's alternative is that a model M, instead of a pair of a domain and a set of relations ($M = \langle D, R \rangle$) is a quadruplet $M = \langle S, O, R, P \rangle$ where S is a set of Scientific Instruments, O a set of Operations, R a set of Results and P a set of Operative Predicates, and every predicate belonging to P is an element of $S \times O \times R$. Let us take an operative predicate as "electrically charged", constituted by the operation "put x on the disk of the electroscope", by the Scientific Instrument *Gold-leaf electroscope,* and from the possible Result "the gold leaves spread apart". The question that apparently comes to mind is what can we refer to with the variable x, given that, by assumption, we have not used a fixed domain of interpretation. We have however on the one hand the universal discourse domain, that precedes any definition of the objects of the theory, and on the other hand we have a theory that creates its own object on the ground of its specific definitions applied to things of which we may informally speak in everyday language (electrons or quarks included). Therefore, the variable can be replaced by everything that is manipulable in a specific way by a specific Scientific Instrument (therefore not, in this case, emotions, moons, toothaches …). The object of the theory is not therefore given, but constructed or selected or filtered from the conditions imposed by operative predicates.

It is not difficult to see the import of such theory from the point of view of a general theory of meaning. The main idea is the necessity of considering something different from intension, but at the same time something which interact with intensions, without loosing the expressive power of intensional logics. Here the three level semantics which we have hinted at in the first paragraph becomes more complex, giving space to an element of procedural aspect which cannot easily be reduced either to sense or to reference. Here however (Agazzi 2012a, 2014: Sect. 4.1) oscillates between a standard tripartition in kinds of expressions (proper names, predicates, sentences) and their sense and reference, and a different kind of analysis where predicates can be considered to have a reference in a "limiting sense", while it is sensible to speak of the *extension* of a predicate. Agazzi (2012a: 12) remarks that "reference and extension are related notions, but are not identical, so that a Fregean and a Carnapian semantics are not really equivalent". Making these further distinctions, he actually follows the suggestion given by Frege in a letter to Husserl, where Frege explicitly claims that for the predicates we need a tripartite division in sense, reference and extension.

Although new suggestions are presented in the general semiotic framework in Agazzi 2014 (especially the distinction between encoding and exemplifying at page 190–191), something seems still to be missing: the specific place of the operational criteria of referentiality, that are implicitly connected with the old discussion on the operational meaning. Here a possible integration of the general structure provided in Agazzi 2014 might impinge upon the relation between the sense and the reference or denotation of a predicate. If we keep the connection with intentional semantics and their notion of intension as functions, we might think that the idea of operational meaning might be developed to perform the role of the specific procedures attached to the intensions, procedures that exemplify particular ways to define the extension of the predicate.[3] A suggestion of this kind might find an implementation in intelligent systems (robotics), where to each mathematical function a specific procedure might help determining individuals and classes of the intended domain, where procedures may be implemented in different kinds of operations connected with specific instruments. While the intension of 'cat' individuates a class and its relation with other classes and individuals, without actually discriminating what is a cat and what is not, some specific procedures given by a video camera with a specific kind of pattern recognition (given through some learning system) may verify if a certain individual meets the requirements that belong in its data base to the set of cats; it may in such a way *recognize* a cat, connecting inferential and referential competence. And most intelligent systems, although without intensional logics, have been used in this direction since the famous "Lunar" by Woods, that were used to recognize and classify Moon stones. Whether the robot also *knew* that what its system recognizes and classifies as a cat or as a stone *is* a cat or a stone is another story. It is what is known as the problem of the Chinese room, to which John Searle gave an answer that looks very similar to what Agazzi begun to define in an old paper of his published in 1967, thirteen years before the famous paper by Searle. Let us then end our presentation of Agazzi's ideas on language and mind on this last point.[4]

4 Intensionality and Intentionality

Since Brentano, and later with Chisholm, there is a deep and privileged relation between intentionality and propositional attitudes like "believing", "knowing", etc., which involves the relation between intentionality (with a 't') and intensionality (with an 's'). Believing and desiring are *intentional states*, that is states directed towards objects of beliefs and desires. Reports of intentional states are typically given in terms of propositions ("x believes *that* p", "x desires *that* p"), where by 'proposition' we mean the intension of a sentence—and since Carnap these reports are classified as *hyper-intensional contexts*, such that the principle of substitutivity *salva veritate* doesn't work for them. Yet reports of propositional attitudes are the main road to the analysis of ordinary language (with a

---

[3]A development of this kind with some historical remarks on the topic is presented in Penco (2013).

[4]Agazzi had presented his paper in English at the "Wiener memorial meeting on the idea of control" that took place in Genoa on October 26–29, 1965 where he also had an interesting discussion with Putnam on the topic of artificial intelligence. For technical reasons the recordings of that conference were seriously damaged and the proceedings never published, Therefore Agazzi published an Italian translation of his paper in the *Rivista di filosofia neoscolastica* (Agazzi 1967). This translation was then reprinted in a volume edited by Paolo Aldo Rossi in 1978.

great variety of subtle alternative theories). However intensional contexts are not *sufficient* for determining intentionality, although they are an indication of intentionality as (Searle 1983, Chap. 7 widely discusses); it looks therefore as if intentionality with a 't' has a original and founding character in respect of intensionality with an 's'.

Searle is well known among the authors who most insisted on the original character of intentionality and on its role in the foundation of meaning, and his argument of the Chinese Room has had a certain role in the disillusion of the perspectives of strong artificial intelligence. The Chinese room experiment suggests us imagining a English speaker inside a room where he receives from outside papers written in Chinese characters. The man has a system of rules of transformation (into English) according to which the Chinese sentences are mapped into other sentences (to questions he will produce answers, and so on) in Chinese characters, although nobody explains him the meaning of the Chinese characters. The man in the Chinese room is like a robot: he has a syntax and his outputs are adequate answers in Chinese, but he does not have a semantics of Chinese; he therefore cannot be said to understand: although we may be impressed by the performance of the Chinese room, we cannot say that the man in the room "knows" or "understands" Chinese because he is only able to manipulate symbols without any idea of their meaning (this is what Searle means by saying that the man has a syntax but not a semantics). Although Searle's idea of "symbol manipulation" in his mental experiment seems a little naïf, his paper has raised a lot of discussion, and has mostly been accepted by many people working in artificial intelligence.[5]

Searle's famous paper has been published in 1980, and its main point is that intentionality is the characteristic feature that makes it impossible to say of a machine that it "thinks" or "understands" as we say it of a human being. Several years before the publication of Searle's paper, Agazzi had presented a similar argument in an Italian journal (Agazzi 1967). He first develops a kind of "intentional stance" as Dennett (1987) developed later. He claims that complex machines impose us to use an intentional language; in fact, if we want to explain a robot's performance on the ground of our knowledge of human activities, it would be "not only possible, but also necessary to illustrate the properties of machines in terms of human psychological predicates" (Agazzi 1967: 14). The use of anthropomorphic predicates or of an intentional language ("the computer has not enough *memory* …", "it has still to *learn* how to make the program run …") are therefore acceptable in a context of advanced robotics. But the use of anthropomorphic predicates cannot make us forgetting the fundamental difference between the *human sight* and the *recording of visual data* from a sophisticated machine. What is a "plus" in human action "is denoted in

[5] Penco (2012) claims that Searle's vision of "symbol manipulation" is very strict, and represents a weak point in Searle's analysis; however the Turing test could be "updated" to problems of understanding language in context.

philosophy with the term of intentionality, with which we refer to the fact that in the knowledge activity of a living being there is a kind of participation or identification of the subject in front of the objects …". Therefore, we are brought to think that we cannot properly say that a machine *sees* an object, "the plus that accompanies the recording of the image in the living being is intentionality or its consciousness" 1967: 16).

Summarizing with a slogan we might say that the *possibility to receive information* is the *common feature* between a machine and a living organism, while the *intentional activity* is *what differentiates* them. Intentionality would become then, to take a term from the Theory of Evolution, *the "missing link" of artificial intelligence* that makes machines incapable to have that specific quality of human subjectivity consisting in *thinking* (see Agazzi 1991). Intentionality however is not translatable into operations of behaviours[6] and its characterization has raised many debates. Searle considers intentionality as a characteristic feature of the brain; is this the best definition of intentionality? Or does it take for granted what one should argue for (that is, that intentionality is typical of living beings)? Agazzi looks for a different kind of definition: as "symptomatic characterization": the most typical symptom of intentionality is the involvement of the living organism in its entirety (and not of subparts of it) in answering the signals that it perceives. We might also imagine to reconstruct a brain and all the details of a nervous system, but this would not amount to have intentionality. Only the activity of the living being in the environment may produce this peculiar feature, and the brain is just a part of it, a necessary but not sufficient part of it.

The definition of the original intentionality is what differentiates Agazzi from Searle (1980); for other aspects, Agazzi's argument is very similar to the one presented by Searle, although Agazzi arrives at a similar conclusion in a different way, through an analysis of the limits of formalisms, and in particular of Gödel theorem (of which Agazzi made the first translation in a language different from German). We can assert the validity of Gödel's formally indecidable proposition (neither provable nor refutable in the formal system) only by metatheoretical considerations that are grounded on the meaning of the Gödelian formula (and on the fact that it is satisfied by every interpretation of the symbols in the universe of natural numbers). These metatheorethical considerations "necessarily pass through the examination of the meaning of the formula that is recognised as valid in the intuitive number theory, although it is not provable in the formalized number theory" (1967: 21–22). The fundamental reason of the limitations of formalisms can be therefore reduced to the presence of a dimension of intentionality, to a kind

[6] See also Agazzi (1991: 239–241). Generally speaking, if we could build a robot that behaves in a way which is undistinguishable from human behaviour, Agazzi, with Putnam, should probably admit that it would be undecidable whether this individual has a consciousness or not (if they were "like us" or they were, following Chalmers, "*zombies*").

of "seeing" (Gödel speaks of a particular form on intuition), and basically to the capacity to confer *meaning* to the formula. It is this theoretical framework that permits Agazzi to present in advance the basic idea of Searle's Chinese room:

> this further step (intentionality) seems to be in a structural way beyond the possibility of a machine, that is always in a condition very similar to the one of a man to whom somebody teaches the grammar and the syntax of a language of which he knows only the alphabet, without having been communicated the meaning of words; he would be able to build correct sentences, or to recognize the correctness of some and incorrectness of others ... but he could not distinguish the true from the false propositions (Agazzi 1967: 23)

Although not framed in the form of a Chinese room, Agazzi's thought experiment is devised to make us imagine somebody endowed with syntax but not with semantics, exactly as Searle's English speaker in the Chinese room. Is this the end of the matter? Difficult to say, and the discussion on the Turing test against which Agazzi and Searle fight, leaves open other possibilities, and Agazzi himself participated to the new discussion on the topic in a recent anthology.[7] But this would bring us too far.

What seems to me clear enough from this short discussion of some of Agazzi's ideas on meaning is twofold: on the one hand the particular consonance of Agazzi's views with some of the most striking solutions to the main problems of contemporary philosophy of language and mind, and on the other hand the attention he pays to some particular features, which makes his contribution always partly different from the standard solutions, suggesting either an update or an alternative to them.

References

E. Agazzi 1966. Riflessioni su alcuni nuovi orizzonti della logica matematica. In *Archivio di Filosofia*: 47-69.

E. Agazzi 1967. Alcune osservazioni sul problema dell'intelligenza artificiale. *Rivista di filosofia neoscolastica* LIX: 1-34. Reprinted in Paolo Aldo Rossi (ed.), *Cibernetica e teoria dell'informazione*: 199-244. Brescia: Editrice La Scuola.

E. Agazzi 1969. *Temi e problemi di filosofia della fisica*. Milano: Manfredi. Reprint Roma: Abete, 1974.

E. Agazzi 1976. The concept of Empirical Data. Proposals for an Intensional semantics of Empirical Theories. In *Formal Methods in the Methodology of Empirical Sciences*, ed. Przelecki et.al., 153-167. Dordrecht: Reidel [reprint in RLF, ch.9].

E. Agazzi 1979. Proposte per una semantica intensionale delle teorie empiriche. In *Studi sul problema del significato*, ed. Evandro Agazzi, 148-166. Firenze: Le Monnier.

E. Agazzi 1989. Logo semantico e logo apofantico. *Epistemologia*, XII: 17-47. Reprint in RLF, ch.3].

E. Agazzi 1991. Operazionalità e intenzionalità: l'anello mancante dell'intelligenza artificiale. In *Intelligenza naturale e intelligenza artificiale*, ed. S. Biolo, 1-13. Genova: Marietti.

[7] See Agazzi (2013).

E. Agazzi 1994. On formalism. In *Philosophical problems today*, ed. Guttorh Fløistad. Vol. 1: 74-135. Dordrecht: Kluwer Academic Publishers, [Reprinted in RLF ch. 1].
E. Agazzi 1998. *George Boole. Filosofia, Logica, Matematica* (editor with Nicla Vassallo). Milano: Angeli.
E. Agazzi 2012a. Meaning between Sense and Reference: Impacts of Semiotics on Philosophy of Science. *Semiotica, Special issue Semiotics and Logic*, CLXXXVIII, 1/4: 29-50 [Reprint in RLF Ch.10].
E. Agazzi 2012b. *Ragioni e limiti del formalismo* [here as RLF]. Milano: Angeli.
E. Agazzi 2013. *The legacy of Turing*. Ed. Evandro Agazzi. Milano: Angeli.
E. Agazzi 2014. *Scientific Objectivity and its Contexts*. Dordrecht: Springer.
Bottani, A. 1997. Schemi concettuali e definizioni operative. In Montecucco 1997: 235-254.
Bridgman, P.W. 1927. *The Logic of Modern Physics*. New York: McMillan.
Carnap, R. 1947. *Meaning and Necessity* (2nd ed.1956), Chicago: Chicago U.P.
Dennett, D. 1987. *The Intentional Stance*. Cambridge (Mass.): MIT Press.
Dummett, M. 1993. *Origins of Analytic Philosophy*. London: Duckworth.
Kaplan, D. 2012. An Idea of Donnellan. In *Having in Mind*, ed. J. Almog and P. Leonardi. Oxford: Oxford Unversity Press.
Marconi, D. 1997. *Lexical Competence*. MIT Press.
Penco, C. 2012. Updating the Turing Test. Wittgenstein, Turing and Symbol Manipulation. In *Open Journal of Philosophy*. 2, No. 3: 189-194.
Penco, C. 2013. What Happened to the Sense of a Concept Word? In *Protosociology* 30: 6-28.
Putnam, H. 1977. Models and Reality. in Putnam H. *Realism and Reasons; Philosophical Papers*, vol 3, Cambridge: Cambridge U.P. 1983.
Putnam, H. 1981. *Reason, Truth and History*. Cambridge: Cambridge U.P.
Russell, G. 2008. *Truth in virtue of Meaning. A defence of the Analytic-Synthetic Distinction*. Oxford: Oxford University Press.
Searle, J. 1980. Minds, Brains, and Programs. *Behavioral and Brain Sciences* 3:417-424.
Searle, J. 1983. *Intentionality*. Cambridge: Cambridge U.P.
Williamson, T. 2003. Everything. *Philosophical Perspectives* 17, 1: 415-465.

Part IV
Anthropology, History and Culture

Philosophical Anthropology

Matteo Negro

Abstract The way in which Agazzi intends to explore the dynamics of knowledge, metaphysics, or language, just to mention some of the many subjects of his researches, is not only inseparable from his idea of human being, but it is in fact its most direct consequence and expression. Ultimately, it is illusory to speak of a scientific image of the entire human reality as it is illusory to assume a complete scientific image of any particular reality. Taking up Wilfrid Sellars' suggestion, Agazzi invites us to verify the validity of the scientific image of humanity and the epistemic compatibility with its 'manifest' image. In particular, according to Agazzi the perspective of the ought, is something we can ascertain as an empirical fact, but it is not limited to the empirical evidence: it refers to a criterion that in itself does not belong to experience but to a 'metaphysical' level; an ideal of perfection, never fully realized, a general requirement of normativity which is present in all aspects of human agency. From this starting point Agazzi's anthropological reflection proceeds to the metaphysical delineation of personal identity, the relationship between mind and body and, ultimately, the dignity of the human person.

1 Introduction

Focusing on the main lines of Evandro Agazzi's philosophical anthropology is an operation which may be conditioned by two objective factors. The first is connected to the vast production of the author in this field: well over one hundred works, including books, articles and contributions, which are inserted fully and organically in an even larger amount of scientific works. The other factor a careful student of the work of Agazzi cannot avoid grasping is the *anthropological value*

M. Negro (✉)
Department of Political and Social Sciences, University of Catania, Catania, Italy
e-mail: negro.unict@gmail.com

M. Alai et al. (eds.), *Science Between Truth and Ethical Responsibility*,
DOI 10.1007/978-3-319-16369-7_15

of his entire production, even of what is most clearly dedicated to very specific epistemological issues (such as philosophy of mathematics, of logic, or of physics). In this sense it is really difficult to separate or isolate the single items of Agazzi's thought, which always have a unitary form. The way in which Agazzi intends to explore the dynamics of knowledge, metaphysics, or of language, just to mention some of his many subjects of research, is not only inseparable from his idea of human being, but it is in fact its most direct consequence and expression.

Agazzi has shown that the sciences are among the highest forms of humanism, that knowledge in its multiplicity is a high expression of the human spirit, as well as an excellent demonstration of its existence. In doing so, he reversed the usual way in which we relate science to persons. Knowledge about people should not be cultivated only by those few areas of culture that make no use of experimental methods; on the contrary, Agazzi has always claimed that the existence and nature of human beings can be demonstrated just by using as a point of comparison the field of experience, accepting the challenge of the sciences on their own ground, discerning somehow in their acquisitions a broader perspective, and defusing potential closures or epistemological reductionisms. Anthropology cannot

> be identified with psychology, sociology, cultural anthropology or linguistics, each of which considers man from a specific and restricted angle. Nor can it be seen as the sum of these particular disciplines, but rather as the attempt to afford a global image of man in which the information provided by these sciences (and several others as well) can be harmonized and receive a *sense*, taking into consideration at the same time other aspects of human reality that are not the subject matter of scientific investigation (Agazzi 2007: 383).

Ultimately, it is illusory to speak of a scientific image of the entire human reality as it is illusory to assume a complete scientific image of any particular reality. Taking up a suggestion of Wilfrid Sellars', Agazzi invites us to verify the validity of the scientific image of human nature and its epistemic compatibility with the corresponding 'manifest' image.

But concentrating here on the works that mainly focus on philosophical anthropology, we will highlight those aspects which are undeniably crucial to this process of reflection.

2 Person and the 'Ought'

In human experience, namely the experience of a being that within the natural world manifests the possession of certain properties which are not shared by other beings, the perspective of action emerges as an original and irreducible datum. Human action is according to Agazzi marked "by an explicit or implicit confrontation with an 'ought'"; "when the notion of the ought is taken in a suitably general sense, we can easily detect its presence in every specifically human action", as is stated in his prominent book, *Right, Wrong and Science* (Agazzi 2004b: 93). Specific human action, in which normativity and intentionality play a prominent role, shows conclusively how there is no immediate coincidence between what people do and what they ought to do, unlike many other operations that they always perform well, because they ultimately do not depend on them, but on their biological nature. The projections of the ought on the different expressions of action constitute *values*, that is, "*ideal models* acting as regulatory parameters for operations, performances, and human actions" (Agazzi 2004b: 95). Action therefore implies a criterion of perfection, which is a synthesis of *idea* and *norm*, i.e. synthesis of the intentional representation both of an ideal state and of the rule of its realization. Therein lies the biggest difference with the pure goal-seeking behaviour which, for Agazzi,

> is so intrinsically, so to speak, and it in no way implies that the agent is capable of presenting the goal to himself in order to attain it. The agent simply follows an internal disposition. This can be modified and improved over time by accidental interventions from the outside world (...) This is why goal-seeking behaviour is often found in machines, plants, and animals, without for all that implying that they actually *intend* to achieve the goal. Such behaviour forms part of their mode of being; it does not correspond to some 'ought'. A value, on the other hand, has the character of a goal which is known and *judged*, that is, one that the agent deems *right* (Agazzi 2004b: 96).

Without reference to an ideal standard, human action would not be adequately understood, since every other attempt to give an account of it would show its inevitable inappropriateness: this applies in particular to that order of discourse which merely describes it in terms of physical causality, expunging the whole dimension of the will and freedom. Agazzi objects to this type of reduction through the recovery of the notion of 'tendency to an ideal perfection' rather than through direct criticism of materialist or deterministic models. He notes:

> The term *ideal*, which seems to indicate only the nonmaterial nature of the model, actually implies more: a reference to something unconditioned and absolute, able to inspire any particular human activity in a manner above and beyond the desire to achieve purely pragmatic ends (Agazzi 2004b: 100).

As long as we strictly adhere to the instrumental or conditional model of the relationship between actions and purposes we will not be able to single out the real leap between being and the ought that underlies human action. Only through the perception of an intrinsic and unconditional prescriptivity, through the recognition of the existence of an absolute value that practice embodies and witnesses—i.e., an end in itself—is it possible to provide with sense the question of human action and its original normativity. Incidentally we can observe that today several studies in the theory of action dedicated to the relationship between freedom and determinism fail to break the deadlock, because they are largely confined to a conditional formalization; hence, to escape this deadlock, in some cases they look with favour on the indeterminist models. In this way, however, they appear to be unable to get rid of the conceptual framework they intend to deny. The intrinsic value and the extent of the autonomy of human action cannot be upheld just by holding that it is either uncertain and unpredictable, or, on the other hand, deterministic.

The perspective of the ought, therefore, according to Agazzi is something that we can ascertain as an empirical fact, but it is not limited to the empirical evidence: it refers to a criterion that in itself does not belong to experience but to a 'metaphysical' level; an ideal of perfection, never fully realized, a general requirement of normativity which is present in all aspects of human agency (individual and social) inducing "persons to respond with specific rules, i.e. with concrete information, to the instances of the ought which are present at each of these levels" (Agazzi 1987: 334). Requirements of the ought taking the form of instances of values to which the freedom of man accedes by virtue of their ideal validity. Also the analysis of legal norms, in particular those that move from the consideration of human rights, shows that the sphere of duty is connatural and primordial. Such rights reflect the perception of absoluteness, the reference of human persons to what is intentionally grasped and conceived as an end in itself, and to which they feel fully engaged. Duty has an absolute character, since "man acts according to duties and is capable of absolute commitment to them" (Agazzi 1996: 50). Agazzi claims that, because of this commitment, the dignity of the person "receives its specifically axiological connotation, for this commitment is by no means a simple *logical* consequence of rationality" (Agazzi 1996: 50). We could say that dignity is not a practical result, the outcome of a practical inference, the product of reasoning or the result of a correct choice of means. Dignity is axiologically prior and manifests itself in natural law and in the framework of duty as a moral duty to do good and avoid evil. Agazzi explains this concept in a very enlightening way in his essay on Maritain:

> The correct way is that of remaining faithful to an *ontological* notion of person, according to what a person is *by nature*: an entity which is able to possess (...) properties, but which does not change its nature for the fact of being *deprived* of them or not being yet able to *show* or to *exercise* them, or no more able to *do* this or that other thing. It is *in virtue* of this ontological nature, and of what it *ought* to manifest if it were fully expanded, that we have the *duty* of respecting any person, and to provide it with the maximal possibilities of expressing the full richness of what is implicit in its nature. This is why we have always spoken of the *capability* of persons to commit themselves in an absolute way, as a foundation for personal dignity and human rights (Agazzi 1996: 53).

3 Risk and Freedom

For the aforesaid reasons Agazzi's invitation to move in the direction of a foundation of human freedom intends to go beyond the footprint of the Aristotelian *phronesis*, or rather to reactivate its foundation, that is the contemplative dimension of the intrinsic ends, 'what is good for a man'. As he himself writes in the Italian edition of his already quoted book:

> This may already be an indication of the road to follow: a road that can no longer be confined to formal analysis, but must measure itself against the consideration of concrete, varied and ever-changing dimensions, which also require *due* recognition and respect. This seems to indicate clearly that what is needed is to resume the discourse on human beings and discover the full range of values inspiring their actions, by acknowledging that their profound freedom consists in the ability to fulfil those values (Agazzi 1992: 171).

Humans, 'condemned' by nature to be free, to quote Sartre, interpret this condition as adherence to what they recognize as valid in itself, as freedom of evaluating the engaging purposes of their life and agency. This sort of dynamics, although originating from an ontological core, is accomplished in the field of experience and presents some recurrent characteristics. One of these is undoubtedly that of risk, which Agazzi analyses in depth. Risk is not simply related to the presence of margins of unpredictability, but it is configured in the first place as a tension inherent in the possibility of human freedom, in a tension which is engrafted in the condition of finiteness, and therefore subject to natural limitation. By risking we take in at once the greatness and uniqueness of the human being, who

> alone can act on the basis of genuine choices, make decisions, propose modifications, project the creation of objects, institutions, and new situations, perfect himself and realize his desires, construct his future and *conceive* his objectives and the options for realizing them (Agazzi 2004b: 147),

together with the perception of the inherent limitations to his ability to choose, to act, to plan and accomplish. But risk, in addition to being a natural tendency, may also be an option of practical reason, and can be used voluntarily to give shape and content to the representation and realization of the meaning of human existence. In this regard Agazzi refers clearly to Pascal's wager. In the frame of the inevitability of choice and uncertainty, risk becomes an important resource because it allows people to invest rationally on the objective most fitting for their fulfilment. So says Agazzi: "at stake is the entire value of the individual's existence. Much may be risked (it is reasonable to risk, says Pascal) with a view toward giving it infinite value, difficult as it is to define this value. One may prefer not to risk, to be content with a finite value" (Agazzi 2004b: 149–150).

The category of risk would therefore be quite ambiguous if it did not reveal itself as an anthropological category. In fact, the various attempts to analyse risk in purely probabilistic or quantitative terms presuppose also a clear anthropological view, which in most cases theorizes the depletion of the human subjects or their mechanization: bodies in space, simply subject to strength and resistance. Paradoxically, the counterpart of this ontological pulverization is, especially in our day, the strong defence of the idea of freedom as a mere 'freedom of action'. Agazzi takes this contradiction very seriously, when he observes that the active exercise of freedom, in its various expressions (freedom of individual and social choice, civil and political liberties) can only be fulfilled thanks to the intrinsic freedom of the human individual, the freedom of the will:

> [T]he common sense of our time accepts as obvious and rightful the claims of the multiple forms of freedom of action, which imply the abovementioned… 'existence of a free will', but it widely shares a view of man according to which the latter freedom is very problematic, and even tends to fade. In fact, the more you spread the scientific interpretations of the human being, the more this will be interpreted as a 'mechanism' whose functioning, although complex, is determined by various factors: physical, chemical, biological, psychological, social, environmental, cultural, and so forth (Agazzi 1999: 5–6).

Freedom is rather an inner principle of a spiritual nature, not assimilable to mere spontaneity, that is, to following the inclinations for which the human person is not really responsible (see Agazzi 2000d). Agazzi, however, also rejects the opposite polarization, the absolutization of the will, which focuses exclusively on a concept of freedom as pure autonomy of the will to the detriment of freedom of action. This perspective is lacking both in not admitting the openness to the action as a genuine projection of the freedom of the will, and in strongly defending self-determination of the will with respect to each supra-individual value. He notes that

> the very concept of self-determination, while preserving an undoubted validity in saying that it is not worthy of man to submit to something he does not approve inwardly, is vacuous unless it is stated 'in view of what' he is self-determining, and it is also noted that not all options are equally valid. Otherwise the concept of moral responsibility vanishes again, and every difference between good and evil disappears, if all options are equally legitimate, and so disappears any reason to speak of responsibility (to quote Nietzsche, we are 'beyond good and evil') (Agazzi 2000c: 5–6).

Freedom is instead capable of self-restraint in adherence to challenging values and goals, but such self-restraint in reality coincides with its exaltation and not with its mortification, since it retains the prerogative to internalize the alternatives and is committed with an operation of affirmation or exclusion which is not subject to gradations or modulations due to external influences (see Agazzi 2000b). This means that the self is always present in the act of will and, in spite of all the cognitive and situational uncertainties implied by the exercise of choice, "our will is free to will, and also to promote the desired actions, in spite of difficulties and deficiencies" (Agazzi 2000a: 5–6).

The traits of Agazzi's anthropological reflection that will lead to the metaphysical delineation of personal identity, the relationship between mind and body and, ultimately, the dignity of the human person, begin to emerge clearly. This analysis is carried out, however, by attempting to probe human freedom and its unconditionality.

4 Human Identity

How does the subject enter the richness of his being? How does the person come into cognitive contact with itself, with the deeper dimension of the self? The task does not seem particularly difficult, since in this case no diaphragm is seemingly interposed between the knower and the object to be known. Agazzi, however, notes very finely that the absence of diaphragms by itself does not eliminate the most bulky obstacle: namely, that any form of knowledge always involves the condition of otherness: that condition is not satisfied in this case. In fact, we would know ourselves in a direct way but we could know oursevelves only as 'others', from which paradoxically there follows the need for an epistemological mediation, namely the need for the subject to be known indirectly, through an 'objective' or 'inter-subjective' knowledge. In this sense the confrontation with the traditional objective knowledge, first and foremost scientific knowledge, which has largely replaced the religious or metaphysical forms, is almost inevitable. But how can scientific knowledge lead us successfully to self-knowledge? How does it make it possible to have access not just to the phenomenal but also to the metaphysical selves, that is, to the very heart of the human person, whose features have been described earlier? If, as noted by Agazzi, "[t]he conceptual space in which [we] encompass all [we] can know is reduced to matter and motion" (Agazzi 1997: 6), how do we achieve this goal? Modernity has not been able to solve this knotty problem, which remains essential. The price paid was a total loss of the unitary image of the human being. Let us add to this the fact that reductionist monism or various kinds of dualism have also reduced the scope of the enterprise of knowledge: the knowledge of the effect would be determined by the impact of the physical world on the doors of perception. The outcome of this shift is certainly no small matter:

> Not being able to count on an *inner* objectivity but only on a *subjectivity* barely offset by the perspective of the 'transcendental', in modern thought persons can be seen as obliged to save their personal individuality, their dignity, their freedom on the basis of pure moral 'certainties', which are postulated outside any genuine 'knowledge' (Agazzi 1997: 8).

The elimination of the sphere of human 'mystery' from the horizon of sciences is a goal which is only apparently achieved by the theories of man-machine interface and by the implementation of mechanical models of various types (chemical, thermodynamic, electromagnetic or cybernetic); these are fragile operations, because they are unsuitable to include in the modelling significant and 'phenomenologically evident' aspects of human reality. By replacing teleology with teleonomy, by suppressing the level of the intrinsic or constitutive finality, we have gradually declined to adequately respond to the question about 'what' the human being is, namely, the question of the essence, although this term seems nowadays obsolete and somewhat elusive. We have given up grasping objectively the 'ontological distance' which is really perceived by us and separates us from any other living being and our own technological artefacts: accordingly we are not allowed to use these as hermeneutic and epistemic paradigms of our own nature. However, Agazzi does not hold a rear-guard position. He indulges in no way in a sort of return to the natural, understood as a primordial, original and pre-technological state, nor does he intend to propose a kind of dualism of natural and artificial. The reasons are almost obvious. The first, very intuitive, resides in the fact that such a condition probably never obtained. Man has always created the artificial, not as a separate universe, but as the expansion of his own capabilities and intrinsic ends. Therefore, on the other hand, as Agazzi rightly states, "the artificial is but the most typical manifestation of human *nature*, which is characterized by the fact that man assures his survival and his progress *adapting* nature to his own needs instead of adapting himself to nature" (Agazzi 1997: 33). The attempt to naturalize the human being in all respects, instead, "reveals itself as a kind of regression to the pre-Socratic philosophical anthropology" (Agazzi 1995: 27), in which the human being is reduced to a mere part of nature.

But what does it mean to be part of nature? Which answer can be given to this question today? The artificial world is or is not our natural world today? The artificial is in effect "a part of nature that proceeds from the peculiar nature of the human being" (Agazzi 2004a: 84), which is clearly characterized by intentionality. But the artificial is also the hermeneutic screen that many have used in their attempt to prove the exact opposite, namely the lack of intentionality in humans, or rather its derivation from underlying physical processes. Nothing is more paradigmatic than the project of the so-called 'artificial intelligence'; that name itself betrays a not inconsiderable philosophical contradiction. If in fact intelligence presupposes thought, then two fundamental questions cannot be avoided:

> either we admit that the (new) machine can perform these [intelligent] tasks as it has an immaterial power, like thought; or it must be admitted that, if a material machine can do smart things, then it is not the case that humans are required to perform such things through an immaterial principle. In both cases there would be a reduction of human beings to a closely *naturalist* level (whereby they lose their specificity with respect to the natural world) and in the second case reduction would also have an explicitly materialist connotation (Agazzi 1991: 2).

Both these different conclusions point to a radical downsizing of human nature; consequently, they have a metaphysical scope and not just a functional one. In fact, in this perspective, the technological machine ideally would have not only an *emulative* but a *simulative* purpose as well: "in order to understand how humans do certain things you try to make a machine that knows how to do them (cognitive goal). In this way the simulative aspect appears loaded with an analogical, heuristic and explanatory intentionality" (Agazzi 1991: 3). By the explanatory claim one means to reduce the ontological gap between human beings and their substitutes by implementing an operational identification. The simulated operationality undoubtedly follows the human transitive activity (the Aristotelian *poiesis*), but it can also trace some aspects of the immanent activity itself (*praxis*), to the extent that the latter employs representational and symbolic systems which can be reproduced. According to Agazzi what remains outside the margins of the simulation is the basic condition of the immanent and transitive activities: intentionality as a power "that allows the internal 'intentional' state to switch to the material symbol that represents it, (…) that allows us to interpret material signs as 'meaning' certain internal states" (Agazzi 1991: 10). This human capacity does not coincide with a particular physical state, but it is the 'inner' condition of the transition from one state to another, as well as the condition of the interpretation of the starting and final states of artificial cognitive systems.

Let us then return to the beginning. With respect to human nature it is not realistic either to imagine a return to the past, or to cultivate the utopian dream of a perfect and total knowledge, without shadows and dark sides, but it is desirable to 'recover the interiority', the genuine wellspring of science and technology. Interiority is not subjectivity, although it encompasses it; on the other hand, according to Agazzi, it may be objectified, even though it constitutively transcends the order set by naturalistic explanations:

We have to win back the idea of an *objective interiority*, because it is a reality that all persons *experience* (even if it is not sensory experience), that they can understand. They can talk to each other about that, they can probe it and they are able to teach other people to analyse it, following the steps which, of course, must be made by each individual (but can you in science do otherwise?) and which allow a certain mutual control of statements (Agazzi 1997: 40).

5 Corporeity and Person

But Agazzi's reflection goes further and delves right into the heart of a matter of great importance and delicacy: the identity and status of the human person. In this regard, he proposes the mental experiment of assuming a conceptual gap between being-human and being-person, and evaluating the theoretical and practical sustainability of such a formal separation. The Boethian definition of a person ('an individual substance of a rational nature'), which he often recalls, historically laid the foundations of an essentialist and substantialist conception of person, and established the referential and coextensive identity of the concepts of human being and person. Today, however, we are witnessing a breakage of this balance, due in large measure to the transition from the substantialist to the functionalist model. Agazzi observes:

> [O]ne has the right to ask why a human individual can, not yet or no longer, be a person, and the answer consists in a *nominal definition* of the person which no longer has a substantialist character, but reduces the concept of a person to a predicate which summarizes a number of *empirically ascertainable functional capabilities*, but without reference to any ontological substrate that implies them. Then we will call functionalist such a conception of a person (Agazzi 2001b: 21–44).

The main limitation of this approach is that although it does not give up the purpose of defining a set of formal or sortal properties, it is unable to arrive at a convincing answer about the referential nexus between these properties and the actual individuals which bear them in a unmistakably unique way. This is because we continue to look at the particular, the real concrete person, without the systemic perspective of totality, that is, with no care about grasping its intrinsic typicality. Agazzi remarks again:

> when the subject matter of a metaphysical consideration is a particular entity, we intend to consider it *as a whole*, that is to give a characterization of it that is compatible with its multiple properties, although not exhausted by any of them. Assuming the point of view of the *whole* is very different from the purpose of knowing the *totality of the real*: the 'whole' is a horizon, is a sort of 'distributive' universal, within which is aprioristically included all that in any way 'is given' to our knowledge, without the pretension of predetermining either the single individual characteristics of the entities that are included, or how they 'give' themselves (Agazzi 2001b: 21–44).

Despite the loss of the idea of substance, for a variety of reasons, some of which may be well grounded, the idea of the totality of the particular as unavoidable metaphysical horizon cannot be given up; its loss would bring about the reduction or fragmentation of being, through the seclusion of properties and propositions including such properties, which can also be true but will not exhaust once and for all the complexity of their reference. For this fundamental reason, never sufficiently understood by today's philosophical anthropology, it is not permissible to assert that the absence of certain properties (such as consciousness) is a sufficient reason to consider them accidental or non-typical. For Agazzi, instead, their absence, understood as contingent deprivation or potential presence, is a full index of personal typicality. Deprivation, in fact, "necessarily entails an ontological reference to the intrinsic essential properties of the *carrier* of the particular privation involved" (Agazzi 1994: 225). Hence, the person exists even if contingently lacking such intrinsic essential properties. The same principle applies to potentiality; it does not affect the possession but only the exercise of certain capabilities or properties, as Agazzi explains: "the *exercise* simply is a kind of external confirmation of this possession, and by no means a transition from potency to act" (Agazzi 1994: 229). But all this brings us back to the enormous problem of how to define the real potentialities of the human person. It is a request for meaning, to which once again one can respond only by a realistic look at the whole of any particular human experience.

Earlier we mentioned the idea of separating, artificially and "contrary to the most immediate content of our existential experience… the *unity* of this experience, in which we do not distinguish soul and body, and in which, in any case, any human being apprehends himself as *one* and not as *two*" (Agazzi 2011: 73); but we saw that this idea has led to an impoverishment of the notion of corporeity. Agazzi has devoted important studies to the so-called 'mind-body' problem, in which he thoroughly examines the contemporary positions and their classical roots, Aristotelian, Cartesian and empiricist. On this occasion we will not dwell, for reasons of space, on the individual clarifications, but we must recognize immediately his merit of having considered Cartesian dualism as the philosophical and cultural antecedent of materialism and spiritualism: these trends of thought are well grounded in that doctrine, but have moved independently towards marking the most substantial distance between our spiritual and material dimensions, and ultimately dealing the coup de grace to the unitary image of the human person. Agazzi correctly guesses that

> the human body itself is a peculiar body indeed, which is so to speak 'oriented' toward the realization of performances which we can still call 'spiritual'. On the other hand, man as a pure spirit does not exist either, as spirit is 'incarnate' in man, and is affected by all the influences of his corporeity. This makes it impossible to say where the body 'ends' and the soul 'begins'; but this happens because man is not *two things*. As we said, man is certainly a complex reality which presents several aspects, but it would be a mistake to understand these aspects as if they were different 'parts' of him (Agazzi 1981: 19–20).

Here emerges Agazzi's view, which inherits significant parts of the Aristotelian philosophy and of Plato's anthropology, reading them in an original way though the systems theory. According to this conception, "the relationship between the body and the person is considered as a relationship between a part and the whole" (Agazzi 2013b: 121). The person must therefore be regarded as a complex whole with certain properties, which do not coincide with the properties of the individual parts or arise from their composition or summation. In this sense, the properties of the parts are related to the properties of the whole without determining them. Conversely, the system as a whole is certainly related to the properties of the subsystems, but it cannot be turned into any of these. The empirical evidence is therefore that of a being which expresses its nature (through thinking and other activities) in the 'functional' multi-level relationship between the system and its subsystems (material or immaterial). In this perspective voluntary action is the peak of the interchange between the whole and its parts, since the whole person acts involving every lower aspect of the system, thus attributing to the temporal flow of various movements and activities a comprehensive and unique identity.

For the same reasons, if we do not consider corporeity as a dimension alien to the whole, and the various kinds of bodily affections (disease, aging, and consequently the experience of suffering or fatigue) play a role that goes well beyond the boundaries of the material or biological. Also in this case failure to take into the right consideration the unity of the system leads paradoxically to a fragmented vision of the body, of its affections and of the treatments that can be offered. As Agazzi realistically complains, today "medicine only recognizes influences entirely interpretable within the framework of causal actions of a *physical* kind, duly understandable on the ground of physical theories" (Agazzi 2001a: 13), but it loses sight of the whole of an ill and suffering person, and even of the very experience that suffering can represent for people (see Agazzi 2013a). Human finitude, in its various manifestations, is therefore not an absolute limit, but an almost necessary condition for the rediscovery of the inner richness, and of what we recognize as supremely good and valid in itself; this condition of finitude is therefore compatible with the sense of perfection and the pursuit of values and existential purposes.

This analysis of the ought, of freedom, of human and personal identity, and of agency and corporeity, has revealed the traits of a lively and fruitful thought, able to communicate with the contemporary world and with scientific knowledge, and at the same time with the great masters of classical philosophy (from Aristotle and Aquinas to Kant), but mainly committed to the rational defence of the originality and irreducibility of human experience.

References

Agazzi, Evandro. 1981. Mind and Body: A Philosophical Delineation of the Problem. *Epistemologia* 4: 3-20.

Agazzi, Evandro. 1987. I fondamenti filosofici dei diritti umani. *Cenobio* 4: 325-338.

Agazzi, Evandro. 1991. Operazionalità e intenzionalità: l'anello mancante dell'intelligenza artificiale. In *Intelligenza artificiale e intelligenza* naturale, ed. Salvino Biolo, 1-13. Genova: Marietti. Rep. 2010. *Natura umana, natura artificiale*, ed. Maria Cristina Amoretti, 63-77. Milano: Franco Angeli.

Agazzi, Evandro. 1992. *Il bene, il male e la scienza. Le dimensioni etiche dell'impresa scientifico-tecnologica*. Milano: Rusconi.

Agazzi, Evandro. 1994. The Human Being as a Person. *Forum. Trends in Experimental and Clinical Medicine* 4: 221-232.

Agazzi, Evandro. 1995. Dimostrare l'esistenza dell'uomo. In *Interpretazioni attuali dell'Uomo: filosofia, scienza, religione*, ed. Evandro Agazzi, 27-38. Napoli: Guida.

Agazzi, Evandro. 1996. Perché gli esseri umani hanno dei diritti? L'educazione civica fra filosofia ed etica. *Nuova Secondaria* 5: 43-53.

Agazzi, Evandro. 1997. *La techno-science et l'identité de l'homme contemporain*. Fribourg: Editions Universitaires.

Agazzi, Evandro. 1999. Libertà d'azione e libertà del volere. *Nuova Secondaria* 2: 5-6.

Agazzi, Evandro. 2000a. La libertà in situazione. *Nuova Secondaria* 9: 5-6.

Agazzi, Evandro. 2000b. I limiti della libertà. *Nuova Secondaria* 8: 5-6.

Agazzi, Evandro. 2000c. Libertà come creatività e autodeterminazione. *Nuova Secondaria* 6: 5-6.

Agazzi, Evandro. 2000d. Libertà e spontaneità. *Nuova Secondaria* 5: 5-6.

Agazzi, Evandro. 2001a. Illness as lived experience and as the object of medicine. In *Life-Interpretation and the Sense of Illness within the Human Condition*, eds. Anna-Teresa Tymieniecka, Evandro Agazzi, *Analecta Husserliana*, 72: 3-15.

Agazzi, Evandro. 2001b. Il significato dell'identità personale. In *Identità personale. Un dibattito aperto*, eds. Andrea Bottani, Nicla Vassallo, 21-44. Napoli: Loffredo.

Agazzi, Evandro. 2004a. L'être humain et la nature dans le discours éthique. In *Le discours bioéthique*, ed. Peter Kemp, 71-85. Paris: Cerf.

Agazzi, Evandro. 2004b. *Right, Wrong and Science: the Ethical Dimensions of the Techno-Scientific Enterprise*, ed. Craig Dilworth. Amsterdam-New York: Rodopi.

Agazzi, Evandro. 2007. Philosophy and Human Understanding. In *The Proceedings of the Twentieth-first World Congress of Philosophy*, vol. 13, eds. Ioanna Kuçuradi, Stephen Voss, and Cemal Güzel, 381-389. Ankara: Philosophical Society of Turkey.

Agazzi, Evandro. 2011.The Scientific Images and the Global Knowledge of the Human Being. In *Rethinking Human Nature. A Multidisciplinary Approach*, ed. Malcolm Jeeves, 70-81. Grand Rapids: Eerdmans.

Agazzi, Evandro. 2013a. Il dolore e la sofferenza umana alla luce della ragione e della fede cristiana. In *Ripensare l'uomo oltre la modernità*, ed. Elio Sgreccia, 47-28. Siena: Cantagalli.

Agazzi, Evandro. 2013b. La persona umana e il suo corpo nella cultura e nella bioetica. In *Ripensare l'uomo oltre la modernità*, ed. Elio Sgreccia, 115–135. Siena: Cantagalli.

Science, Historicity and Complexity

Giuseppe Gembillo

Abstract In this contribution the author intends to reconstruct Evandro Agazzi's reflections on the development of contemporary science after the elaboration of the Theory of Systems and the emergence of the perspective of Complexity. Reflections that convinced him that Reality is historical and that, in order to understand it, an interdisciplinary approach and a new definition of Science are needed. As is well known, classical science and its language were incompatible with historicity; Galilei and Descartes claimed to discover the eternal laws that characterize the mind of God, the ontological structure of Nature and the "form" of our intellect; up to Fourier, Darwin and Mach the objectivity, universality and eternity of scientific laws represented a metaphysical assumption that quickly become an undisputed dogma. In 1883, the same year in which Wilhelm Dilthey was still opposing "Sciences of Nature" and "Sciences of the Spirit", Ernst Mach published *Mechanics Exposed in its Historical-critical Development*. In his work, summarizing the results emerging from thermodynamics, electromagnetism and evolutionism, Mach inserted historicity into science using history not only as a reconstruction of the event-science but also as the only way for its theoretical comprehension. By so doing, Mach aroused endless debates involving several non specialized scholars, including, for example, Lenin, who understood very well that the crisis of scientific determinism initiated by Mach would eventually involve "objectivity" and the process of realization of Marxism, and reacted consequently. What that followed has strongly affected the Italian scientific and philosophical debate, marking indefinitely its physiognomy.

Keywords Systemicity · Complexity · Historicity · Interdisciplinarity · Scientificity

G. Gembillo (✉)
Department of Human and Social Sciences, University of Messina, Messina, Italy
e-mail: gembillo@unime.it

M. Alai et al. (eds.), *Science Between Truth and Ethical Responsibility*,
DOI 10.1007/978-3-319-16369-7_16

1 The Specificity of the Italian Situation

In effect, when first introduced in Italy by Giovanni Vailati and Benedetto Croce, Mach's thought caused fierce debates. Vailati didn't suffer their consequences in a direct way, since he had a protector in his master Giuseppe Peano, and he died too young. On the other hand Croce—who unlike Vailati also accepted Henri Poincaré's conclusions—was harshly criticized at once, and from then on he was accused of wanting to "underestimate" science, becoming a victim of a die-hard prejudice until now. This reaction can be explained in some way if we bear in mind the specificity of our (Italian) situation. In Italy when it comes to science, also in a moderately critical key, we live, so to speak, "on the verge of a nervous breakdown". The never completely metabolized condemnation of Galileo has always been seen as condemnation of science as such, and every time it has been called into questions, a sort of indignant defense against "resurgent obscurantism" has sprung up. This situation has also been supported by a typical inclination of the Italian scientific culture, characterized by the lack of great scientists-philosophers. From Galileo Galilei to Carlo Rubbia we had great experimental physicists, but no one with the predisposition or the interest to question science methodologically or epistemologically. This led to the paradoxical case of Enrico Fermi, who in 1934 was unable to understand what he had done, and he thought he had discovered a new element, without realizing he had instead caused the fission of the atom.

Also as concerns historiographical reconstruction, in Italy the history of science has been, in general, the history of the approach to the Galilean revolution, based on the assumption that as soon science was born, there would be no longer "history" but, indeed, only science, or in other words definitive acquisition. Something similar occurred about the way philosophy of science was introduced "officially" in the Italian University by Ludovico Geymonat. Indeed, it has been affected by two background prejudices, a theoretical and a political one. Geymonat introduced logical neo-positivism in Italy not as a philosophical movement worthy of being known on the historiographical side, but as an alternative theoretical proposal opposing the Crocean underestimation of science, without taking into account that—in the meantime—the neo-positivists themselves had already considered their initial program obsolete. Furthermore, he extremely politicized it, overstating a connection with Marxism that, as a matter of fact, had represented only the personal choice of some neo-positivists. A clear testimony of all of this, limiting myself to just one example, is the way he structured his *History of philosophical and scientific thought*, in which he emblematically dedicates very little space to Mach and Croce, but around forty pages to Louis Althusser. In accordance with these choices, the Italian scientific debate on the subject has developed in a generic and sometimes superficial way that essentially remained the same up till now if, even after more than a century, some people still talk about Crocean "underevaluation" of science in allusively and quick way, without accurate arguments presented on scientific papers or specific monographs, but just through rapid and vague hints, expressed in general terms, as we feel compelled to do with what is "obvious" (cfr. Gembillo 2006). In other words, a large part of the Italian philosophy of science, before opting for bracketing the problem and identifying itself almost completely with the scientific procedures, manifested itself as a direct or indirect polemic against Mach and Croce. Most certainly there have been important and honorable exceptions, one of which is represented in an eminent manner by the thought of Evandro Agazzi.

2 Teoreticity and Historicity in Agazzi's Thought

As a matter of fact, Evandro Agazzi has always maintained a great balance in evaluating the various theoretical conceptions. This was due to the fact that he always based his findings on a rigorous and careful historical reconstruction that allowed him to enunciate theoretically relevant judgments, fitting the specific nature of the subject he was gradually taking up. Above all, he thought over and reconstructed with great attention the history of post-Galilean science, going upstream against the Italian tradition. To this purpose he wrote several works on the history of physics, he edited two volumes in the history of science, he translated James Clerk Maxwell's *Treatise* (Maxwell 1973) with an interesting and well-documented historical introduction. In other words, his originality lies in the fact that each of his theoretical works has always been grounded expressly on a comparison and on a solid historiographical basis, and this basis covers everything that has happened historically and theoretically from Galileo up to the present day. Since his first work in this sense, *Temi e problemi di Filosofia della fisica* (1969), he has dedicated, for example, wide historiographical reconstructions and epistemological discussions to quantum physics and to the relativity theory.

Given the impossibility to talk about all the works he wrote about the relationship between science, historicity and complexity, I will limit my discussion to some thoughts expressed in one of his most theoretically mature and clearer texts in terms of argumentation and training: *Le rivoluzioni scientifiche e la civiltà dell'occidente* (2008), that specifically includes the issue of Complexity. I will focus specifically on the problem, which he discussed in detail, "of the widening of the scientific landscape in the 20th century".

After recalling quickly some of the most significant developments that characterized science and technology in the twentieth century—namely quantum physics, space travels, the moon landing, the discovery of DNA, the techno-sciences, etc.—Agazzi makes an interesting reference to the methodological and theoretical aspect that more deeply marked the difference between classical science and that of the twentieth century, highlighting that the reference to the twentieth-century science is not enough

> to give us the measure of the new intellectual model that characterizes in its entirety contemporary science, which is distinguished from modern science especially by the abandonment of a strictly deterministic and analytic perspective in the name of a fair appreciation of the complexity and globality (Agazzi 2008: 152).

From this observation he draws the motivation for a historical-theoretical reconstruction of the disciplines, which were decisive to impose the methodological turning point that he mentioned, a turning point that refers "to the theme of complexity, systems theory, cybernetics" (ibid.) From these new sciences, Agazzi draws inspiration to support and transform the realism that has always characterized his thought. Now the concept of "form" in the active sense of *Gestalt* suggests him that "here emerges the most notable feature of the modern dynamic complexity compared with the previous types of complexity: the forms generated by the nonlinear dynamics, are not just 'beautiful', but often also remarkably similar to those of natural objects, especially those that characterize living organisms" (ibi: 160).

In this concrete horizon it is possible to get the meaning of the self-development that characterizes all the living forms, or more clearly that "shows a spontaneous tendency toward the growth of complexity itself, even starting from situations that did not seem to prepare it (a phenomenon called self-organization), just as it happened many times in the course of natural evolution" (ibid.). Agazzi knows that these considerations—starting from the second half of the twentieth century—have been increasingly corroborated at all levels, and he notes that this is "why someone has started to think (and in some cases it has been shown) that at the heart of many—if not all-of these extraordinary structures there are natural dynamics of this kind" (ibid.). He avoids carefully the temptation to prefigure or endorse some new form of biological determinism, and to this purpose he specifies first and foremost that "with that we would not have a surreptitious return to reductionism" (ibid.).

However, he seems to attribute the impossibility, in these cases, to make exact forecasts, to our inability to follow the complex evolution of an organism, and not to its intrinsic impossibility. In his opinion, that is,

> this 'creative' aspect of complexity never suppresses the irreducible limit to our ability to calculate. Therefore, even if we could determine the dynamic that governs the genesis of a certain structure, as the structure generates itself, our ability to forecast will decrease, and therefore we couldn't know in any way what will be the global properties of the structure itself, except by letting it evolve and then seeing the result (ibid.).

We couldn't know it also because the developments are not determined by an abstract law, but by the concrete historical and evolutionary Reality.

> In other words, it will never be possible, in general, to predict the behavior of a structure that is the result of a nonlinear dynamic starting from the law that governs this dynamic; to get a prediction we will identify some new law, which must be obtained from the analysis of the direct reality and not deduced from the previous law (ibid.).

In other words, contrary to what was supposed by classical science, the top level is not explained by the lower level, and therefore not reducible to it, and "this applies to all levels of reality: atoms with respect to molecules, molecules with respect to cells, cells with respect to the body, and so on" (ibid.).

For this reason there is no direct affiliation between the different disciplines, since for each of them the main reference is always concreteness, in the sense that "every time you go one level up it is necessary, so to speak, 'to return to reality', looking for it in its own objects, according to specific methodologies, with the very important result of ensuring again the autonomy of the different sciences, which are therefore, each in its own order, all equally 'fundamental' as physics" (ibid.). In this way, the classical hierarchy is replaced by the network interactive relationship where what counts are equal relations, and not hierarchical dependencies. This is what emerged from today's complexity theory and the historically complex processing systems.

3 From Atomism to Complexity System, the Cybernetics

Agazzi unites properly the concept of Complexity "to the idea of an organized, structured system, different from a pure multiplicity consisting of elements essentially of the same type" (Agazzi 2008: 163). In order to better clarify this concept he uses a particularly effective comparison, based on a fundamental shift suffered by the mathematics of the late nineteenth century (Cantor), noting that "the relationship between the components of an object or the complex process and the object itself considered as a unit is not the relation of 'membership of an element in a set', but that between 'the whole and its parts'"(ibi: 164). He intends to further clarify the differences between set-theoretical approach and complex approach by noting first of all that

> in set theory the axiomatic construction can be done using as the only primitive notions that of element and the relationship of membership of an element to a set. This means that *the nature of the elements* is completely irrelevant, and that they are not supposed to be provided with structures or internal relations. We could say that the ontology of set theory (i.e. the type of entities that it presupposes) is *atomistic* (ibid.).

Understanding perfectly the meaning of the perspective of complexity, he notes that

> vice versa, if we consider, for example, a living organism, it is naturally conceived as consisting of 'parts' and not by 'elements', and these are the various 'organs', all different from each other, characterized by specific structures and functions and at the same time connected and cooperating to form a unity, namely a *whole*, provided with its overall properties as well (of which the most notable is precisely life). Moreover, these parts are themselves a 'whole' compared to the other parts that constitute them. The ontology underlying this concept is holistic (ibid.).

To be more precise and to achieve a better comprehension, and with different arguments, of the difference between the atomistic and the holistic approach, it should be duly noted that "in the holistic perspective the structure and functions of the parts are of considerable importance, and the relationships between parts and whole are not of mere membership, but of correlation, interaction, and also lie on other relations established between the different parties" (ibid.). The difference between the two perspectives is really radical since, unlike the set theoretic approach, within the systemic-relational approach "as a result of all these relationships, it is said that each whole *is more than the sum of its parts* in the sense that it has properties that are not typical of any of its parts, and it does not even appear to be a sort of sum or combination of them" (ibid.).

Agazzi rightly insists on "the feature of *originality* or *creativity* that characterizes the complex systems", and observes that, considering the problem from a more rigorous historical perspective,

> the holistic point of view has always been implicitly subtended to the life sciences, but only in the twentieth century it emerged with strength and explicitness, casting a new light on many problems and resulting in refined conceptual elaborations that can be rightfully considered as clarifications and developments of the notion of complexity (ibid.).

In a sense, this point of view has been integrated by System Theory and by Cybernetics, issues on which Agazzi develops a broad, structured and rigorous discourse, which in this context cannot be examined in depth, but cannot be completely omitted, either. I will briefly mention it, beginning with the "General System Theory", highlighted by Agazzi both for its methodological originality and because—precisely for this reason—it has encountered many obstacles and aroused fiery controversies. As a matter of fact, although the concept of system has always accompanied philosophical and scientific thought, actually

> systems theory is relatively young, since it achieved success in the second half of the twentieth century, being at first the object of animated debates and controversies. This is not surprising, since any 'new' discipline has to face a more or less controversial debate in order to establish its titles of scientificity (ibi: 171–172).

Agazzi prevents the objection that such a finding could not always be true, and there are several examples of disciplines that have developed peacefully: he grants that this

> is true, but everything depends – so to speak - on the degree of intensity with which are combined the claims of scientificity and those of novelty: in other words, the instinctive reaction of the scientific community, in front of a new discipline, seems to be the statement according to which, if it is really new it is not science, and that if it is really science, it is not new (ibi: 172).

To avoid misunderstandings we should clarify that we need to be careful and, for example, point out that throughout the history of Western scientific culture it usually happened that

> when novelty comes not in the sense of 'a new development' of an existing framework, but in the more challenging sense of being able to make claims that 'do not fit' in the previous framework, in that case the most common reaction is to state that the new claims are actually 'unscientific' extrapolations or digressions (ibi).

For this reason, Agazzi rightly notes "systems theory has been the object of controversy, not for its content, but because it presented itself as a way to conceptualize and theorize that wanted to be 'new' compared to the already tested and established scientific criteria, considered too narrow" (ibi). In other words, although the concept of system has been historically used to describe coherent theoretical philosophical constructions or, for example, even those of Newton and Linnaeus, the new theory inspired by it aroused skepticism "since the concept of system was used in a partly new and even alternative sense, compared to its traditional sense" (ibi).

Connecting these observations to those expressed earlier about the idea of complexity, Agazzi—to avoid any misunderstanding or possible confusion—reminds that the classical system "corresponded to the need for *unification*, *reduction of multiplicity to one*, about which we have already discussed, which had found its fullest expression in the mechanical determinist reductionism" (ibi: 172–173). As is clearly shown by all that has been said, "vice versa, the new systems theory would be an attempt to understand the role of *diversification* and the peculiar type of structured and ordered unity that it entails" (ibi: 173). Moreover, if we face the problem without bias and with the necessary calm, we can agree that, on closer view, it

> gave a framework and a rigorous and harmonic characterization to a number of concepts, such as the one of ordered whole, of functionality, of hierarchical structure, of organism, of development, of adjustment, of interrelation between individual and environment, of centralization, of self-preservation, of finalized process, which are frequently used and indispensable in several sciences (from the biological ones to the psychological and social ones), but that had been used with an almost exclusively intuitive meaning, or with a meaning just a little more specific than in common language (ibid.).

The most common reaction to this attitude was, as Agazzi precisely emphasizes, a kind of "temporary tolerance", waiting and hoping that the approach would adjust to the traditional scientific standard. But complex systems theorists, starting from the founder, Ludwig von Bertalanffy, were not disposed to align to this wishes, and defined in a more precise way their methodological and ontological approach, converging in outlining a "generalist" perspective whose "aspect of interest and originality lies in 'breaking' the traditional scheme, and thus introducing new and prolific perspectives in many disciplines, not only belonging to the natural sciences" (ibi: 176).

This is, essentially, what Agazzi intended to say about the new approach to the Real. However, "to better understand the meaning of systems theory we should consider in any case another almost coeval and strongly linked discipline, cybernetics, and also the contribution that systems theory itself gave to a new scientific methodology, interdisciplinarity" (ibid.). To coherently realize the latter, one didn't need only cybernetics, but also information theory, soon named computer science, and the life sciences. After a due reconstruction of the essential events related to these new approaches, Agazzi rigorously explains why the relations between disciplines should be investigated in all their complexity, drawing conclusions on which we shall now linger to enucleate their most innovative aspects.

4 The Meaning of "Interdisciplinarity"

At this point Agazzi, facing the issue of interdisciplinarity directly, declares himself convinced that

> starting from what has been exposed above, it is clear that systems theory, information theory, cybernetics, complexity theory are subjects whose concepts, methods, and principles find direct application in many fields of scientific research, technological achievements, production structures, social institutions, and the organization of various services in the contemporary civilization (Agazzi 2008: 185).

This means that they have “interdisciplinary” character. In this regard, however, for the sake of clarity he states that this is a sort of “transdisciplinarity”, that should not be confused with the “multidisciplinarity”, defined by the exterior combination of different disciplines that do not really come into interaction. As he exactly explains,

> what characterizes interdisciplinary as something different compared to the simple ‘co-existence’ of various disciplines (which is often referred to as *multidisciplinarity*) is the fact that the interdisciplinary perspective aims to achieve a convergence, a coordination, a mutual exchange between different disciplines in order to solve a particular problem (or certain problems) of cognitive or practical nature. In this sense it is an effort of “unification” which is the opposite of the *reduction* (ibi: 185-6).

In other words, we are dealing with something substantially opposite to the traditional approach, since

> the unity we try to propose is not considered as a reduction of the complex to the simple, of the differences to the uniformity, but as the comprehension of the *complexity* through the coordination and the synergy of the different *parts* of a *whole* that is holistically comprehensible, since each part contributes specifically to the operation of the whole and to the formation of its irreducible features (ibi: 186).

He insists on the close connection between this approach and the one that emerged in systems theory and complexity theory, and adds that the need for interdisciplinarity has to be understood also in the light of the practical need to make decisions in complex situations. Also, he points out that this approach should not be interpreted as a kind of substitute for specialization, but it should represent an appropriate integration to it. If it is true that the inclination towards what Ortega y Gasset rightly called “barbarism of specialization” (Ortega 1930) is very negative, it is also true that without specialist proficiency the integration of knowledge would suffer of over-generalization. By specifying this, Agazzi appropriately reaffirms the distinction between *Interdisciplinarity*, *Multidisciplinarity* and *Transdisciplinarity*, and once again underlines that only the first “has the acceptance of the complexity as a prerequisite, and aims to understand it (Agazzi 2008: 188). But to be understood it needs to be based on the interaction between rigorous and precise experiences in the sense that

> the needs of interdisciplinarity cannot be considered satisfiable by a suppression of specialization. This is certainly a consequence of the increase of knowledge, but it is also one of its *conditions*, so that giving up the specialization would mean to deprive ourselves of increasingly reliable and advanced knowledge for our practical achievements, apart from giving up an increase of knowledge (ibi: 188-9).

Furthermore, an equally important aspect, “every science is provided with its own *methodological criteria* and inferential processes (i.e. *logics*)” (ibi: 189). For this reason it seems appropriate to mention, in line with what Agazzi is highlighting here, the “polilogics of complexity” (Gembillo 2008). From this originates the coherent conclusion that true interdisciplinarity

> is built *starting from the disciplines* and respecting the seriousness and commitment of their technical investigation. Secondly, it is clear that the impossibility to dominate many specializations is not linked to the inability to contain an excessive mass memory of

> knowledge, but rather to the inability to adopt simultaneously many perspectives, many mental habits, many different methodological settings and master also the related concepts and operational methods (Agazzi 2008: 190).

Another reason is that, as discovered by Humberto Maturana, our brain develops in a certain direction, and from that direction it "*looks at reality*" (Maturana 1995). But since, as Goethe said, "everything has been given to man both as a treasure and as a curse", also in our case,

> all this represents undoubtedly an obstacle, but also constitutes a richness: once again, not because accepting the plurality of disciplines you know *more* of the real, but because you know it *better*; in other words it is possible to get more aspects, exploring more deeply its richness, since we are able to appreciate more adequately its *complexity* (Agazzi 2008: 190).

Achieving this is difficult, since "the unity of the thing does not guarantee by itself the unity of knowledge" (ibid.), because who assumes the task of understanding the complexity of the events must have an unordinary capacity of synthesis and be fully aware of the fact that

> the truly interdisciplinary discourse is established when one admits that, in front of a complex problem, different optics are *necessary* for its comprehension, and one makes an effort to consider it under each views, comparing them and trying to see how each one contributes to the comprehension of the whole. (ibi: 191).

We could say, at the end of the short *process* outlined until now, that

> the successful completion of interdisciplinary work is the achievement of a certain *synthesis*. This however cannot be understood as the achievement of a unique and definitive *image*, but rather as the overcoming of the one-sidedness of the individual views, through the awareness of their differences, their partial mutual translatability, the existence of interconnections, homologies, and analogies, that increase our comprehension of the studied reality, without *exhausting it*. The unity of the 'thing' corresponds, by now, also to a certain unity of our knowledge of the thing, but the inherent complexity of the latter is never completely captured (ibi: 192).

If this is true we cannot deny that "interdisciplinary work therefore is never fully accomplished: it is reasonable to stop when one has reached the desired *objectives*" (ibid.). Given that all this is based on the clear recognition of the historical and temporal structure of each existing thing at any level, then, "claiming that this work comes to a definitive and absolute end would correspond to the pretension of being able to ultimately know the truth, which instead is an endless challenge for the human being" (ibid.). And this happens not only because the individual subject cannot possess the ability to dominate all the possible points of view on reality, but because, as we have pointed out elsewhere, Reality is constantly changing and so it is never subject to a "full" and complete description of it (Gembillo and Anselmo 2013).

Agazzi's reasoning culminates ideally and theoretically with the consideration that what emerges from all this is the need to redefine the concept of science starting from all the objects of which it is responsible. To this purpose he takes as an example three sciences: "Cosmology, climatology, and the science of new materials, since they are significant examples of these new settings, linked to interdisciplinarity, complexity, systemic vision, that characterize the new spirit of contemporary science" (Agazzi 2008: 193). Of these three sciences I will focus only on the one that, having a longer historical tradition, allows not only for an adequate theoretical analysis, but also a concrete comparison between old and new conceptions of scientificity. I am obviously talking about that "science" that strives to understand the Cosmos.

For this purpose, Agazzi first underlines a serious limitation, wondering:

> The fact that in practice modern cosmology studies the universe using as sources only cognitive theories of physics (general relativity and quantum theory) is not perhaps in contrast with the complexity that the universe obviously possess (from a certain point of view, it should be the most complex object of all)? (ibi:166).

This and many other questions of the same content imply, as we will see now, just a new idea of scientificity, which does not escape the analytical sensitivity of Evandro Agazzi.

5 A New Idea of Scientificity

The new idea of scientificity emerges starting from Cosmology, the discipline that Kant defined structurally unscientific, because inherently unable to draw on possible experience. On the contrary, for Agazzi, after the scientific revolutions of the twentieth century, "the most interesting aspects for us originate from the fact that this discipline, which is now recognized as a science, and moreover a physical science, can be considered so only by widening the usual scientific criteria of natural science" (Agazzi 2008: 195–196). Obviously we should not hide the difficulties, nor is it appropriate to face the problem superficially, because we have to take into account that the issue is complicated by the fact that "it is difficult to identify the object of cosmology" because in order to do so correctly you must be able to answer the question "What kind of object is the universe?" (ibi: 196). Clearly the most difficult problem is that it does not occupy a visible space in its entirety, nor has boundaries defined with precision, since it expands in all directions. Precisely for this reason, "it cannot be empirically denoted as a well-defined system of things, neither can it be characterized as a structured set of certain attributes or properties (as in other sciences)" (ibid.). Therefore we cannot omit the fact that, apart from what noticed so far, "compared to the natural sciences (to whose context it purports to belong) cosmology does not fulfill the requirement of empirical controllability, i.e. the ability to prove its theoretical hypotheses" (ibid.).

Therefore its scientificity lies somewhere else, in a meeting point that until some decades ago seemed inadvisable, because it was linked to the already mentioned contraposition, between science and history. On the other hand, Agazzi, inheriting the best of what in another occasion I have defined as "complex neo-historicism" (Gembillo 1999), expressly states that

> difficulties of this type can be overcome by recognizing that cosmology - since in essence it makes an effort to reconstruct a history of the universe - has the right to claim the conditions of scientificity that are recognized for historical sciences, which in fact are not of

> experimental nature, and cannot avail themselves at will of the past data, even though they are "empirical" (i.e. they cannot avoid to be based on factual data), yet they try to offer not just descriptions, but also interpretations and explanations (Agazzi 2008: 196).

This does not mean that cosmology should belong to a different "context", abandoning the field of classical sciences to enter among the human sciences, since an important aspect links it to the former context, namely how it implements "interpretations and explanations", since

> to provide the latter, cosmology draws on the physical sciences and their theories, so it would seem that it is able to adopt that 'nomological-deductive' model of scientific explanation that (at least according to the current epistemology) characterizes mature science and physics in the first place (ibi: 197).

Agazzi rightly believes that, however, this is not the correct answer to the problem of the scientific status and the sense of Cosmology, since it goes beyond the methodology of classical sciences. He adds that "you cannot say that either," and he justifies his considerations by inviting us to reflect that "the laws of physical theories—confirmed now—serve to explain phenomena that take place *within* the universe, but there are no laws to explain the phenomenon *of the* universe taken as a whole" (ibid.). Therefore it is no longer possible, either in this specific case or in general, to appeal to any form of reductionism, since it is now clear that the Whole cannot be explained on the basis of its individual parts, and

> this fact blocks the way to a possible analytical-reductionist solution, as seen above, that would consist in showing that the properties of a 'whole' result from the composition of the properties of its parts, in other words, that the laws that govern the parts allow to deduce the laws of the whole (like, for example, some people think that the properties and laws studied by biology can be derived from those of chemistry) (ibid.).

This dream, already secular and difficult to erase completely despite its already clear illusoriness, still does not take into account the specific nature of that "whole" represented by the Universe in its entirety, nor the fact "that there are no laws or properties of the universe as a whole that can be established with a minimum of explicitness, in order to show later how they are derived from the laws of physics" (ibid.). The solution to the problem lies in the idea of commonality that has already emerged clearly, for example, in scientists like Ilya Prigogine, and here appropriately stressed by Agazzi, who notes that

> these, that in the eyes of many scholars are among the objections that one can address to the recognition of cosmology as a science, actually lose almost all their strength if we stress again that the epistemological characteristics of this discipline are very close to those of the historical sciences, and it represents a noticeable example of a *historiographic* natural science (ibid.).

Agazzi with his usual grace argues in this way an issue that has aroused fierce controversies and sharp contrasts among the most important scholars of the twentieth century. On his part, he adds that

> therefore, as the 'scientific' historian may use sectorial knowledge related to various spheres of human history in which perhaps some 'laws' or at least 'regularities' can be traced, without thereby being forced to recognize the existence of 'laws' of history as a

> whole, so the cosmologist uses different contributions of theories provided with laws concerning certain aspects of the evolution of the universe, in order to reconstruct the lines of such a development, and to tell such a story as *objectively* and *rigorously* as possible (ibid.).

The guarantee that everything respects the parameters that make an approach rigorous is provided by the fact that it must be recognized that "this research for objectivity and rigor is already the necessary and sufficient mark to qualify as scientific a cognitive endeavor (even if, of course, its results should be judged and evaluated in the light of the requirements of rigor and objectivity actually achieved)" (ibid.).

These conclusions were made possible precisely by the methodological turning point occurred thanks to systems theory and the perspective of Complexity that have shifted the attention from the elements to the whole, from the parts to the whole. Agazzi does not fail to highlight it, stressing that what said so far

> might be enough to report the interest and the peculiarity of this new science that, in particular, has been set up thanks to the legitimation of the 'holistic' point of view that characterizes contemporary science, which has completely overcome the narrow overspecialization of the strictly analytical approach (ibi: 198).

Having said so, and to support all the arguments carried out so far, it should also be noted that "cosmology is an interdisciplinary science in the right sense that we have already made clear, namely in the sense that it seeks 'contributions' not from a wide array of different disciplines, but just from those few disciplines that are really needed to investigate the problems it studies" (ibid.). In this case it is important to insist that "in the essence physics, astronomy, astrophysics, mathematics" (ibid.) are the ones whose interaction provides methodological credibility to Cosmology at the moment when it presents itself as a rigorous "historical reconstruction" of an inherently historical entity as the Universe turned out to be. The reference to these few disciplines might seem restrictive for the interdisciplinary character that Cosmology presents, but

> watching things a little deeper we notice that it simultaneously uses two fundamental physical theories that, until now, could not be 'unified', namely the theory of relativity and quantum theory, but it uses them avoiding their insurmountable collisions, in the sense that relativity essentially serves to determine the choice of a 'model' of the universe (for example the expansion model rather than the steady state one), while quantum physics, even in its more advanced parts regarding the elementary particles, is used to 'fill in' the model and actually write the different chapters in the history of the universe (ibid.).

Naturally, Cosmology can pursue its proper goals not in a vague and extemporary way but "only thanks to a sophisticated mathematical processing of selected models, and by using the tools and results of astrophysics and astronomy, which can provide those few but significant empirical findings upon which it can rest" (ibid.).

A similar debate should be carried out for the other sciences that have characterized the twentieth century, making it a fundamental theoretical turning point. However, even limiting myself to the considerations taken into account so far, I think I gave a basic idea of how Evandro Agazzi managed to get in tune with the ensuing scientific and epistemological issues and to show the way for a deep renewal of science in our time. To wit, a renewal that finally leads to heal the wound that science itself opened at the moment of its birth and then had to heal. Today this is happening, or at least this is the direction taken by those who have had the taste and the courage to question the activities that they themselves were helping to strengthen and renew. It is not easy to take note of all this for those epistemologists who bet everything on the perspectives proposed by the old science, that seemed too good to be abandoned. Agazzi, however, fully belongs to those who understood that being practical and realistic does not mean believing in a static and abstract reality, but conforming oneself to its "perennial historicity".

References

Agazzi, Evandro. 1969. *Temi e problemi di filosofia della fisica.* Milano: Manfredi.
Agazzi, Evandro. 2008. *Le rivoluzioni scientifiche e la civiltà dell'occidente.* Milano: Fondazione Boroli.
Gembillo, Giuseppe. 1999. *Neostoricismo complesso.* Napoli: ESI.
Gembillo, Giuseppe. 2006. *Croce filosofo della complessità.* Soveria Mannelli: Rubbettino.
Gembillo, Giuseppe. 2008. *Le polilogiche della complessità. Metamorfosi della Ragione da Aristotele a Morin.* Firenze: Le Lettere.
Gembillo, Giuseppe, Anselmo, Annamaria. 2013. *Filosofia della complessità.* Firenze: Le Lettere.
Maturana, Humberto. 1995. *La realidad: ¿objetiva o construida?* Guadalajara: ITESO.
Maxwell, James Clerk. 1973. *Trattato di elettricità e magnetismo*, ed. E. Agazzi, 2 vols. Torino: UTET.
Ortega y Gasset José. 1930. *La rebelión de las masas.* Madrid: Revista de Occidente.

History of Science, Epistemology, and Ontology

Flavia Marcacci

Abstract In his historical works, Agazzi explicitly examines some methodological perspectives. As a matter of fact, according to him, the history of science needs methodological perspectives in order to clarify its own contents. Similarly, epistemology needs the history of science to find realistically itself. These are, respectively, a top-down and a bottom-up aspect of the relationship between history and epistemology. Thus, history of science can be used not as a mere erudition exercise, and epistemology can concretely improve any reasoning about science. As a consequence of these considerations of Agazzi's, at least two different ways to practice history of science are determined. On the one hand, a historic history of science; on the other hand, an epistemological history of science. But as is well known, the methodology of history is a delicate question: historical events are contingent and often unique; they have causes, which allow to study them scientifically, but they cannot be predicted either deterministically or statistically, for their causes are too many and complex. Many philosophical questions are opened: for instance, whether the history of science reports just a gallery of images about science and about reality, or it reports some knowledge about the ontology of scientific objects. This paper supports the latter point of view, by inquiring in which sense even history can be considered to have an ontic space.

1 Introduction

In the first chapter of his latest book, *Scientific Objectivity and Its Contexts*, Agazzi examines the substitution of the notion of truth by that of objectivity in modern science through a deep (historical) analysis of the philosophical sources. As a consequence of

F. Marcacci (✉)
Department of Basic Sciences and Foundations,
University of Urbino 'Carlo Bo', Urbino, Italy
e-mail: flaviamarcacci@gmail.com

M. Alai et al. (eds.), *Science Between Truth and Ethical Responsibility*,
DOI 10.1007/978-3-319-16369-7_17

this substitution, the historical development of modern philosophy was characterized by a dualistic epistemological approach. Thus, science was conceived as a "cognitive" enterprise, but phenomenalistic epistemology was unable to explain how science could relate to the real world.

The radical crisis of contemporary science, which occurred at the beginning of the twentieth century, revealed that this belief was deceptive. These masterly pages by Agazzi skillfully and harmoniously integrate the historical-critical and epistemological instances in a way useful for both the history and the philosophy of science.

Historical analysis is in fact used to show the frequent confusion, even of the terms themselves, between the epistemological and the ontological approach. However, Agazzi insists on one point in particular: the passage from Galileo to Kant was a shift from a conception of science based on careful observation of phenomena—without however falling into dualism—to a conception of science and knowledge based on phenomena in a way that radically excludes what lies behind the phenomena. Kant's intention was to promote a positive conception of "appearance" (the "affections" of Galileo) in contrast to the negative Greek conception. However, from then on—according to Agazzi—science had either to defend its ability to describe reality, or to take refuge in conventionalist or instrumentalist positions and use a phenomenalistic epistemology. Therefore, science is considered by some as «knowledge without truth, yet still deserving to be considered knowledge» (Agazzi 2014: 9).

If Agazzi is right, as I believe he is, then the instrumentalist and phenomenalistic philosophy of science has produced a history of science that is simply a gallery of representations of reality that has nothing to do with reality itself. The aim of this paper is to investigate how the history of science ought proceed to integrate the philosophy of science and clarify which philosophical claims it is able to respond to.

To accomplish this, we must begin with Agazzi's work as a historian of science: his prestigious edition of the two-volumes *Storia delle scienze*, published by Città Nuova in 1984, and the important critical edition of James C. Maxwell's *Trattato di elettricità e magnetismo* (Maxwell 1973). References to history and its problems have never been lacking in his writings and reflections, augmenting their weight and providing concrete examples to the more purely theoretical dimension of the philosophical speculation for which he is so well known and respected. Perhaps for this reason alone it would not be enough to speak of Agazzi as a 'historian' of science, but rather as an epistemologist of the history of science. In an explicit (at times also implicit) way Agazzi allows history to be oriented towards speculative reasoning and *viceversa*. This is currently a very relevant attitude, if we consider that on the international scene we are still largely pondering the relationship between the history and the philosophy of science, to the point that there are organized philosophical paradigms such as *historical epistemology* and *historical ontology*.

We will return to this important aspect at the end. We will begin now with a digression guided by Agazzi's own works.

2 On History as Science and Science as History: In Search of a Methodology

Writing the history of science a science is not an easy job. The *Storia delle scienze* opens with a long and deep methodological introduction by Agazzi meant to raise some of the many and very real problems involved in constructing a history of science:

- Discussion of the "criteria of demarcation" between science and the non-science.
- Analysis of the meaning of the term "science" and the difficulties of finding an agreed-upon meaning. (*philosophia naturalis*, *episteme*, *techne*, *metaphysics…*)
- Determining a list of disciplines called "scientific", about which the history of science should be written.
- Analysis and choice of criteria for historical accuracy.

If these are problems every historian must confront, Agazzi points out two potential risks: reading the past in light of today's scientific language and concepts, and not at all "historically"; or, worse, "evaluating" the past in the light of contemporary scientific ideas, reducing it to a "retrospective" of discovery and error that, again, in no way grasps its historical meaning.

> If, in fact, a "result" or a "discovery" is thought of as a simple and pure component of ideal scientific knowledge, which is identified and collocated once and for all like a brick in a wall, it takes on an ahistorical coloring and its comprehension is reduced to that of its role within the logical-empirical context of science to which it belongs. Instead of a historical process, science would appear to undergo an "internal growth", not dissimilar to that of a tree developing according to the preordained genetic plan of its seed. (Agazzi 1984, vol. I: 9).

Thus informed of the risks, we must not overlook the fact that our contemporary perspective on history could represent an opportunity. It is important to escape the false myth of a "complete historicization" with the pretense of understanding an era by rigorously using only what belonged to that era. Rather, it is the presence of "supra-historical" elements that allow us to get close to the past. Agazzi asserts the importance of pinpointing elements of permanence of past science in present science, and cites Euclidean geometry as an example: when a "value of knowledge" (ibi: 10) persists, the history of science is not used as mere erudite cultural satisfaction.

If it is difficult to do history as if it were a science, it is true that science is also history. The historical components present in a scientific theory must always be identified, isolated and then correlated within the theory itself. In this sense it is useful to evaluate these components in and of themselves, and not confuse the "value of knowledge" (or even the scientific tool) which makes the theory useful even today. Various examples could confirm Agazzi's thesis. Let me cite only one: the early modern approach to astronomy, entirely astrometrical, was integrated into and then replaced by observational astronomy and astrophysics,[1] which use today many more methods of analysis and instruments to observe and measure, beginning with electromagnetic telescopes and continuing straight up to the "neutrino telescope". Nevertheless, astronomy and celestial mechanics, the oldest components of astronomy, are today a less central component of the science of the cosmos. And yet they are an important part. On December 19, 2013 the spacecraft Gaia went into orbit as part of the European Space Agency's space astrometry mission.

Returning to our discourse, it would be incorrect to suggest that studying the astrometry of the Seventeenth Century helps us with that of today. Such a study has a historical value which may rather be of interest to the philosophy of science if it helps describe the scientific revolution in a new or unexpected way. This, however, does not imply a ban on using current knowledge to better understand the past. Proof is given by the widespread use of calculating programs to evaluate archaeoastronomical data (for example, to verify a past eclipse) or to validate the accounts of past astronomers in the history of experimental science (for example, confirmation of observational data gathered by astronomers like Galileo.)

If science *today* is interested in confronting theories (on a synchronic level), then the history of science is interested in the diachronic dimension of science in order to understand theories in various periods. Even if this proves difficult, if we want to develop a coherent and ordered discourse on the history of science we need to stress a few fundamentals of historical research:

- Evaluation of the cultural context (e.g. instruments, cultural background, the role of science in society…);
- Functionality of a scientific theory in its historical context (such as the astrometrical techniques of our example);

[1]Modern science was born in the 1600s through changes wrought by astronomy: we all know heliocentrism triumphed and Galileo wrote the *Dialogue Concerning the Two Chief World Systems* (Imprimatur 1632, Index of Prohibited Books 1633), which caused a furor in the Catholic Church. The scientific context of Galileo's reflections is less known, however. After the work of Tycho Brahe, nobody believed any longer in the Ptolemaic system. The attention of professional astronomers was focused on the tension between the Tychonic, or quasi-Tychonic, and Copernican systems. The title page which tells of this contest is famous: it is from the *Almagestum Novum* (1651) by Giambattista Riccioli (1598–1671), which presents itself as a powerful attempt to summarize this debate. Riccioli thoughtfully 'dismantles' the observational data and new astronomical findings (the phases of Mercury and Venus, the moons of Jupiter, sunspots) and debated theories (tides) in order to show the plausibility of a semi-geocentric system. This is to point out that with the scientific instruments of that time it was not possible to decide on the 'true system' (even if Riccioli would propose a semi-Thychonic system with spiralling planetary orbits). Why then did we eventually pass over to a heliostatic one, and one different from that envisioned by Copernicus had? Not for astronomical reasons, but rather for physical reasons: Newton was to explain the physical behavior of the planets through gravity in 1687. Not until well into the second part of the 17th century there were enough clear reasons for choosing one system over the another, even if many scientists were convinced of heliocentrism through their scientific 'instinct'. It was not a choice defensible by definitive astronomical proof, though. That would arrive later, e.g. with Giambattista Guglielmini (1760–1817) in 1789–92 for terrestrial rotation, and with Giuseppe Calandrelli (1749–1827) in 1806 for terrestrial revolution.

- Supra-historical durability of data (e.g. trigonometrical calculations of the 1600s, still valid today, although in a different formalism);
- Study of the formation and conception of scientific objects over the course of history (e.g. the change in meaning of the word "planet").

In light of these considerations, we can understand why Agazzi is against the partial and unilateral reduction of the history of science to an "internal" or "external history". The historian's job is to assemble these two instances completely and harmoniously. Depending on the particular and actual needs of his or her research, the historian will have to use elements of both approaches. In Agazzi's work we can perceive a faint tendency to give more room to "external history", but such a preference is quickly rectified in favor of an ordered and scientific consideration of historical research. When a historian has to narrate events connected through an interdisciplinary approach to construct a credible and intelligible version of them, his or her originality is inescapable. In such cases, the use of current scientific terminology is not excluded.

Let's now attempt to summarize Agazzi's suggestions by identifying two ways of writing the history of science: a *historical* history of science and an *epistemological* history of science.

1. The historical history of science would principally be the history of science written using the standard methods of historical research: attention to sources and documents, textual and contextual analysis, the appeal to philology, etc. We might call this the *bottom-up* approach, as it begins with historical data in order to reconstruct an adequate epistemological picture of events. This is in part the work done by the "new philosophy of science". This level of analysis has inevitably influenced the philosophy of science in its epistemological origin.
2. In the history of science what matters more are the theoretical perspective and the conceptual keys orienting the historical reconstruction. This approach might be called *top-down,* as it looks at history from a theoretical point of view, which is initially independent from the historical data. For example, this happens when we evaluate past theories using present-day concepts: in the example of astrometry of the 1600s, we might speak of the underdetermination of competing astrometrical theories, although at the time no philosopher could have come up with the idea of 'underdetermination'. In light of such an idea, however, many problematic relations among theories can be explained. This level of analysis implicates epistemological problems (as how we know the reality of which scientific theories speak) even before ontological ones (what reality is behind the theories).

These two ways can and must be methodologically integrated at various levels: given a specific scientific theory, we study its historical content, make an epistemological analysis and give an evaluation to summarize it. At each of these levels, the historical and epistemological sensibilities come into continuous contact with each other (Fig. 1).

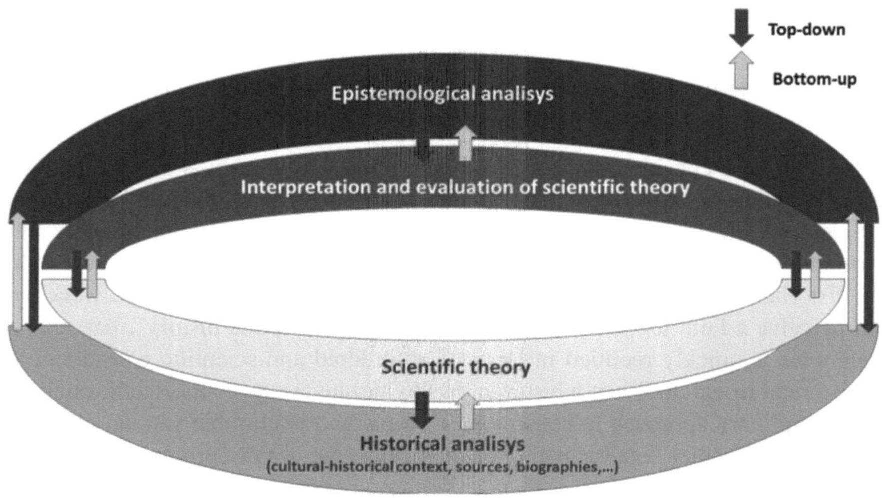

Fig. 1 Interaction between historical analysis and epistemology

3 An Ontic Space in the History of Science: The Unpredictable Contingent

If we have moved into methodological territory, what philosophical work is there left to do? It would seem that it is a contest between two tendencies: absolutizing history and relativizing knowledge (historicism) or relativizing history and submitting it to knowledge (idealism). Rightly, Agazzi writes:

> We can write history since not everything is historically determined, since certain supra-historical elements exist which, present in our day, give us the possibility of using them as a guide to read the past and likewise to understand which different collocations and functions they were able to receive in that past (Agazzi 1984: 9).

That is, science shows something that, although partial, is real: partial with regard to the historical context, real with regard to that which it attempts to describe.

One might think that, as certainty is never guaranteed, science constructs interpretative pictures which inasmuch as possible are *plausible*. This would be the reason for which, throughout history, no matter what Laudan says in his famous list of false but tenacious arguments (Laudan 1981), those theories that paint a realistic picture of reality have a better chance. And yet Putnam wrote

> The assertion that 'the Earth is flat' was, without doubt, rationally acceptable for three thousand years. Today it is absolutely not. And yet it would have been wrong to hold that such a statement was true three thousand years ago, as that would mean the Earth had changed shape since then." (Putnam 1981: 55).

Putnam's observation is useful to us not so much for the questions it raises about the relationship between science and reality, which we cannot deal with here, but rather about the relationship between science and scientific reality, in its "historical" characterization. In effect, historical facts carry a great deal of weight in science. Consider, for example, that there is really no need for scientists to agree on which tools to use in their work in any given era: the agreement has already been made for them first during their education and then in their working years. But such an agreement is a symptom of *historical determinism*, as are the use of certain instruments and not others, the accidental nature of certain discoveries and the refinement of certain concepts. These are *historical facts*, then, and not merely theoretical necessities (Agazzi 1992).

Yet, they are *not only* historical facts:

> In itself, it was certainly a good thing to introduce historical and social awareness to the understanding of science, and it is also useful to submit the scientific enterprise to sociological study: the information we get is always interesting and enlightening. It is another thing, though, to insist on reducing scientific knowledge to nothing more than a social product. This is the error of much sociological epistemology, which has never really been able to show the causal link between the social conditions in a given place and time and, for example, the shape of the natural laws expressed there, in addition to the inability to explain the cross-cultural acceptance of the contents of scientific knowledge (whose validity, therefore, doesn't seem "relativized" by the social circumstances that produce it) (Agazzi 1992: 39).

We can use Agazzi's words to pose a problem: how to pinpoint Agazzi's cognitive value in the history of science and make it useful in addressing reality? Adopting the mentality of classical metaphysics often followed by Agazzi, we might ask how the history of science can contribute to investigating *the thing which is* and not only its representation? A balanced view of history's role in science which doesn't reduce everything to *history* is therefore *epistemologically* important in order not to relinquish saying *something* ontic, not to give into the idea that everything is mere representation, and not to reduce knowledge to an entirely relativistic sociocultural matter.

In other words, the historical dimension of science shows us how contingency plays an important role in the growth of scientific knowledge. This fact has an interesting philosophical dimension which forces us to ask ourselves what is to be understood by the word "science", if it means not just formal analysis and experiment, but also contingency and fortuity. In his book *Epistemologia e scienze umane* (Agazzi 1979) Agazzi looks into the concept of "scientificity" itself with the aim of piecing together a constructive interdisciplinary definition of it. It is undeniable that certain disciplines such as physics and mathematics are dominant, when it comes to establishing criteria for scientificity, because of their ability to present themselves as highly persuasive fields of knowledge, and have inevitably become models for other types of knowledge. But "science can have many meanings," wrote Aristotle.

What we need to do is to propose a kind of model of the concept of science, one which is flexible enough not to be taken prisoner by a single example yet at the same time avoids emptiness. That is, namely, to avoid the outcome in which any and every subject may in the end call itself scientific.

As with scholasticism, analogic is that which is a halfway to "univocal" (indicating a single type of reality) and "equivocal" (vague and multipurpose enough to be attributed to many realities). Agazzi pauses here to discuss two aspects which are more or less attributed to scientific discourse: rigor and objectivity. Without going through Agazzi's entire reasoning process, I will look at his conclusions:

- As regards the concept of rigor, it is suggested that all types of knowledge that wish to call themselves "sciences" must be able to show the deductive rigor of their reasoning based on the elucidation of their own premises and the logical connections between premises and conclusions, as well as the ability to return to the facts that are to be proved. This doesn't mean that the initial hypotheses can't be modified, but simply that such modification must be done openly so that one's reasoning may come under scrutiny.
- As regards the concept of objectivity, it is not reducible to either that of mathematization or that of quantification. If anything, it is reducible to that of intersubjectivity, as something which does not depend exclusively on a single subject. In effect, "objective" should be that which is inherent in the object much more than "something which is not inherent in the subject." But science, before it is a discourse on being, is a discourse on being known. In this sense science shouldn't settle for an intersubjective agreement "of awareness" (risking the reduction of knowledge to awareness) or of "perception" (risking the reduction of knowledge to perception), but should rather focus on an agreed use of a given predicate. If in this regard the exact and natural sciences have amassed a great literature, the same cannot be said of the humanities.

Agazzi proceeds suggesting that knowledge cannot be considered scientific only in relation to the object under investigation: any object can in fact be the object of various scientific disciplines. Even the objects of the exact sciences are susceptible and can be dealt with through "non-exact" approaches (Agazzi 1979: 74–75). To illustrate this point, we shall think of a hot air balloon: this object that can be studied in the fluid dynamics, or in the history of the means of transportation, or be the subject of a poem like the famous poem by Vincenzo Monti *Al signor Montgolfier*. The history of science moves ahead precisely along this ridge between the humanities and the exact sciences, since it employs languages, concepts and reasons from each of these areas. Thus, a philosophical approach to the history of science needs to consult and integrate all those other approaches, in order to offer a complete and objective account. This is why history of science also raises profound philosophical questions.

A wide-ranging theoretical response to the issues Agazzi has raised would be desirable, for these issues are vital also to the philosophical and scientific disciplines. The recent suggestion of a "historical epistemology" appears to move in this direction, and so do some other proposals: the idea of an "applied metaphysics" (concerning the conditions that make an idea "thinkable", Daston 2000); the study of the material preconditions of science (with direct impacts also on the debate between realism and antirealism in science); or the conception of historical epistemology as a historically-based theory of the long-term developments of scientific knowledge, supported by an established, empirically-based epistemology (e.g., the cognitive sciences) (Renn and Hyman 2012: 20).

Trying to discern a common trend among these suggestions, we might hypothesize a historical analysis, which questions the transformations of the "objects" (including concepts, laws, or theories) which populate science. A trivial example already mentioned is the semantic slip of the word *planet*, but we might also consider terms like *epigenetics*, *probability*, and many others. Hacking speaks of a "historical ontology" by asserting that there are objects that begin and cease to "exist", and historical ontology should investigate the causes of their "birth" and "death". That is, we should investigate how the various scientific entities were introduced or rejected in the course of history (Hacking 2002). Similarly, we might speak of a "historical ontology of science" as the completion of a "historical epistemology": i.e., of the study of the processes by which theories themselves (or paradigms, or similar meta-entities) and their objects appear and disappear in the history.

Of course, a "historical ontology" cannot by itself decide which entities have some value which is not purely historical, i.e. which ones *really* exist or don't. To be sure, this is a question we cannot avoid asking, but which must be answered in the light of the best presently accepted theories.

So, why should a historical ontology of science interest us? Of course, in order to explore the various kinds of critical approach, cultural backgrounds, and practical or political interests orienting the research, etc. Eventually, also the question of the truth conditions of historiographic theories themselves and the criteria for ascertaining their truth-values becomes relevant and inescapable. All of this is surely very interesting for epistemology, too.

Secondly, a historical ontology helps to find out whether something "true" and "real" remains despite scientific change (e.g., the trigonometry used in the Ptolemaic astronomy). This is significant for two reasons: showing that "truth can resist tenaciously", and that science, while having an intrinsic historical nature, can nonetheless represent reality. The history of science is not just a story of theories that merely *imagined* reality, but of theories that somehow grasped at least some authentic part of reality; the image we get from the history of science is not that of a merely conventional enterprise, but of how reality is progressively understood in a deeper and deeper way.

Nevertheless, if we wish to speak of a "historical ontology of science", the presence of unique events and their role in the construction of the scientific theories has to be justified and included. This is not of course a question that we can even initially take up here. Rather, I just want to notice that one of the main problems we encounter in this discussion is the peculiar nature of historical events: they are often unique and unrepeatable, totally casual and accidental. According to the classic metaphysical tradition, there cannot be a science about this kind of facts. In order to be an object of science, an event or an object must be predicated necessarily and universally (*An. Pr.* I, 13) or, at least, usually (ὡς ἐπὶ τὸ πολύ, *ut frequenter*: *An. Post.* I, 11).[2]

[2]In fact, there are events happening necessarily and always (e.g. the sunrise), and events happening usually (e.g. the generation of a plant). But a science of what is unique and unrepeatable, of what is totally casual and accidental is impossible (e.g. *Fis.* II, 8, 198b10–199a8).

Even setting aside the interpretative problems concerning this complex issue in Aristotle, it is clear that in his view frequent events can be studied scientifically because what happens frequently has a constant (although not necessary) *cause*. Nonetheless, it should be noticed that even when historical events are unique, apparently casual and accidental, still they have causes. So they too can be studied scientifically. No doubt, they are contingent, for their causes are themselves contingent; and they *seem* accidental (i.e., they cannot be predicted exactly, and never happen in the same way) because since they typically depend on an open plurality of always changing causes.

Theories, paradigms, discoveries, etc., are precisely historical entities in this sense, unique and unpredictable, but amenable to (meta) scientific study. So, a "historical ontology of science" must be very clear about this when it seeks to find out the *truth* about them, and to provide an ontology for them. Even history has an ontic space to be understood.

4 Conclusion: The History of Science Between Epistemology and Historical Ontology

How do we organize an epistemology of the history of science which does not reduce it to a mere narrative of representations whatever, but captures its ontological scope as well? The top-down/bottom-up methodology can offer us a decent methodological recommendation. Keeping Agazzi's two suggestions together—that the humanities furnish themselves with working definitions and that we adopt an idea of science as "analogy"—we might even hope that the historical knowledge of science equip itself with a "formal" epistemology which is able to make the chosen assumptions visible in every era and in light of particular theories. In any case, the direction we need to take is fourfold:

- The history of science can be considered a science, even if this involves raising a number of epistemological questions about it.
- Science is historical, and what in it is historical and contingent, nonetheless has scientific and philosophical value.
- History of science is the history of theories about *the reality we know*. Ontology and epistemology change along with these theories.
- History of science does not speak only of a succession of theories as if they were a series of imaginary representations of reality, but rather of theories that have some kind of ontological meaning.

The first two points focus on a methodology according to which epistemology and history of science have to work together in two directions: bottom-up and top-down. The last two points focus on the distinction between *historical ontology of science* and *historical epistemology of science* within the framework of the history of science.

It seems to me that we can assert that all four points convey and respect Agazzi's lessons, who as a philosopher of science has not forgotten the importance of verifying the historical foundation of every theoretical supposition. Vice versa, as a historian he has understood the importance of identifying those theoretical junctions in history, which clarify the present and help us to consider future developments in science. In addition, he does all this in the hope that philosophy and history really can question reality and not reduce it to a mere representation.

References

Agazzi, Evandro. 1979. Analogicità del concetto di scienza. Il problema del rigore e dell'oggettività nelle scienze umane. In *Epistemologia e scienze umane*, ed. Vittorio Possenti, 57–76. Milano: Massimo.
Agazzi, Evandro. 1984. *Storia delle Scienze*.2 voll. Roma: Città Nuova Editrice.
Agazzi, Evandro. 1992. *Il bene, il male e la scienza. Le dimensioni etiche dell'impresa scientifico-tecnologica*. Milano: Rusconi.
Agazzi, Evandro. 2014. *Scientific Objectivity and its context*. Cham: Springer.
Daston, Lorraine. 2000. *Biographies of Scientific Objects*. Chicago: University of Chicago Press.
Hacking, Ian. 2002. *Historical Ontology*. Cambridge (MA): Harvard University Press.
Laudan, Larry. 1981. A Confutation of Convergent Realism. *Philosophy of Science* 48.1: 19–49. http://philoscience.unibe.ch/documents/TexteFS11/Laudan1981.pdf. Accessed 21 January 2015.
Maxwell, James Clerk. 1973. *Trattato di elettricità e magnetismo*, ed. Evandro Agazzi. Torino: Utet.
Putnam, Hilary. 1981. *Reason, Truth and History*. Cambridge-New York: Cambridge University Press.
Renn, Jürgen, and Hyman, Malcolm D. 2012. The Globalization of Knowledge in History: An Introduction. In *The Globalization of Knowledge in History*, ed. JürgenRenn: 15–44. Berlin: Edition Open Access.

Contributions to Latin-American Philosophy

Lourdes Velázquez

Abstract Philosophy is universal and there is apparently no sense in dividing it geographically. Yet this happens very often and not just for expository reasons. Agazzi has defended the thesis that there is no unique model for philosophy (i.e. that of the Western tradition), since philosophy is in a deep sense the "self-consciousness of cultures" and for this reason it reflects the deepest intellectual and spiritual grounds of a given culture. This applies in particular also to Latin America, where several intellectuals have tried to make explicit the specific features of its philosophy. Agazzi has had a tradition of personal relations of collaboration and friendship with them, and has promoted (owing to his position in international philosophical institutions) several opportunities of mutual encounter and cooperation. His action was also decisive in promoting Spanish as an official language of the World Congresses of Philosophy, and he has also succeeded in having many Latin-American philosophers elected as members of the Steering Committee of the International Federation of the Philosophical Societies (FISP).

1 Introduction

Geographical connotations of philosophy are notoriously a quite delicate and controversial issue, since philosophy is considered as a universal discipline, or even as a specific kind of investigation characterized by a certain intellectual style (that might be equated with a critical comparison of rational arguments) and also

This paper was supported by the Research Project FF12012-33998, from the Ministry of Economy and Competitiveness of Spain.

L. Velázquez (✉)
Faculty of Bioethics, Universidad Anáhuac, Huixquilucan, Mexico
e-mail: luluvela@prodigy.net.mx

L. Velázquez
Universidad Pontificia de Mexico, Mexico City, Mexico

M. Alai et al. (eds.), *Science Between Truth and Ethical Responsibility*,
DOI 10.1007/978-3-319-16369-7_18

by a certain list of fundamental *problems*, that constitute the subject matter of its recognized branches. Therefore, what really counts in philosophy seems to be the accurate delineation of a given issue, its careful analysis, the evaluation of the arguments advanced in order to defend or oppose certain solutions, and all this independently of the individual thinkers who have concretely proposed these doctrines or pieces of doctrine. Their names are often mentioned as a kind of mnemonic tool, more or less in the same way as we associate a proper name to a given theorem in mathematics or to a certain law in physics (speaking, for example, of the "theorem of Euclid" or the "law of Ohm"). If already the name of the individual person seems of little interest, even less important should appear his/her nationality. All this may be true, but on the other hand when we move our consideration from what we could call "systematic philosophy" to the history of philosophy, we immediately realize the paramount importance of linking philosophical ideas and doctrines with particular persons, institutions, cultural environments and traditions. In particular we find perfectly natural to speak, for instance, of a French philosophy (in which we include thinkers such as Descartes, Comte, Bergson, Sartre and Ricoeur) distinct from a German philosophy (in which we include Kant, Fichte, Schelling, Hegel, Nietzsche, Husserl, Heidegger), without excluding by this that the thought of a certain philosopher might have been influenced by that of a different philosophical tradition (like Sartre's existentialism, e.g., with regard to Husserl and Heidegger). Of course, there are cases in which such a national attribution is problematic for different reasons (e.g. because a philosopher has emigrated to a country different from his own original one and has even acquired a new citizenship), but, as usual, exceptions do not eliminate a rule.

What are, hence, the conditions that justify the recognition of a national or regional philosophy? It does not seem reasonable to reduce them to a pure contingent matter of fact, such as that of being taught at Mexican rather than French, German or Italian universities (indeed, it is likely that the contents of such teaching are quite similar, or that an excellent specialist, e.g., in German philosophy is a professor at a Japanese university). For a similar reason it would not be sensible to make the national attribution strictly dependent on the nationality of a given philosopher, not only owing to the already mentioned cases of change of nationality, but also to the more substantial fact that a thinker may have contributed significantly to the philosophy of a country very different from his own original one (think e.g. of Lévinas, who has certainly greatly contributed to contemporary French philosophy but was born in Lituania). In order to significantly speak of a "regional" philosophy we must be able to single out not a whole bulk or a system of doctrines but at least a few *original contributions* that were proposed within the context of a certain country or geographic region and developed as efforts for answering certain cultural issues typical of that region, in such a way as to constitute—if not a school in a proper sense—at least a rather well defined and recognizable community of intellectuals and a tradition of research. In the case of Latin America we can affirm that such conditions were realized.

2 Three Domains of Original Contributions of Latin American Philosophy

It is possible to single out three sectors in which philosophical reflection and research has demonstrated original and specific traits in Latin America, they belong to philosophical anthropology, philosophy of history and the philosophy of liberation (they have also received a kind of standardized denomination in Latin America as the "ontological", "historicist" and "liberationist" trends).

2.1 The Ontological Trend

The denomination "ontological" must not induce one to believe that this trend belongs to ontology understood in its classical traditional sense of the general theory of being, of what exists, but rather in the more modern sense that we find, for instance, in Heidegger's work *Being and Time,* where "being" refers to human existence. Therefore, this trend is inspired by the idea that Latin-Americans have something peculiar in their own nature, in their constitution, that characterizes them in comparison with other humans belonging to different cultures. This is why we have said that this ontological trend can perhaps be referred to philosophical anthropology, but even such an attribution would be debatable, as we shall see, because the deepest motivations of this human ontology are of a social-political nature. If we wanted to find an analogy, we could refer to thinkers (like Fichte or Gioberti) who, in the first decades of the nineteenth century, nourished the spirit of countries such as Germany or Italy that were starting the struggle for their national independence: they stimulated the national pride by presenting an alleged specificity and superiority of their people in comparison with other nations.

Indeed the birth of the first trend (that is also known as "philosophical Americanism") can be traced back to the first decades of the twentieth century in Mexico, as a product of the nationalist milieu that had promoted the Mexican revolution of 1910, with its traits of nationalism, anti-imperialism and anti-oligarchism. This mixture of ideas had produced some reflections on the "being" of the Mexican and Latin-American people, that had found expression in several literary writings with philosophical pretension such as, for instance, *La raza cósmica. Misión de la raza iberoamericana* (1925) and *Indología: una interpretación de la cultura iberoamericana* (1927), both authored by José Vasconcelos Calderón. It is only with the book by Samuel Ramos, *El perfile del hombre y la cultura en México* (1934), however, that one can appreciate the first configuration of a real project of a philosophy about the Mexican. Equally important was the creation of the "*Grupo Hiperión*" that included philosophers such as Emilio Uranga, Jorge Portilla, Luis Villoro and Joaquín Sanchez McGregor. The most important work produced within this group was *Análisis del ser mexicano* (1952) by Uranga.

Starting from Mexico, the *philosophical Americanism* produced a display of works in the whole of the continent during a temporal span of about four decades (1930–1970). A particular mention deserve the following books: *La seducción de la barbarie. Análisis herético de un continente mestizo* (1953) and *América profunda* (1962) by the Argentine Rodolfo Kusch; *América Bifronte. Ensayo de ontología y filosofía de la historia* (1961) also by an Argentine author, Alberto Caturelli; *Pueblo continente* (1937) by the Peruvian Antenor Orrego; *El problema de América* (1959) by the Venezuelan Ernesto Mayz Vallenilla; *El sentimiento de lo humano en América* (1951) by the Chilean Félix Schwartsmann; *La invención de América. Investigación acerca de la estructura histórica del nuevo mundo y del sentido de su devenir* (1968) by the Mexican Edmundo O'Gorman; *La filosofía de lo mexicano* (1960) by Abelardo Villegas.

All these works produced a lively discussion in the whole continent concerning the existence or non-existence of an original Latin-American philosophy, discussion whose elements are present, for example, in works like *Filosofía Argentina* (1940) by Alejandro Korn; *Sobre la filosofía en Iberoamérica* (1940) by Francisco Romero; *Hay una filosofía iberoamericana*? (1948) by Rizieri Frondizi; *Cuales son los grandes temas de la filosofía latinoamericana*? (1958) by Victoria Caturia de Bru; *El problema de la filosofía hispánica* (1961) by Eduardo Nicol: *Filosofía española en América* (1967) by José Luis Abellán; *La filosofía iberoamericana* (1968) by Francisco Larroyo.

2.2 The Historicist Trend

Also the second line was started in Mexico and was initially prompted by the influence of the Spanish philosopher José Ortega y Gasset through the mediation of his disciple José Gaos who landed in Mexico at the end of the 1930s as a refugee from the Spanish civil war. Following the historicist thesis of his teacher, Gaos outlined the project of reconstructing the history of ideas as the ground for the elaboration of a *Philosophy in Spanish language*, as sounds the title of his most important work published in 1945. Nevertheless the great figure of the Latin-American historicism is certainly Leopoldo Zea, a direct disciple of Gaos, who proposed and developed a systematic reflection on the history of the ideas in that continent as a necessary precondition for the creation of a native philosophy. Starting with his doctoral dissertation *El positivismo en México* (1943), passing through *América en la historia* (1957), *El pensamiento latinoamericano* (1965) and *Dialéctica de la consciencia americana* (1976) up to his very original *Filosofía de la historia americana* (1978), Zea has realized a trajectory that has made of him the most powerful promoter of the project of a Latin-American philosophy.

The pioneering work of Leopoldo Zea had continental impact and counted with important followers among which four most salient figures should be mentioned: the Uruguayan Arturo Ardao, the Peruvian Francisco Miró Quesada Cantuarias, and the Argentine Arturo Andrés Roig and Horacio Cerutti Guldberg. The contribution of these scholars concerns mainly their methodological reflection on the problem of the history of ideas. Of Ardao is particularly noteworthy his seminal paper *Historia y evolución de las ideas filosóficas en América Latina* (1979), and of Miró Quesada two excellent books: *Despertar y proyecto del filosofar latinoamericano* (1974) and *Proyecto y realización del filosofar latinoamericano* (1981). Arturo Andrés Roig has developed an extraordinary work of reflection on the history of ideas in his books *Teoría y crítica del pensamiento latinoamericano* (1981) and *Rostros y filosofía de América Latina* (1994). On his side Horacio Cerutti, already known in the 1970s for his criticism of the project of a philosophy of liberation, has published important reflections on the history of ideas in *Hacia una metodología de la historia de las ideas (filosóficas) en América Latina* (1986) and *Filosofar desde nuestra América* (2000).

The legacy of the Latin-American history of ideas has been received in several countries whose authors we briefly mention: Yamandú Acosta in Uruguay; Hugo Biagini, Adriana Arpini, Clara Alicia Jalif de Bertanou and Dina Picotti in Argentina; Joao Cruz Costa in Brazil; David Sobrevilla in Peru; Carmen Bohórques and Javier Sasso in Venezuela. In Cuba is noteworthy the work of Pablo Guadarrama of the University of Santa Clara while in Colombia has been important the constitution in 1977 of the Grupo de Bogotá on the initiative of a few professors of the Saint Thomas University. In Mexico is prominent the activity of Mario Magallón in the Centre for Latin. American Studies of the Universidad Nacional Autónoma de México. As to other countries, deserve mention the work of José Luis Abelán in Spain and of Jorge Gracia, Ofelia Schutte and José Luis Gómez Martinez in the USA.

2.3 The Liberationist Trend

While the first two trends originate at the Northern border of the continent (Mexico), the philosophy of liberation was born in its most Southern part, that is, in Argentina. It was there that, at the beginning of the 1970s, moved its first steps a philosophical movement that was sensitive to the concerns already expressed in other sectors of the Latin-American intellectual life, such as the sociology of dependence and the theology of liberation. An important preparation was the publication in 1968 of the book by the Peruvian author Augusto Salazar Bondi in which the thesis is advocated that the authenticity of a Latin-American philosophy will come as a consequence of the self-consciousness of the situation of alienation and dependence in which that continent had remained submitted. It can be said that the foundational events of liberation philosophy were the Second National Congress of Philosophy celebrated in Cordoba (1972) and the publication, the same year, of the book *Hacia una filosofía de la liberación latinoamericana* where made their appearance the figures that started this movement: Enrique Dussel, Mario Casall, Carlos Cullen, Horacio Cerutti, Julio de Zan, Daniel Guillot, Juan Carlos Scannone and Oswaldo Ardiles. All were philosophers of distinct origins and orientations who, however, agreed upon the necessity of a philosophy committed with the processes of political, social and cultural emancipation of Latin America.

The persecution realized by the military dictatorship in Argentina compelled the philosophers of liberation to a massive exodus around the 1970s. In Mexico settled Enrique Dussel, who was to become the most outstanding figure if this movement, and with whom are strictly associated its most salient theoretical developments. He wrote there his programmatic book, *Filosofía de la liberación* (1973), and from there started the "continentalization" of the movement. In Mexico was signed in 1975 the famous "Declaration of Morelia" in which there is a convergence of philosophers belonging to the three trends just mentioned: Abelardo Villegas, Leopoldo Zea, Francisco Miró Quesada, Arturo Andrés Roig, Enrique Dussel. The tireless prolific work of Dussel, whose scope is comparable only with that of Leopoldo Zea, had made of the liberation philosophy a movement known worldwide. It is sufficient to recall the dialogues hold in the 1980s with philosophers of the level of Karl-Otto Apel, Richard Rorty, Paul Ricoeur and Gianni Vattimo. Among the many works of Dussel those that could deserve a special mention are: *Filosofía ética latinoamericana* (1973), *Método para una filosofía de la liberación* (1974), *Introducción a la filosofía de la liberación* (1977), *1492:el encubrimiento del otro. Hacia el origen del mito de la modernidad* (1992), *Etica de la liberación en la edad de la globalización y la exclusión* (1999), *Política de la liberación* (2008).

Liberation philosophy has found diffusion in various countries. In Brazil one must note the work of Hugo Assman, Roberto Gomes and Sirio Lopez Velasco; in Colombia that of Jaime Rubio Anguio and Germán Marquínez Argote; in Costa Rica has been very important the contribution of Franz Hinkelammert, an original German thinker known for his books *Crítica de la razón utópica* (1984), *La fe de Abraham y el Edipo occidental* (1990) and *El grito del sujeto* (1998); in Bolivia are notable the contributions of Juan José Bautista, a disciple of Dussel and Hinkelammer.

The legacy of Latin-American philosophy, in its three mentioned trends, has been substantially transformed at the beginning of the twenty-first century thanks to the work of three chief figures: the Cuban Raúl Fornet-Betancourt (Raúl Betancourt), who wrote *Crítica intercultural de la filosofía latinoamericana actual* (2004); the Ecuadorian Bolivar Echeverría, who authored *La modernidad de lo barroco* (1998); and the Colombian Santiago Castro-Gómez, author of *Crítica de la razón latinoamericana* (1996) and *La hybris del punto cero* (2005). Fornet-Betancourt proposes an "intercultural turn" of the liberation philosophy that could promote it to being a privileged tool for the dialogue between distinct philosophical traditions; Echeverria's work could be considered as a critical development of the ontological trend along the patterns of the philosophy of culture, especially in those texts where he characterizes a "baroque ethos" of Latin America at variance with the capitalist rationality of European modernity; Castro-Gómez appears in line with the historicist trend, revisited according to the genealogy of Michel Foucault and of the Latin-American post-colonial studies.

2.4 General Considerations

The cursory but also rather detailed sketch we have given of Latin-American philosophy should be sufficient to give an idea of its peculiarity: it cannot be equated simply with the philosophical studies and institutions present *in* Latin America (that are comparable with those realized in other regions of the world, and especially of the Western world). Neither can it be identified with the philosophical work of thinkers born in a country of Latin America. For example, Mario Bunge was born in Argentina, but his academic activity took place at the McGill University of Montreal, so that he can be qualified as an Argentine-Canadian philosopher; his thought, however, constitutes a systematic, self-contained and widely articulated system that has acquired a reputation as a significant element of contemporary philosophy as such, and to qualify it as an expression of Latin-American philosophy might sound restrictive. The same applies in a way to Francisco Miró Quesada: we have mentioned him as one of the most significant thinkers of Latin-American philosophy, but at the same time one must recognize that his philosophical work and reputation have been much broader than the thematic field of Latin-American philosophy as we have considered it thus far (let us mention simply his contributions to logic, philosophy of science, philosophy of law). Finally, there are many philosophers, born in Latin America or active in Latin America, who have done a professionally excellent and influential work in different fields of philosophy without having devoted special attention to the topics of the Latin-American philosophy as we have described it here (let us simply mention Juliana González in México). Therefore, what can be considered *specific* of Latin-American philosophy is that it has been characterized by certain rather precise *thematic issues* and, at the same time, that it has been elaborated by *professional philosophers* and not, for example, as something like an "implicit philosophy" couched in ideological doctrines, literary works, popular customs or traditions: it is precisely this aspect that will help us understand why Evandro Agazzi has had some special merits with respect to Latin-American philosophy understood *in a broader sense*, a sense that put it on the same level as other "regional philosophies" quite independently from its most peculiar products.

3 Evandro Agazzi's Contributions

3.1 Reputation and Visibility in Philosophy

A peculiarity in Agazzi's way of considering philosophy is his effort to find certain objective criteria for evaluating the "reputation" of a philosophical school, association, institution, tradition. This term does hardly occur explicitly in his writings, but its sense has inspired the attitude and the concrete activity that he has developed during the many years he has spent especially as a top officer of the

International Federation of the Philosophical Societies (FISP), of the International Institute of Philosophy (IIP), of the International Academy of Philosophy of Science (AIPS). The objective ground of such a reputation should obviously be the *philosophical quality* of the work performed, but here difficulties immediately surface since (as Agazzi has often noted with disapproval) in contemporary philosophy we often see that the partisans of a certain school simply reject as "non-philosophical" the work of the representatives of a rival school, despite the fact that all of them belong to the philosophical "profession". Such judgments are based sometimes on methodological reasons (an alleged lack of rigor), and sometimes on the alleged philosophical irrelevance of the subject treated. The joint effect of these two criteria has led for a long while to restrict the genuine domain of philosophy to the Western tradition, considering other traditions, at best, as containing "implicit" philosophical elements within an essentially religious context, or some "pre-philosophical" intuitions in the context of a certain popular "wisdom". It is clear that such an intellectual attitude was not particularly suitable for evaluating the admissibility of associations and even of individual persons into institutions such as FISP and IIP that are constitutionally committed to be internationally open and representative. But then an additional and more serious difficulty must be faced, that of the linguistic barriers. Indeed the production of a philosopher, the activity of an association, the global level of the philosophical teaching in a country may be of a very high standard but, if the language of that country is scarcely known abroad, this good philosophical quality remains "invisible" from the outside, and the philosophical life and activity within that region is condemned to a status of practical isolation. Agazzi has succeeded in reducing significantly both difficulties.

3.2 Philosophy as Self-Consciousness of Cultures

Regarding the first issue, Agazzi relied on his general way of conceiving philosophy as an organic reflection on the "world of Life" whose roots are present in the rational nature of every human being and gradually expand to form general views shared by larger and larger communities. These views regard the sense of life, the relations of humans with the rest of nature, the moral duties, the right social order, the possible existence of an ultra-mundane reality and life and so on. When, in the history of a given community, certain persons appear who are gifted with a special intellectual acumen, an aptitude for critical reflection and a capability of organizing these views in some systematic way, we can say that a philosophy is emerging, though the record of the names of these philosophers might be lost for different reasons. These ideas were presented by Agazzi in a seminal lecture on *Philosophy as self-consciousness of cultures* delivered at the conference of the International Society of Metaphysics held in Nairobi in 1981, that had called the attention of several outstanding participants (I was present at that conference and remember, in particular, the expressions of appreciation of Francisco Miró Quesada. Ioanna

Kuçuradi, Odera Oruka). When the proceedings of that conference were published in 1983, Agazzi's paper attracted the attention of Leopoldo Zea who published its Spanish translation in his journal *Escritos de filosofia* the same year. This circumstance was not accidental, since Agazzi's thesis was a lucid way to make explicit the ideas on which (as we have seen) rested the "ontological trend" of Latin-American philosophy. Agazzi himself has remained faithful to this approach up to the point of resuming and expanding it (with the title *Philosophy as self-consciousness of cultures and as condition for intercultural understanding*) in a lecture delivered at the meeting of the International Institute of Philosophy of 2008 in Seoul, devoted to the theme "Comparative and intercultural philosophies".

One must note, however, that the strongest stimulation to elaborate this thesis had come to Agazzi from another special circumstance, that is, from having started in 1979 his activity of professor at the Swiss University of Fribourg, where he attracted several African students, some of which wanted to write a doctoral dissertation under his guidance. The easiest solution was to let them work on some topic or author of the standard Western philosophy; the most challenging was to engage the best of them to do a research on the philosophy of their own culture. Agazzi adopted both practices, but the second imposed him a serious reflection on the very idea of an "African philosophy". The topic was by no means new, and its discussion had moved between two poles: on the one hand certain scholars maintained that it is possible to extract from indigenous cultural expressions an implicit philosophy by using the categories of traditional Western philosophy. This is essentially the approach of *La philosophie bantou* (1945) by the Belgian priest Placide Tempels, and has remained (though with a much more elaborated linguistic and ethnological sophistication) in the "etnophilosophie" of the Rwandan philosopher Alexis Kagamé. On the other hand, a position closer to the "ontological" trend of Latin-American philosophy is to be found in Léopold Senghor, who has insisted on the "negritude" as a peculiar racial characteristic and has also linked his literary and philosophical production with concrete social-political commitments and activity (indeed he was even President of his country, Senegal, from 1960 to 1980).

It lies outside the scope of the present contribution to give more details on this particular aspect of Agazzi's intellectual life, but the mention of the above circumstances (as well as of others that induced him to be among the founders of the Afro-Asian Philosophy Association) is useful in order to understand why he was so appreciative of Latin-American philosophy at moments when even many philosophers working in Latin America did not appreciate very much that philosophy.

The analogy with Latin America, however, must not be pushed too far: almost all African countries (and also several Asian ones) had been subjected to colonial dominance until the second half of the twentieth century, and therefore had remained deprived of universities and in general of all those educational institutions that constitute the "infrastructure" for the development of the forms of "high culture", including in particular philosophy. In short, no "professional philosophy" had existed in the African countries during the colonial period, and this explains quite well why the question of the very existence of an African philosophy could be raised (and is still raised) by many scholars. Latin-American countries, on the contrary, have gained their political independence in the first decades of the nineteenth century, and have had the opportunity of establishing a more or less efficient system of high education (moreover inspired by the Iberian models of the colonial period, in which philosophy was present). In addition, the fact of having received several European intellectuals escaping racial and political persecutions just before the second World War contributed to an enrichment and to a recognition of the philosophical life of certain countries, especially Argentina and Mexico, and this is reflected in the fact that, since the foundation of FISP in 1948, an Argentine and a Mexican philosopher have constantly appeared among the members of the Steering Committee of this Federation (the firsts of them were Francisco Romero and Samuel Ramos, respectively). In addition, one must not ignore that one of the most brilliant, versatile and dynamic Mexican philosophers, José Vasconcelos, had delivered a series of very successful lectures at various universities and institutions of the USA during his voluntary exile after his political defeat in the Mexican presidential elections (1929), and had also visited several European countries. Therefore, it was rather natural that Vasconcelos (who had returned to Mexico in 1940 and had recovered his philosophical leadership in his own country) was invited as a main speaker at the World Congress of Philosophy that was held in Venice in 1958, where he also extended the invitation to hold the next World Congress in Mexico. The young Agazzi, by the way, had the opportunity of personally meeting Vasconcelos on that occasion.

The 13th World Congress of Philosophy did actually take place in Mexico in 1963, but it could not be presided by Vasconcelos (who died in 1959) nor by Samuel Ramos (who died even a couple of months before him). It was organized and presided by Francisco Larroyo who then automatically became President of FISP (as it was tradition at that time). That congress was very successful and certainly contributed to a better knowledge of the philosophical life in Mexico by the international community. Nevertheless it was only with the World Congress of 1978 in Düsseldorf that a Mexican was invited again as a main plenary session speaker: he was Leopoldo Zea, who had been included in the program following the strong suggestion of Agazzi, who was a member of the Scientific Committee of that Congress.

3.3 The Study of Latin-American Indigenous Thought

The particular view of Agazzi, strictly relating philosophy with cultural traditions even outside institutionalized structures, very naturally induced him to be interested in those philosophical conceptions that were present in Latin-American cultures before the "conquest" realized by the Spaniards. Scholars doing important work in this direction existed especially in Mexico (Miguel León Portilla, Alfredo López Austin, Mercedes de La Garza) and Peru (María Luisa Rivara Tuesta), and Agazzi established friendly relations with them, and especially a strict cooperation with Carlos Viesca Treviño who was the head of the Institute for the History and Philosophy of Medicine of the Mexican UNAM. This cooperation traduced itself concretely in the constitution of two research teams, one in Italy and one in Mexico (directed respectively by Agazzi and Viesca), that organized a series of meetings with regular alternation in the two countries on historical-philosophical issues of medicine. The papers and the proceedings volumes of these meetings (edited by Agazzi and Viesca in Italian and Spanish) have been a valuable output of such an initiative, whose merit has been double. On the one hand, it attracted the interest of European scholars on this aspect of the Latin-American cultures, that were usually studied mainly from an ethnological or archaeological point of view (in particular. Agazzi has hosted several articles on pre-Hispanic Latin-American philosophy in the international journals of which he is the editor). A significant demonstration of this fact was that the author of the present contribution obtained her doctor degree in philosophy at the university of Genoa with a thesis (redacted and published in Italian) on *Philosophy and medicine in ancient Mexico.* On the other hand, it testified also in Mexico that "Mexican philosophy" was not only the contemporary one, but included also the indigenous pre-Hispanic philosophy (actually the author of the present paper teaches "Philosophy of Mexico" at the Pontifical University of Mexico City and includes pre-Hispanic philosophy in the subject matter of her classes).

3.4 Personal Contacts and Cooperation

The care for personal relations has always characterized Agazzi's activity also in the domain of intellectual life and research, and for this reason he never accepted the role of an armchair intellectual whose chief work consists in reading (and writing) books and papers. He has given much importance to direct dialogues, real discussions, concrete encounters and for this reason has never underestimated meetings, conferences, congresses, tours of lectures of which he has organized a great number in all the continents. It is on the occasion of such initiatives that he was able both to deepen his knowledge of the philosophical work of persons and communities, and to offer to several philosophers the opportunity of knowing or better knowing one another and coming to a significant collaboration. This in particular has happened with regard to Latin-American philosophy. We have sketched above a rather systematic presentation of this philosophy, articulating it into three fundamental trends, each one including a great deal of authors and titles, with a wide chronological and geographical dissemination. This was obviously a historiographic systematization without the pretension of reflecting the existence of "schools" in a strict sense. Indeed many of those authors did not have a mutual personal acquaintance, and at times did only a partial reference to their respective works. In the span of less than two decades Agazzi attained a quite deep knowledge of Latin-American philosophy and, in particular, established links of personal acquaintance, and often of sincere friendship with many of the most salient representatives of this philosophy, to whom he also offered the opportunity of concretely coming together and confronting their views in conferences and congresses that he was able to promote also thanks to his functions as an officer of FISP, IIP and several other philosophical institutions. Leaving aside Mexico for the moment, we note that particularly frequent have been his visits to Brazil, Peru and Argentina, but also Venezuela, Ecuador, Colombia and Chile have known his active presence, so that it is not exaggerated to say that Agazzi has had certain merits in the progressive consolidation and international recognition of the *specific* contribution of Latin-American philosophy. He has been a sincere friend of Miró Quesada, Leopoldo Zea, Mayz Vallenilla, Mario Bunge, Alberto Caturelli, Enrique Dussel, Tarcisio Padilha, Miguel Reale, Newton Da Costa (to mention just a few of them of different countries and orientations), and it was in a way symbolic that the last opportunity in which Miró Quesada, and Zea joined together was at Agazzi's home in Mexico City where his wife offered a dinner to the participants of the meeting of the IIP in 2005. It is certainly significant that Agazzi received five doctoral degrees honoris causa by Latin-Amerucan universities.

3.5 The Privileged Links with Mexico

The first visit of Agazzi to Mexico was on the occasion of an international philosophical symposium in Ixtapan de la Salle in 1979, where he gave an invited lecture, and he returned then to this country very frequently, either in connection with meetings and conferences organized by institutions (such as FISP, IIP, the Interamerican Philosophical Society, the International Academy of Philosophy of Science, the International Association of Christian Philosophers, the Mexican Association of Philosophy), or on invitation of numberless universities and academic institutions to give lectures, courses, seminars in several parts of the Mexican Republic. This closer contact with the "normal" practice of the philosophical activity induced Agazzi to go a step further in the appreciation of Latin-American philosophy by recognizing its *good quality standard*. It is a judgment that he has quickly extended to the forms of "high culture" in general, and that he also expresses by reproaching Mexican intellectuals for often considering themselves as belonging to the "third World". According to him, the level of the Mexican culture is objectively comparable with that of the European countries and with the average of the USA. Of this last country Mexico has imitated several things and, among them, the model of a "free market" educational system, in which one finds a rather restricted number of excellent (and expensive) universities and colleges, beside hundreds of second-rate or even poor (and cheaper) universities and colleges. In Europe, where education is still principally promoted and regulated by the State or other public authorities, the situation is much more uniform. One cannot underestimate the importance of this second kind of contact Agazzi has had with Latin-American culture: the personal acquaintance with the outstanding personalities whom we have mentioned above, and the knowledge of their work, could have given him the impression that they were just isolated peaks in a low-level landscape, but coming into direct contact with the standard activity of (good) Mexican universities, both public and private, offered the evidence that the general landscape was not that low-level at all. Agazzi had the possibility to confirm this diagnosis also for other Latin-American countries.

The most significant consequence of all this is that, after retiring from he universities of Fribourg and Genoa (of which he is emeritus professor), he has become full professor in Mexican universities, public (such as the Universidad Autonoma Metropolitana, UAM) and private (such as the Universidad Panamericana). In both of them he has given lectures and seminars of no lower level than those he gave in Europe or in the USA, but simply relying on his pedagogic expertise to make accessible even certain "difficult" pages of the classic authors to young students endowed with a poor educational background (totally adhering in such a way to the ideal of social promotion that must inspire higher education). Therefore, as one can speak of an "Italian period" and of a "Swiss period" of Agazzi's career, one must now speak of a "Mexican period" that is characterized not only by his activity in Mexican universities and institutions (he is a member of the Mexican System of National Investigators, and of the Mexican Academy of Sciences) but also by the significant fact of having acquired the Mexican citizenship. In such a way Agazzi is concretely contributing today to the life and recognition of Latin-American philosophy in which, however, he had been present for many years through his work: it is sufficient to remember that the Spanish edition of his *La lógica simbólica* has been the basic textbook on which logic has been studied by many generations of students (including the author of the present paper) throughout Latin-America.

3.6 The Battle for Spanish as Philosophical Language

What I have said, however, does not still concern an even more important contribution that Agazzi has brought to the promotion of Latin-American philosophy, the one that we have qualified above as the question of *visibility* and is strictly related with the existence of "linguistic barriers". At first sight it sounds very strange to speak of a linguistic barrier in the case of Spanish that is the second spoken language of the world (after Chinese and before English itself). Nevertheless the question is different and regards the status of *international language* that a given idiom has outside the domain of its native speakers, at least from certain points of view (as it was the case for Latin until the end of the eighteenth century). As regards philosophy, in particular, it is certain that Spanish was not an international language in the first half of the twentieth century, and this reflected itself in a palpable isolation of the philosophical life of the Spanish-speaking countries, an isolation that could not be really overcome by a few representatives of that philosophy who were fluent enough in French,

English or German and could hold talks or write papers in those languages. In short, it is really difficult for a philosopher to think or express his thoughts in a foreign language. This matter of fact was made particularly evident to Agazzi (who did not experience it personally, being fluent in several languages) by an accidental circumstance. He was trying to convince an outstanding Italian philosopher to be an invited plenary session speaker at the Düsseldorf World Congress of 1978 and, in order to overcome his denial, he visited him at home and received the following confidential confession: this eminent personality (who was at ease in reading several languages) did not afford reading out loud a paper and participating in a public discussion in a language different from Italian. Until that moment, French and English were the official languages of FISP (the Federation that is in charge of organizing the World Congresses of Philosophy), whereas German as well was admitted as a third official language just for the World Congresses (evidently owing to the great weight of German philosophy in modern times). From that moment on Agazzi started a work of persuasion within FISP in order that also Spanish and Russian be admitted as official languages for the World Congresses (in addition also to the language of the host country). The resistances were very strong, but a first significant success came at the Congress of the Interamerican Philosophical Society (Tallahassee 1981) that approved a motion moved by Agazzi in favour of admitting Spanish as an official language for the World Congresses of Philosophy. In the meanwhile Agazzi obtained the sufficient support within FISP Steering Committee and a change in the Statutes of the Federation (admitting Spanish, Russian and the language of the host country as official languages of the World Congresses) was approved by FISP General Assembly at the World Congress of Montreal (1983). The positive affects of such a measure were soon visible through the great increase of Spanish speaking participants (in particular of Latin-American philosophers) in the World Congresses.

One cannot conclude this survey without mentioning the policy in favour of a greater weight recognized to Latin America that Agazzi pursued in his position of Secretary general and then of President of FISP. Under his impulsion the Latin-American philosophers who were elected to the Steering Committee of the Federation attained the highest level (Argentina, Chile, Peru, Venezuela, Brazil, Mexico had one representative each), and the crowning of this policy was represented by the election of Agazzi's successor as President of FISP at the Moscow World Congress in 1993. This successor was Francisco Miró Quesada who, proposed by Agazzi, obtained a great majority of votes (the tradition of the automatic succession had been discontinued after the Montreal Congress of 1983). This event may be considered a symbol of the full dignity and visibility that Latin-American philosophy had finally acquired and for which Evandro Agazzi had long committed himself. Today the fact that a Latin-American woman (namely, the author of the present paper) holds for the first time the office of Vice-President of FISP continues this tradition.

References

Agazzi, Evandro. 1967. *La logica simbolica*. Barcelona: Herder (Spanish translation by J. Pérez Ballestar of *La logica simbolica*. Brescia: La Scuola, 1964).

Agazzi, Evandro. 1978. *Temas y problemas de filosofía de la física*. Barcelona: Herder (Spanish translation by J. Vidal of *Temi e problemi di filosofia della fisica*. Milano: Manfredi. 1969. Reprint Roma: Abete, 1974).

Agazzi, Evandro. 1996. *El bien, el mal y la ciencia*. Madrid: Tecnos (Spanish translation by Ramón Queraltó of *Il bene, il male e la scienza*. Milano: Rusconi. 1992).

Agazzi, Evandro. 2008. *Scienza* (interview with di Giuseppe Bertagna). Brescia: La Scuola.

Agazzi, Evandro. 2011. *La ciencia y el alma ade Occidente*. Madrid: Tecnos.

Basave Fernandez Del Valle, Agustin. 1958. *La filosofía de José Vasconcelos, el hombre y su sistema*, Ediciones Cultura Hispánica. Madrid: 451–454.

Cacciatore, Giuseppe. Una filosofia per l'America Latina: Leopoldo Zea in *Cultura Latinoamericana*, Annali 2003, n. 5. Salerno: Pagani. Oédipus. 2004. pp. 431–453.

Colonna, Roberto. 2008. *Filosofía sin más. Leopoldo Zea e i "Cuadernos Americanos"*. Firenze: Le Cáriti.

Various Authors. 2012. America Latina e Occidente. In *Pagine Inattuali*, 1. Salerno: Arcoiris Edizioni. In this text are devoted to the work of Leopoldo Zea, in particular, the papers by Mario Magallón Anaya (*Validità del pensiero filosofico-politico del messicano universale: Leopoldo Zea*), Giuseppe Cacciatore (*Un profilo di Leopoldo Zea*) and Alberto Filippi (*Leopoldo Zea e i labirinti dell'Occidente*).

Vasconcelos, José. 1925. *La raza cósmica, misión de la raza iberoamericana. Notas de viajes a la América del Sur.* Madrid: Agencia Mundial de Librería. Reprinted 1927. Also published by: Calpe Argentina, Buenos Aires 1948; Aguilar, Madrid 1966; Asociación Nacional de Libreros, México 1983; Porrúa, México 2001. English edition: *The cosmic Race,* translated, with an introduction, by Didler T. Jaén. Baltimore, MD: John Hopkins University Press, 1997.

Zea, Leopoldo. 1957. *América en la historia*. México: Fondo de Cultura Econòmica. Italian translation by D. Pastine: *America latina e cultura occidentale.* Milano: Silva, 1961. English translation by Sonja Karsen: *The Role of Americas in History.* Lanham, MD: Rowman & Littlefield 1991.

Zea, Leopolodo. 1968. *El positivismo en México*. México: Fondo de Cultura Economica. English translation by Josephine H. Shulte: *Positivism in Mexico*. Austin: University of Texas Press 1974.

Zea, Leopoldo. 1988. *Discurso desde la marginacion y la barbarie*. Barcelona: Anthropo. Italian translation by E. Sfondrini: *Discorso sull'emarginazione e sulla barbarie.* Roma: Bulzoni 1988.

Zea, Leopoldo. 1969. *La filosofía americana como filosofía sin más*. México: Siglo Ventiuno. Italian translation by A. Pierini, R. Ciucci, A. Salvini: *Filosofia latinoamericana.* Lucca: Pacini Fazzi 1993.

Part V
Metaphysics, Ethics and Religion

Metaphysics and Ontology

Paolo Musso

Abstract The metaphysics of Evandro Agazzi is strictly related to his epistemology. In his metaphysical reflection three phases can be distinguished. In the first one (1975–1983) Agazzi is mainly committed to the task of demonstrating the legitimacy of metaphysics, here taken only as the knowledge of the suprasensible, by showing its substantial continuity with natural science. In the second phase (1983–2002) Agazzi refines his vision by focusing on metaphysics taken as the knowledge of the most general features of reality, so modifying the way of conceiving its relationships to science, from a mere "division of labour" to a positive feed-back. In the same period Agazzi also investigates the philosophical causes of the contemporary antimetaphysical attitude, coming to the conclusion that it ultimately depends on the oblivion of the correct meaning of intentionality, and on the gnoseological dualism which, from Descartes on, has derived from it. In the third phase (1997–2011) Agazzi shows that metaphysics is in a relationship of substantial continuity and positive feed-back also with logic. Precisely from this reciprocal interaction he derives the proof of the existence of various different ontological levels of reality, even in a greater number than it has been traditionally recognized.

1 Introduction: The Originality of Agazzi in the Context of Contemporary Philosophy

Among the many aspects which make the position of Evandro Agazzi so original and almost unique in the context of contemporary philosophy,[1] the most relevant one is very likely the fact of his being a philosopher of science who defends the

[1]For a complete and systematic synthesis of Agazzi's philosophy, including also his metaphysics, see his recent *Scientific objectivity and its contexts* (Agazzi 2014), which unfortunately was published too late to be taken in consideration here. For a shorter synthesis see Musso (2004, Chap. 11).

P. Musso (✉)
Department of Applied and Theoretical Sciences, University of Insubria, Varese, Italy
e-mail: paolo.musso@uninsubria.it

M. Alai et al. (eds.), *Science Between Truth and Ethical Responsibility*,
DOI 10.1007/978-3-319-16369-7_19

possibility of a metaphysics of "classical" kind, i.e. *cognitive* and taken not as a mere reflection about the most general aspects of physical reality, but also about the *suprasensible*: as is well known, indeed, and as Agazzi himself has written in the first line of his first article explicitly devoted to the problem of metaphysics, «a substantial part of contemporary philosophy qualifies itself as antimetaphysical» (Agazzi 1975: 4). But this is not all: while today the (few) supporters of metaphysics usually see science as something tendentially hostile or, at best, irrelevant, on the contrary in Agazzi metaphysics is never opposed to science, nor is merely *juxtaposed* to his epistemological reflection, but arises *from its inside*, and it is strictly related to it, in what we could call a relationship "of positive feed-back", where they help, reinforce and complete each other. The best proof is that the essential elements of his metaphysics had already been sketched by Agazzi long before than he started to explicitly focus his reflection on it, in the context of works apparently devoted to "pure" philosophy of science and even formal logic.

2 The Roots of Agazzi's Metaphysics: Formal Logic and Scientific Method

When in 1975 Agazzi writes his first "metaphysical" article, he has already published his most important books: *Introduzione ai problemi dell'assiomatica* (1961), *La logica simbolica* (1964) and *Temi e problemi di filosofia della fisica* (1969), where he sketches the fundamental framework of his thought in the fields of logic and epistemology. They are three works very deep and rather technical, nonetheless, as we have said, they contain practically all the elements that successively will turn out to be essential to his metaphysics.

First of all, in his two logical works Agazzi immediately introduces a concept which may well be considered the real "fil rouge" (cf. Musso 2007) of his whole philosophy (and therefore also of his metaphysics), which here he calls, according to the modern custom, «intentionality». However, in the last years Agazzi has started to speak more and more explicitly of «intellectual intuition» , in order to underline that he takes intentionality in a particular sense, very close to that of the medieval philosophers and rather different from that generally accepted nowadays, when it is either completely rejected or in any case tendentially reduced to a mere subjective psychological state or to a socially acquired skill.

In *Introduzione ai problemi dell'assiomatica*, an exposition and commentary (in my opinion, still the best of all) of the famous Gödel's Theorem, all the discourse about intentionality is contained in very few but nonetheless fundamental pages. Here Agazzi shows that the undecidable proposition *G*, expressly constructed by Gödel to prove the incompleteness of arithmetic, is nonetheless *true*, since it says of itself precisely that it is undecidable, but nevertheless «it is possible to proof it only by means of metatheoretical arguments and not by deducing G from the axioms of the system P (indeed we know that such deduction is impossible, since G is indemonstrable)» (Agazzi 1961: 186). It follows that

> the set of true propositions is wider than the set of demonstrable propositions, i.e. "theorems", and then also that what human thinking can understand to be true goes necessarily beyond the scope of what can be demonstrated, what is the same than saying that maybe the true distinctive feature of human thinking is not its discursive activity but its capability of "seeing" the truth. In other words, [...] human thinking cannot be replaced by a thinking machine which could execute all the logical operations that it can do (Agazzi 1961: 199).[2]

Later, in *La logica simbolica* Agazzi extends this seminal intuition to the whole problem of the foundations of logic, by questioning the very widespread thesis of the complete conventionality and therefore of the complete meaninglessness of formal systems—what in any case «mathematical logicians had never really stated, while it has become instead the favorite slogan of some philosophical circles» (Agazzi 1964: 356). Furthermore, he also shows that even «the so-called "pure calculus" can be considered well established [...] only if it is "based" on some intuitive evidence» (Agazzi 1964: 355). Consequently, the said capability of "seeing the truth" in an intuitive way is the real basis of *all* systems of modern mathematical logic, which, in turn, are neither arbitrary nor meaningless, but are based, ultimately, on reality, or, at least, on some properties of reality, in a way which is not very different, after all, from that of scientific theories, as we are going to see immediately.

In *Temi e problemi*, indeed, Agazzi explains, in a synthetic but substantially complete way, his famous theory of scientific objectivism, firstly clarifying that the *objects* of scientific theories are not "things" as a whole, but only some *properties* of theirs, identified through standard *operations*, which allow the establishment of an intersubjective agreement among the different observers. This kind of "weak" objectivity, limited to the *ascertainment* of the existence of such agreement, without saying anything about its objective ground and therefore without any ontological commitment, is the unique generally accepted by contemporary epistemology. But Agazzi stresses that, as a matter of fact, it can be separated by "strong" objectivity, which instead is based on the relationship with reality, only basing on «epistemological dualism [...] that consists in conceiving the *real object* as something situated beyond the *known object*, so that we can never reach it» (Agazzi 1969a: 364–365). Now, it is well known that this presupposition is the basis of modern philosophy, but nonetheless

> it must be rejected not only because it is dogmatic, but also because it is self-contradictory. Indeed, to say that, beyond the object that I know, there is another that I do not know, I need to have at least ascertained *that it exists*, but this means already to know it (Agazzi 1969a: 365).

Moreover, if theories

> were mere agreements, we may decide (that is, precisely, "agree") to never modify them, even in the case that new experimental results contradict them. The fact that this never happens and, instead, all recognize that in such cases we must modify the theory, is a proof that, in reality, nobody is ready to seriously admit such conventionality (Agazzi 1969a: 369).

For this reason, a theory, when adequately confirmed, may surely be said *true* (once again against what is maintained by the overwhelming majority of contemporary

---

[2]This topic will be discussed in a more systematic way in Agazzi (1967, 1991a).

epistemology), although never in an absolute sense, since it is always «true or false *of* a given universe of objects» (Agazzi 1969a: 369) or, as Agazzi likes to say, it is «absolutely true relatively to its objects».[3] So, at last, we can find also a space for philosophy and, particularly, for metaphysics, since

> a truth in an absolute sense should be nothing else than a truth which holds for *all possible objects*, i.e. a truth which, holding for all possible kinds of objectivation, refers to reality not as objectivated, but *as such*, and therefore, as we have seen, exceeds the field taken in consideration by science and is instead related to philosophy (which, typically, when wanting to assume a cognitive task, aims to investigate *reality as such* and therefore appears as *metaphysics*) (Agazzi 1969a: 369–370).[4]

3 From the Inside of Science: Experimental Method and Metaphysics

It is just this reflection about the ontological implications of logic and science, nowadays usually questioned, if not even explicitly denied, what has allowed Agazzi to overcome the contemporary antimetaphysical bias in his peculiar way, i.e. by making the legitimacy and even the need of metaphysics emerge *from the inside* of logic and natural science.

Nonetheless, in his first metaphysical works Agazzi is dealing only with the latter, while, at least for the moment, he thinks that «the case of mathematics is a little peculiar, so for the purposes of our discourse we can ignore it» (Agazzi 1975: 5). Indeed, as he says both in *Scienza e metafisica oggi* (1975)[5] and in *Considerazioni*

---

[3]Cf., e.g., Agazzi et al. (1989: 189). So, Agazzi goes on, «in this sense truth is even supra-historical, in the sense that, *relatively to its referents*, a true discourse remains eternally true (thus, e.g., [...] the Pythagorean theorem is eternally true with respect to its referents, i.e. the objects of Euclidean geometry, while it may be no more true with respect to other referents, as is absolutely natural). In this way it is possible to conciliate a certain way of conceiving the absoluteness of truth with its relativity, without denying the capability of science of achieving a certain degree of *definitivity* in its various fields» .

[4]In a sense, Agazzi's objectivism may be seen as an original synthesis of Aristotle's theory of abstraction, Galileo's prescription of not investigating the "intimate essence" of the things, but only "a few affections" of them, and Einstein's methodological lesson that all the physical concepts must always be operationally defined.

[5]Even if here he says that «beside [...] some motivations of specifically philosophical nature, in the context of the antimetaphysical debate also the authority of science is very often alleged» (Agazzi 1975: 4), where such "motivations of specifically philosophical nature" are identified with «reactions against real or imaginary forms of "idealism"» , «the problematic attitude shared by most contemporary philosophy» and «more generally [...] the radical distrust in the synthetic use of reason» (Agazzi 1975: 4), but then he does not deepen his analysis. The other two metaphysical papers of this first period (chronologically the second and the third one), *The role of metaphysics in contemporary philosophy* (Agazzi 1977) and *Science and metaphysics in confrontation with nature* (Agazzi 1978), do not present substantial novelties with respect to *Scienza e metafisica oggi*. The last important text I am going to refer to in the following, *Scienza e fede* (1983), in a sense can be seen as a link to the next phase of Agazzian metaphysical reflection.

epistemologiche su scienza e metafisica (1981), as well as in other writings of this period (see, e.g., Agazzi 1969b), in his opinion the main cause of the widespread antimetaphysical attitude in contemporary philosophy is to be identified, more than in internal causes, in the conviction that modern science «can be established under the condition, both necessary and sufficient, of the exclusion the mediation of experience» (Agazzi 1975: 5), i.e. of any attempt of pushing our knowledge beyond what is certified by our senses.

But the truth is that, on the contrary, the mediation of experience is needed *also for science*, which progresses thanks to the construction of *theoretical hypotheses* able to *explain* the empirical facts. Indeed, contrary to what has been generally maintained from Descartes on, «what satisfies the empiria is the requirement of the ascertainment» (Agazzi 1981: 318), while «the logos is not involved in the ascertainment process, but only to account for what is already sure: it is just this attitude that gives rise not only to philosophy, but also to science» (Agazzi 1981: 315–316), which, not for nothing, Agazzi has often defined as «the invention of the "why?"». Now, since what can account for experience is, by definition, something that *is not given* in the experience itself (otherwise it would not be necessary searching for it, because experience would already possess their own reasons), it follows that «the zone where it happens such "giving the why" implies the mediation of experience. Thus, also in science is implemented what represents the basic method of metaphysics» (Agazzi 1981: 319).

It is well known that for a long time neopositivism has attempted to demonstrate that the explanatory process is only apparent, because in reality, thanks to the analysis of language carried on basing on formal logic, theoretical propositions can always be reduced to a mere "abbreviated description" of a set of empirical propositions. But just the failure of such attempt has shown once and for all that this is impossible, as Agazzi explains in details, particularly in *Considerazioni epistemologiche*, where he develops arguments already partially developed in *Temi e problemi*, even though this, «in contemporary literature, is not well clarified in its reasons, because [...] it gives the impression that the problem depends only on the limitations of the logical techniques, so that it is only a matter of finding more powerful logical tools» (Agazzi 1981: 315–316), while in reality it is a genuine in-principle impossibility (cf. Agazzi 1981: 316–317). Equally false is the idea, commonly accepted for a long time, that theoretical propositions can be derived from empirical facts thanks to a logical reasoning, inductive or deductive (cf. Agazzi 1981: 320–321). On the contrary, «a hypothesis is a fruit of the synthetic use of reason, [...] since the logos does "intention" a hypothetical abstract construct, which is not directly into our experience, even if then it has to be *related* to experience» (Agazzi 1981: 322).

One could object that, although we may admit that up to this point science and metaphysics have used analogous methods, just this last feature distinguishes the *kind* of mediation of experience typical of the two, by showing that only the first one is legitimate, since it is the only one that remains faithful to experience. In other words, science surely uses a *metaempirical* mediation of experience, but this is not yet, in itself, a *metaphysical* mediation in the strict sense: the first feature, indeed, is a *necessary* but not yet *sufficient* condition for the second (cf. Agazzi 1983: 142). But we can reply by repeating the same reasoning previously made about scientific method. Indeed,

> the determination of the operational predicates of a given science is equivalent to the determination of its investigation field or, as we prefer to say, of the *whole* of that science, in the sense that any statement incapable to be related, directly or indirectly, to this set of predicates, automatically falls out of it. [...] Pushing this reasoning to the limit, we can say that the set of all possible empirical operational criteria defines the whole of the experimental science *tout court* or, if we prefer to see the matter from the point of view of what is "pertinent", we can say that science in a broad sense has as its field of objects the *whole of experience*. [...] The metaphysician [...] simply is one who wants to investigate the *whole* without any further qualification, not claiming to know a priori that it transcends the whole of experience, but also not accepting to exclude a priori that it actually does (Agazzi 1975: 11-12).[6]

In order to know if the whole as such actually transcends the whole of experience, first of all we need to establish

> whether the analysis of experience can be carried on by using predicates which, although being applicable also to experience, do not have as their necessary denotation experience itself. Now, there are some cases in which some famous "starting points" of the metaphysical discourse does not seem to enjoy that privilege: among them there is, e.g., the *Cogito*, which cannot be intentioned out of the precise experiential situation of self-consciousness. In other cases, instead, we are allowed to think that such requirement is actually present (Agazzi 1975: 18).

This is the case, e.g., of the concept of *being*, since, although it is undoubtedly able to refer also to empirical objects, its intentionality «does *not* contain in itself the reference to experience. At the level of the semantic logos (i.e. of the pure *meaning*), when I state that something *exists*, I do not mean that I am perceiving it» (Agazzi 1981: 328). Therefore, as it uses concepts which, despite having been born within empirical experience, do not necessarily refer to it, to be faithful to experience the metaphysician is fully legitimated not to "come back" to it at every moment, just because it is experience itself that, so to speak, has "shown the way" which leads beyond itself.

At this point, in order to construct a cognitive metaphysics only one step more is needed, i.e. demonstrating that using such concepts to go beyond sensible experience is not only *legitimate*, but also *necessary*, what can be achieved only by demonstrating that avoiding to do this would be contradictory. It is precisely from this that follows

> the intrinsic unavoidability of the apparent apodicticity and presumption of the metaphysical discourse, which just for this reason very often seems so disagreeable to those who

[6]So, this is «a *criterion of demarcation* of a methodological kind; indeed, the limitation of the discourse to the whole of experience, regardless the way it is formulated, is a choice, an option, which is preliminary to the foundation of the scientific discourse, since it is even its definitory postulate» (Agazzi 1975: 12). In fact it is the *unique* possible criterion of demarcation, as proved by the failure of all the others which have been proposed, all of merely formal nature (what proves, once again, that scientific method is not a matter of logic).

> are looking at it from the outside and is easily mistook for dogmaticity. [...] A scientific explanation, indeed, although it obeys to rather precise and rigorous criteria, is never a matter of consistency or inconsistency. The rigorous metaphysicians, instead, [...] are condemned to "wanting to be right" in an absolute way; they cannot be satisfied with saying: "things are so and so, but they may be also different", because in this case they would not be in a condition under which refusing to transcend experience implies a contradiction, and so all their efforts would fail (Agazzi 1975: 19-20).

Agazzi has never fully committed himself to the enterprise of actually building a cognitive metaphysics in all its amplitude (for this he has always made reference to the work of his master Gustavo Bontadini). Nonetheless, he has always pointed out that any attempt of denying the legitimacy of such operation would be in turn «a metaphysical demonstration; indeed, in order to demonstrate that the whole as such has a given nuance, a given property (that of coinciding with the whole of experience) it is necessary to assume "the point of view of the whole as such"» (Agazzi 1981: 327). So, we have demonstrated that at least the *horizon* of metaphysics can never be eliminated, although its specific content still remains to be determined: however, having some metaphysics, at least implicit and so to speak "negative", is *unavoidable*, since, for the above reasons, even materialism turns out to be a form of metaphysics.[7] Anyway, at least one time, precisely in *Scienza e fede* (1983), Agazzi, after having repeated, without substantial novelties, the same arguments that we have just exposed, has gone so far as to sketch at least the *general features* that any cognitive metaphysics should possess. First of all,

> the "contents of knowledge" of such a metaphysics are few in number, but of immense value: we can say that they do not go much beyond the demonstration of the existence of an absolute being of non-sensible nature and maybe[8] also the existence of a spiritual dimension in the human being. But the quantitative poverty of these contents is counterbalanced by the exceptional acquisition of a "conceptual space for transcendence"» (Agazzi 1983: 152).

This means that

> the world of transcendence may at least appear as inhabited by beings which we can "understand" (even if in a limited way) through "concepts" and about which we can develop a discourse, brief, poor, but testable (even if through a kind of test which is not purely empirical), whose existence can be "known" (thanks to the combined efforts of both experience and *logos*), and not merely "imagined" or "postulated"» (Agazzi 1983: 154).

Obviously, this does not mean that cognitive metaphysics may hope to *exhaust* or even approximate all the richness of such a world, since to go beyond the few certain outcomes sketched above

> it necessarily uses "analogical" methods, which transcend the rigor of pure logical non-contradiction and get closer and closer to the methods of [...] "hermeneutics". [...] But

[7]As well as it is unavoidable to have a religious faith, at least implicit and "negative", given that, for reasons absolutely analogous to those explained above with respect to metaphysics, also atheism is a kind of faith (cf. Agazzi 1969b: 178 and Musso 2011, § 9.5).

[8]As it will be shown at the end of Sect. 6, some years later Agazzi has gone beyond this "maybe", at least implicitly.

> what is important is that such further construction may rely on solid anchorage points [...], whose absence would risk to make swinging our whole construction in the field of the mere subjective opinion (Agazzi 1983: 155).

Finally, Agazzi concludes his reflection by questioning the thesis, nowadays very common, that in metaphysics there is no progress as in science, but only a succession of theories based on so heterogeneous principles, that they are unable to communicate with each other. First of all, indeed, even rejecting its exaggerations, which have come to the point of completely denying the very existence of an accumulation of knowledge in the scientific field, speaking of a mere succession of "paradigms" or "conceptual schemes" incommensurable to each other (just as it is believed to happen in metaphysics), Agazzi points out that contemporary epistemology has undoubtedly demonstrated that the merely "cumulative" conception of scientific progress is too superficial, since during scientific revolutions also deep conceptual changes take place, depending on which also the previous knowledge, although preserved, is understood in a partially new way. Secondly, on the metaphysical side, from one hand, also here «it exists [...] this *cumulative* aspect inside *each* metaphysical discourse» (Agazzi 1981: 334), while, from the other, «without any doubt, nowadays we are analyzing reality from the point of view of the whole through richer and more penetrating tools [...] because in some way we take in account also the "outcomes" of the philosophies of the past» (Agazzi 1981: 334), although not *in the same way* of science, in which, instead, the old theories are taken in account *only* in the light of the new theories, which must incorporate all the true aspects of the former ones. Not even the fact that metaphysics is often in dialogue with a faith (usually but not necessarily religious) makes any essential difference: also in science, indeed, we always start from "believing" in a given hypothesis, which only later and step by step is submitted to a rational control. Thus, «the situation of the "knowledge inside a belief" or the "knowledge inside a faith" is absolutely general» (Agazzi 1981: 334) and the only difference between science and metaphysics consists in the *different kind of questions* they attempt to answer.

4 From Science to Metaphysics and Back Again: A Mutual Positive Feed-Back

Just starting from *Scienza e fede* we can see an important turn in Agazzi's metaphysical reflection, which until that point, as we have said, had been essentially aimed at defending the possibility and legitimacy of metaphysics in the strict sense, i.e. as the rational knowledge of the suprasensible, what could not have another outcome than a sort of "division of labor" and therefore a substantial reciprocal extraneousness.[9] On the contrary, starting from the publication of the

[9]Not for nothing, Agazzi had concluded *Scienza e metafisica oggi* by stating that science «is intrinsically *a-metaphysical*, [although] not anti-metaphysical» (Agazzi 1975: 21), what is surely true with respect to metaphysics taken in the second sense, but not in the first.

essay *Science and metaphysics: two kinds of knowledge* (1988), Agazzi begins to propose a more complex idea of metaphysics, by distinguishing in it, on the footsteps of Aristotle, two fundamental meanings:

(a) «the science of "reality as such", i.e. of the most universal *features* of reality» (Agazzi 1988a: 12);
(b) «the science of those *dimensions* of reality which overstep its empirically ascertainable level (or, to put it briefly, [...] the science of the "suprasensible")» (Agazzi 1988a: 12).

While about the second aspect there are no substantial novelties,[10] the focus put on the first aspect, until that point only discussed in passing, leads Agazzi to a partially different conception of the relationships between science and metaphysics, which pass from a state of mere "non belligerency" to a state of «mutual dynamism» (Agazzi 1988a: 22), as expressly stated by the title of one of the sections of *Science and metaphysics*. But this can be seen in almost all the further Agazzian writings, but particularly in *Metafisica e razionalità scientifico-tecnologica* (2000), in the item *Realismo* of the *Dizionario di scienza e fede* (2002) and in *Metaphysical and scientific realism* (2002).

Indeed, as Agazzi shows through both a brief overview of the history of philosophy and an analysis of our perception, «the individual can be "known" only within the framework of a universal model» (Agazzi 1988a: 14), regardless the fact that it is conceived in the sense of Plato's Ideas, Aristotle's forms, Kant's categories or modern psychology's *Gestalt*.[11] Science has often forgotten this aspect, because it does not start from zero, but from ordinary knowledge, of which it represents essentially a deepening and, so to speak, a "specialization" (cf. Agazzi 1985: 188): as such, science usually finds its basic data already "done" and therefore it can avoid reflecting on what makes its own existence possible. Nonetheless, the need for a universal model is unavoidable for *any* kind of knowledge and therefore also for scientific knowledge, in even three different senses:

> First of all, the atoms are not given prior to the unit, but may be singled out by an *analysis* of the whole *Gestalt*, of which they appear as constituents. Secondly, this *Gestalt* may serve to organize other and different atoms, and in *this* sense it is universal. Thirdly, the atoms themselves may be "identified" because they have in turn a certain *Gestalt* (which enables us to say that they are *the same* atoms – the Platonic "recognizing" – when they are organized in different structures and units). In conclusion, there is no moment in which our knowledge can dispense with the universal, be it because we need the "unity of the multiplicity", be it because we must be able to grasp "the permanent under the mutable" (Agazzi 1988a: 15–16).

This is the reason why in the past it was even believed that it is possible to deduce from some universal properties all the particular aspects of physical reality. The

[10] They will appear only in the first essays devoted to the relationships between logic and metaphysics (see Sect. 6).

[11] "Regardless", of course, only from *this* point of view. For other, not less important aspects, instead, such a difference is decisive, as Agazzi himself has shown in those same years (cf. Sect. 5).

extraordinary intuition of Galileo has been precisely to understand that such a "deductive" method in the case of natural science does not work, This is why sometimes it is believed that science is in itself antimetaphysical: but this is nothing but a misunderstanding, since

> the "affections" Galileo is speaking of are not at all Kantian "phenomena", taken as "pure appearances", but some particular "accidents" of natural "substances". [...] Therefore, committing ourselves to study these "affections" does not mean renouncing to know the things of the physical world, but studying precisely their objective properties» (Agazzi 2000: 101).[12]

Therefore, «if we take metaphysics in the first of its two basic meanings, [...] metaphysics appears as the unrolling of the general conditions of *intelligibility* of reality, and in this sense it is unavoidable» (Agazzi 1988a: 21): as such, it is inextricably interlaced to science itself, even if very often only implicitly and unconsciously (cf. Agazzi 1988a: 25),[13] but nevertheless *really*. However, it is true that metaphysical theories do not interact with scientific ones in an automatic and mechanic way: on the contrary,

> the relationship between science and metaphysics is analogous to the relation between experiments and theories in science. Experiments presuppose a theory, since they are designed and performed by using the concepts, laws, methods of a certain theory, and with the view of answering "questions" asked within it. In this sense they "depend" on the theory. However, their outcome does not depend, and it inevitably introduces a modification in the theory. If an experiment is successful, it not only "confirms" or "corroborates" the theory, but actually *enriches* it, by bringing in an additional detail to the *Gestalt* of the domain of objects which the theory is about. If an experiment shows a "negative" result, the theory must be modified, its proposed *Gestalt* proves not to be fully adequate, and it may even happen that it has to be abandoned and replaced by another. [...].What theories are with regard to experiments and empirical data, metaphysical frameworks are with respect to scientific theories. They are *Gestalten* of a higher order, within which theories take shape: therefore theories "depend" on these more general criteria of intelligibility, but are not "deduced" from them and interact with them in a *feed-back* loop, which in any case produces modifications (of different importance) in the metaphysical background» (Agazzi 1988a: 22-23).[14]

Therefore, not only science is not an enemy of metaphysics, as we have already seen, but now it appears even to be its best friend, if correctly understood, so that in *Dizionario di scienza e fede* Agazzi goes so far as to say that

[12]For a detailed explanation of this fundamental point see Agazzi (1994) and Musso (2011): 135–146.

[13]It must be recognized that, from Popper on, this tenet has become rather common in contemporary epistemology, but the difference with respect to Agazzi is that, according to the nowadays ruling antirealist attitude, such "metaphysics" is usually taken as a mere set of "beliefs" or "general ideas" about reality, accepted by most scientists in a given historical moment basing on conventional and/or social reasons, so that they have no real cognitive scope.

[14]For some very significant examples of such a complex but substantially "virtuous" interaction between science and metaphysics, see Agazzi (1986, 1988b, 1991b).

if we recover scientific realism, we also recovered the *cultural* condition needed to establish metaphysical realism, since scientific realism can be recovered: by overtaking epistemological dualism, by recognizing the role of intellectual intuition, by accepting the synthetic use of reason in the mediation of experience. These are the conditions needed to build a cognitive metaphysics, in the double sense of an investigation of reality as such and of a knowledge of the suprasensible. The differences from science are not eliminated, but are reduced to the fact that metaphysics assumes "the point of view of the whole" without any restriction» (Agazzi 2002a: 1188).

5 At the Core of the Problem: Epistemological Dualism and Intentionality

Still in the Eighties and still in the same essays, Agazzi starts to deepen also another important issue he had begun to deal with in *Scienza e fede*. In his first metaphysical essays, indeed, he had preferred to let on the background the causes of the contemporary antimetaphysical attitude *internal* to philosophy, having preferred to focus almost exclusively on the problem of its relationships with science. Now, instead, he begins to deeply reflect also about them, once again by carrying on various considerations he had partially made in *Temi e problemi*.

The most important outcome of his analysis is surely that of clearly identifying the source point of the said antimetaphysical attitude in the already mentioned "epistemological dualism", i.e. the

> radical change emerged from a tacit and gratuitous presupposition that characterized "modern" philosophy (conventionally inaugurated by Descartes), according to which what we *immediately* know are our representations or *ideas*, and not "reality" (Agazzi 2002b: 37),

so forgetting the correct meaning of

> the *intentional identity* of thought and reality: in a perception or in an intellectual intuition our cognitive capacities "identify" themselves with the objects, though remaining ontologically distinct from them. [...] The representation, from this point of view, simply is "the way of being present" of a given thing to our cognitive capacities, and "depends" in an ontological sense on both, but not in the sense of being "produced" by either of them (Agazzi 2002b: 38).

Such misunderstanding is the basis of the «radical change [that] happens with Kant» (Agazzi 2000: 99), which leads to the rejection of the possibility itself of a cognitive metaphysics and «is the direct consequence of two presuppositions of his "critical" philosophy: the thesis of the unknowability of the "thing-in-itself", and the negation of the possibility of an intellectual intuition» (Agazzi 2000: 99),[15] which, from then on, has become a true prohibition, or, more precisely, an unquestionable dogma. Indeed,

[15]It is just starting from this article that Agazzi begins to use explicitly and with increasing frequency the expression "intellectual intuition" instead of "intentionality", although without completely abandoning the latter.

> in modern philosophy, both realism[16] and idealism suffer from a common disease, i.e. the fact of having ignored the true nature of intellectual intuition. Empiricists reject it in an absolute sense, while rationalists admit it as a capability of our intellect of knowing its own abstract contents, but also in this case it is only a matter of an intuition of essences, and not of that abstractive intuition which is able to see the intelligible inside sensible reality, thanks to the said intentional identity (Agazzi 2002a: 1185).[17]

However, in the long run such an attitude has ended up by questioning not only the possibility of metaphysics, but also of *science*, as proved, from one side, by the paradoxical antirealist drift which has involved the overwhelming majority of contemporary epistemology, and, from the other side, by the fact that «it was just Kant who provided the first antirealist interpretation of natural sciences» (Agazzi 2002a: 1185). Just for this reason, the defense of the true nature of intellectual intuition seems to Agazzi the crucial point for a correct understanding of both science and philosophy and, particularly, metaphysics.

6 From Logic to Metaphysics: Intentionality and Levels of Reality

The above considerations have paved the way to the last phase of the Agazzian metaphysical reflection. From the end of the Nineties up to now, indeed, Agazzi significantly shifts the *focus* of his metaphysical reflection towards his never repudiated "first love", i.e. logic, which he had been working on also during the former years, but never, if not very briefly, in his metaphysical papers.

First of all, indeed, we find, above all in *On the criteria for establishing the ontological status of different entities* (1997) and partially also in the already mentioned *Metaphysical and scientific realism* (2002), as well as in *Idealization, intellectual intuition, interpretation and ontology in science* (2007), a strong defense of realism and therefore of the ontological scope of scientific theories, which here is made even more detailed and precise, by adding to the previous arguments a careful examination of the objections coming from logic and philosophy of language, which, as we know, represent the privileged point of view of modern epistemology. Firstly, Agazzi notes that

> the founder of modern semantics, i.e. Gottlob Frege, had already distinguished the *sense* of a linguistic expression (*Sinn*) from its *reference (Bedeutung)*. Therefore, we can safely continue to say that the task of semantics is the study of meaning, provided we recognize that meaning articulates itself into *two* different (though interconnected) aspects: *sense* and *reference*, and, moreover, that the criteria for assigning meaning are different from the criteria for assigning reference. We must, in other words, keep faithful to a *three-level semantics*: the level of the *sign* (the linguistic expression), the level of the *sense* (what is meant by the sign), and the level of the *referent* (the object about which the sense is predicated). Unfortunately, this has often been forgotten by many semantical theories of our

[16]In the sense of "empiricism", as is immediately clarified in the following.

[17]See also Musso (2004, 2011).

> century, that have been typically *two level* semantics. Some of them have identified meaning with reference (following the model of the "extensional semantics" largely developed in mathematical logic); some others have identified meaning with sense and, for that reason, have found great difficulties in assigning *existence* to entities spoken about in language, since existence [...] is paradigmatically related to referents» (Agazzi 1997b: 41).

Now, «intentional states may happen to be directed towards *abstract objects* "encoding" certain properties, even when there are no *physical objects* "exemplifying" these properties» (Agazzi 1997b: 42), which represent their sense, which, therefore, can exist also *without* a referent. It follows that «we should attribute a particular kind of existence (let us call it, e.g., *intentional existence*) to the abstract objects, without equating it with the *physical existence* of other objects» (Agazzi 1997b: 42). Therefore, although scientific objects are, so to speak, «clipped out of things» (Agazzi 1997b: 43), nonetheless they do not *coincide* with them, which instead represent their *referents*, since «a thing [...] *does not encode* any property, but may *exemplify* many properties» (Agazzi 1997b: 44). Indeed,

> a thing is a "potentially infinite bunch of objects", meaning by this that a thing may be considered under potentially infinite points of view, and in such a way be also considered as endowed with potentially infinite properties. But exactly for this reason it would be arbitrary to say that a thing is *totally characterized* by any particular set of properties (which is the proper meaning of "encoding") (Agazzi 1997b: 44),

as happens, instead, in the case of abstract objects. However, this does not mean that scientific theories are not true and that their objects do not correspond to anything real, since the correspondence between them and the things which exemplify their properties can always be established through suitable referential procedures (cf. Agazzi 1997b: 51). But this means, in turn, that not only the objects which directly correspond to the physical world are real, but also those which «are the referents of true sentences, which are recognized as true on the basis of theoretical considerations and arguments» (Agazzi 1997b: 55). As we have seen, indeed, a proposition is always true or false "of" something: thus, should it not have a real referent, «a true sentence would be true of nothing, that is, not true at all» (Agazzi 1997b: 54). Therefore, the procedures which represent the operational criteria of referentiality are also

> *criteria* for truth, and we are actually recognizing that, in the case of empirical or "factual" knowledge, they play the role of *fundamental criteria*. This does not simply mean that they are sufficient for granting truth *immediately*, but this also means that it is through these criteria that truth is so to speak "injected" in the discourse of empirical sciences, whose theoretical tools would never be able *by themselves* to produce any sentence having referential purport (Agazzi 1997b: 55).

This allow us also to solve the old problem of the distinction between theoretical and factual statements: the latter, indeed, do not differ, as it has been maintained, neither for being "absolutely simple", nor for being "purely empirical", nor for any other logical aspect, nor even «from the point of view of the *sense* [...] The point of discrimination is that a factual statement is, in addition to having a sense, also directly *referential*» (Agazzi 2007: 313), where the key word is "directly", since we have just seen that *all* true statements are referential.

However, from our point of view the most important point of the above reasoning is that it allows Agazzi to get to the bottom of the problem (already sketched in *Temi e problemi* and partially carried on in *Scienza e fede*) of establishing *what kind of reality* corresponds to the referents, which may be different depending on the different kind of theories and, more generally, on the different kind of discourses, so recovering «the traditional thesis of the "analogical" meaning of being, which goes back already to Aristotle» (Agazzi 1997b: 41). Indeed, «determining further this kind of reality amounts to assigning an *ontological status* to these objects» (Agazzi 1997b: 55) and from what we have said above it follows that «this status is entirely determined by the criteria of referentiality through which a given science (but in general a given discourse) recognizes its *data*, or immediately true sentences» (Agazzi 1997b: 55). Now, it is clear that in the case of natural sciences (those mainly considered by Agazzi in this essay) all referents have a physical nature. However, since what we have said holds for *any* kind of discourse in general, in other cases things may be different: so, according to the kind of criteria of referentiality, it may become necessary to consider other kinds of reality. For example,

> if these criteria are the reading of a literary text, the ontological status of the objects is that of characters in a novel, or in a poem; if these criteria are the reconstruction of a dream, the corresponding objects are particular psychic states; if these criteria are simply mathematical calculations, these objects are only mathematical constructions, etc. (Agazzi 1997b: 55).[18]

Also these referents are "real", indeed, not less than physical objects, although *in a different way*, since, according to the fundamental and never forgotten teaching of Bontadini, «"real" is what is "different from nothing"» (Agazzi 1997b: 41).[19]

Agazzi deepens further the logical side of this topic especially in *Logic, truth and ontology* (2004) and then, in a more technical form, in *Consistency, truth and ontology* (2011), thus showing firstly the inescapable ontological commitment of logic, since «logic cannot be disconnected from truth, but truth in turn cannot be disconnected from ontology» (Agazzi 2004: 42). It is true, indeed, that the validity of logical laws is independent of any particular model, but this is not to indicate the disconnection from ontology as such, but rather «a transition from regional ontologies[20] [typical of the individual sciences] to *general ontology*» (Agazzi

[18]The correctness of the Agazzian approach has been at least partially demonstrated by the recent discoveries in the field of chaos and complexity, which have proved the in-principle (and not only practical) impossibility of constructing a unified scientific theory of the whole reality (cf. Musso 1993 and1997, Arecchi 1985, Arecchi and Arecchi 1990). Be aware that this has nothing to do with the construction of the hypothetical Theory Of Everything (TOE), which would be only a unified theory of the fundamental physical forces and therefore is perfectly possible. About truth in mathematics and the reality of mathematical objects, cf. Agazzi (2011). See also Musso (2013).

[19]However, as Agazzi had already clarified previously, it does not mean that «it is impossible not to be realist [...]: we call realist a discourse which *intends* to speak of *a certain kind* of reality and succeeds in it» (Agazzi 1985: 187).

[20]The expression "regional ontologies" is taken from Husserl.

2004: 14). Just for this reason logical laws are neither "tautological", nor "empty", nor "meaningless", as it is usually maintained: «on the contrary, they are more properly considered as "always true", and this does not mean "true in no model" but "true in whatever model"» (Agazzi 2004: 14). Therefore, by briefly resuming the analysis already developed in *La logica simbolica*, Agazzi describes the various kinds of logic (intuitionistic logic, logic of entailment, modal logic, epistemic logic, deontic logic, quantum logic, inductive logic, logic of confirmation, paraconsistent logics, dialectic logic, minimal logics, artificial intelligence, non-monotonics logics), different from classical formal logic, which have been developed in the last decades, whose

> multiplicity [...] mirrors the fact that correct arguments, that is, truth-preserving arguments, are applied with different modulations according to the different ontological regions where they are applied and this confirms that logic is inevitably *ontologically sensitive* (Agazzi 2004: 23).

This not only makes evident the existence of a strict relationship between logic and metaphysics taken in the first sense, i.e. as the science of the most universal *features* of reality, but it also shows that the latter is much richer and complex than it was traditionally believed. But even metaphysics in a strict sense, i.e. taken as the science of the suprasensible, receives from logic a further legitimation. Metaphysics indeed cannot attain, by its nature,

> an *immediate truth*. It might be possible only as a form of *truth by argument*, that is, as a truth that can be attained as a *logical consequence* of already attained truth. In the sciences we have plenty of examples of such a way of proceeding [...] and this happens because we use logical tools that are *general* in the said ontological domain and their application is not restricted to the observable parts of that domain (Agazzi 2004: 24).

Now, analogously, passing from regional ontologies to the general one,

> if, by using [...] general arguments and general ontological principles, we can correctly infer from the consideration of *empirically true* sentences certain *true* propositions of a non-empirical character, we must say that the entities to which these true propositions *refer* really *exist*, despite not being endowed with certain ontologically *particular* features (such as that of being perceivable through the senses). This is simply the consequence of the ontological commitment of logic (Agazzi 2004: 24).

In this way, the same result that in the first phase of the Agazzian metaphysical reflection had been achieved by reasoning on the concepts and the method of science, it is now achieved also by reasoning on the laws of logic. What is common to the two ways of reasoning is still the fact of making emerge the legitimacy of metaphysics from the inside of those which are usually considered its worst enemies, which at the end of the Agazzian reasoning turn out to be its best allies.

Finally, Agazzi focuses his attention on one in particular of the many "kinds of reality" he has identified so far, which is of a very special importance for metaphysics: that of thinking. He starts by claiming the full reality and irreducibility of the *entia rationis*, which we have already partially spoken of, since a clear example of them is represented just by the objects of scientific theories, whose non-empirical nature has been convincingly demonstrated by Agazzi in the above mentioned essays. However, he develops a more systematical discussion of the whole issue in *Thought and ontology* (1997), where firstly he clarifies that «thinking is a bipolar activity; one pole being the thinking subject's mind, and the other pole being that which is thought» (Agazzi 1997a: 13). Therefore, thought is the "content" of thinking, i.e. that «towards which this activity is *intentionally* oriented» (Agazzi 1997a: 15). Thought, in turn, has as his object "external" reality. Now,

> if external and internal are not meant according to a naive pictorial-spatial characterization [...], this would oblige us to say that, while thinking and thought are *distinct* but not *different* (since they share the same mental *nature*), reality and thinking (or thought) are *different* (they do not have the same *nature*) (Agazzi 1997a: 16).[21]

On the other hand, being different from nothing, both thinking and thought are "something", and so are real: but such a reality has, precisely, a different nature with respect to material reality. In this way it turns out to be demonstrated also the second of those "contents of knowledge which are few in number, but of immense value", i.e. «the existence of a spiritual dimension in the human being» (Agazzi 1983: 152),[22] about which in *Scienza e fede* Agazzi had only said that "maybe" it is demonstrable. And, once again, the key has turned out to be intentionality.

7 Conclusions: The Interest for Metaphysics Today

At the end of *Scienza e metafisica oggi* Agazzi proposed a reflection which is still valid, maybe even more than at that time. After having wondered why, despite the demonstrable compatibility between science and metaphysics, *in fact* it has been precisely scientific progress what caused the crisis of metaphysics, he answered that the reason is essentially «not theoretical, but pragmatic» (Agazzi 1975: 22), since

> the increasing intellectual interest for sciences has gradually made decreasing the *interest* for the other intellectual activities, among which also that for philosophical research. In its ambit, the discipline which is farthest from any empirical interest, i.e. metaphysics, was fatally destined to suffer more (Agazzi 1975: 22).

Then, he asked the crucial question, which is still such also for us, that is: so standing the matter, «how could such an interest arise again nowadays?» (Agazzi 1975: 22). His answer was that

[21]This is precisely the error of idealism, which not only believes (correctly) that thinking and being are *coextensive*, but also (incorrectly) that they are *identical* (cf. Agazzi 1997a: 16). And this happens, once again, because idealism misunderstands intentionality, by taking «the identity of being and thinking [...] not as an intentional identity, but as an ontological identity (i.e. by reducing being to thinking)» (Agazzi 2002a: 1182).

[22]About this issue it is also important the reasoning developed in Agazzi (1967, 1991a) about Artificial Intelligence.

> if it is true that the growing of science has led to a weakening of the interest for metaphysics, it will be a reflection on what science, after all, cannot provide us, that could once again give rise to the interest for metaphysics. It is undeniable that there exists a whole set of problems which, just due to their non-empirical nature, are not suitable to be discussed by science: they are essentially problems about how to "give a sense" to the world and the human life, that is, problems on which human beings risk their life in a way instead of another one. [...] If such problems exist, and we want to use our reason in order to understand and try to solve them, as far as we may hope to succeed in it, it is only metaphysics that can help us even today» (Agazzi 1975: 22).

I think this is true, but I would like to add a further point to reflect on. Indeed, this can be taken not only in the sense of a capability of metaphysics of going beyond science, but also in the sense of its *continuity* with science, i.e. as its capability of accomplishing the task of the search for the ultimate sense of reality, to which science itself tends, although it cannot get it with its own strength, even if sometimes it gives us at least a glimpse of it. This is evident especially if we consider the problem of the sense of reality not from the point of view of its drama (which is real and important, of course), but from that of its *fascination*, which is also real and not less important, and is shown to us, at least to a large extent, just by science. Basing on my personal experience, I have to say that very often this approach is more convincing than the first one, above all if the goal is to build a metaphysics which could be rational and cognitive, and not merely "hermeneutic", because in this way it is more emphasized the aspect of substantial continuity with science, despite the diversity of their objects and methods.

But, of course, all that holds only under the condition that science is considered, precisely, *knowledge*, and, more precisely, *true* knowledge. If, indeed, it were a mere socially determined convention, as nowadays is maintained by the overwhelming majority of the philosophers of science, its fascination would be, in turn, a mere social product without any objective value, unable to justify the search for something "beyond" itself which could represent its ultimate ground. In this sense, it would not be so strange if some day, looking at the whole matter from a historical perspective, we had to reach the conclusion that the greatest contribution of Agazzi to the defense of the value of metaphysical knowledge has been precisely his courageous, passionate and rigorous defense of the value of scientific knowledge.

References

Agazzi, Evandro. 1961. *Introduzione ai problemi dell'assiomatica*. Milano: Vita e Pensiero.

Agazzi, Evandro. 1964. *La logica simbolica*. Brescia: La Scuola (revised and extended edition: 1990).

Agazzi, Evandro. 1967. Alcune osservazioni sul problema dell'intelligenza artificiale. *Rivista di Filosofia Neoscolastica* 59: 1-34.

Agazzi, Evandro. 1969a. *Temi e problemi di filosofia della fisica*. Milano: Manfredi. Reprint: 1974. Roma: Abete.

Agazzi, Evandro. 1969b. L'ateismo e la scienza. In *Ateismo sfida ai cristiani*, ed. Giuseppe Lazzati, 164-180. Milano: Vita e Pensiero.

Agazzi, Evandro. 1975. Scienza e metafisica oggi. In *Studi di filosofia in onore di Gustavo Bontadini*, ed. Francesco Carlomagno, I, 3-22. Milano: Vita e Pensiero.
Agazzi, Evandro. 1977. The role of metaphysics in contemporary philosophy. *Ratio* 19(2): 162-169.
Agazzi, Evandro. 1978. Science and metaphysics in confrontation with nature. In *Man and nature*, ed. George McLean, 3-14. Calcutta: Oxford University Press.
Agazzi, Evandro. 1981. Considerazioni epistemologiche su scienza e metafisica. In *Teoria e metodo delle scienze*, ed. Carlo Huber, 311-340. Roma: Università Gregoriana Editrice.
Agazzi, Evandro. 1983. *Science et foi. Perspectives nouvelles sur un vieux problème / Scienza e fede. Nuove prospettive su un vecchio problema*. Milano: Massimo.
Agazzi, Evandro. 1985. La questione del realismo scientifico. In *Scienza e filosofia. Saggi in onore di Ludovico Geymonat*, ed. Corrado Mangione, 171-192. Milano: Garzanti.
Agazzi, Evandro. 1986. Considerazioni epistemologiche sul principio antropico. *Nuova Secondaria* 9: 34-37.
Agazzi, Evandro. 1988a. Science and metaphysics: two kinds of knowledge. *Epistemologia* 11(1): 11-28.
Agazzi, Evandro. 1988b. Evolution and teleology. In (eds.) *Person and God*, eds. George Mclean and H. Meynell, 275-286. Lanham, New York, London: University Press of America.
Agazzi, Evandro. 1991a. Operazionalità e intenzionalità: l'anello mancante dell'intelligenza artificiale. In *Intelligenza naturale e intelligenza artificiale*, ed. Biolo Salvino, 1-13. Genova: Marietti.
Agazzi, Evandro. 1991b. The universe as a scientific and philosophical problem. In *Philosophy and the origin and evolution of the universe*, eds. Evandro Agazzi and Alberto Cordero, 1-51. Dordrecht, Boston, London: Kluwer Academic Publishers.
Agazzi, Evandro. 1994. Was Galileo a realist? *Physis* 31(1): 273-296.
Agazzi, Evandro. 1997a. Thought and ontology: the meaning of a correlation. In *Thought and ontology*, ed. Mark Sainsbury, 13-22. Milano: Franco Angeli.
Agazzi, Evandro. 1997b. On the criteria for establishing the ontological status of different entities. In *Realism and quantum physics*, ed. Evandro Agazzi, 40-73. Amsterdam, Atlanta: Rodopi.
Agazzi, Evandro. 2000. Metafisica e razionalità scientifico-tecnologica. In *Annuario di Filosofia 2000. Corpo e anima. Necessità della metafisica*, ed. Evandro Agazzi, 97-124. Milano: Mondadori.
Agazzi, Evandro. 2002a. Realismo. In *Dizionario interdisciplinare di scienza e fede*, eds. Giuseppe Tanzella-Nitti, Alberto Strumia, II, 1181-1189. Roma: Urbaniana University Press.
Agazzi, Evandro. 2002b. Metaphysical and scientific realism. In *The problem of realism*, ed. Michele Marsonet: 35-63. Aldershot: Ashgate.
Agazzi, Evandro. 2004. Logic, truth and ontology. In *Logic and metaphysics. Proceedings of the international conference of Genoa 2001*, eds. Michele Marsonet and Margherita Benzi, 33-58. Genova: Name.
Agazzi, Evandro. 2007. Idealization, intellectual intuition, interpretation and ontology in science. In *The courage of doing philosophy. Essays presented to Leszek Nowak*, eds. Brzeziński Jerzy, Klawiter Andrzej, Kuypers Theo A.F., Łastowski Krzysztof, Paprzycka Katarzyna, Przybysz Piotr, 303-314. Amsterdam, New York: Rodopi.
Agazzi, Evandro. 2011. Consistency, truth and ontology. *Studia Logica* 97(1): 7-29.
Agazzi, Evandro. 2014. *Scientific objectivity and its contexts*. Cham, Heidelberg, New York, Dordrecht, London: Springer.
Agazzi, Evandro, Fabio Minazzi, and Ludovico Geymonat. 1989. *Filosofia, scienza e verità*. Milano: Rusconi.
Arecchi, Fortunato Tito. 1985. Caos e ordine nella fisica. *Il nuovo saggiatore* 1(3): 35-51.
Arecchi, Fortunato Tito, and Iva Arecchi. 1990. *I simboli e la realtà*. Milano: Jaca Book.
Musso, Paolo. 1993. *Rom Harré e il problema del realismo scientifico*. Milano: Franco Angeli.
Musso, Paolo. 1997. *Filosofia del caos*. Milano: Franco Angeli.
Musso, Paolo. 2004. *Forme dell'epistemologia contemporanea. Tra realismo e antirealismo*. Roma: Urbaniana University Press.

Musso, Paolo. 2007. Il filo rosso dell'intenzionalità. In Filosofia, Scienza e Bioetica nel dibattito contemporaneo. Studi internazionali in onore di Evandro Agazzi, ed. Fabio Minazzi, 293-298. Roma: Presidenza del Consiglio dei Ministri.
Musso, Paolo. 2011. *La scienza e l'idea di ragione. Scienza, filosofia e religione da Galileo ai buchi neri e oltre*. Milano, Udine: Mimesis.
Musso, Paolo. 2013. Las matemáticas, Dios y la inmortalidad del alma. *Quaerentibus. Teología y Ciencias* 1(2): 141-158.

The Autonomy of Science in a Systems Theoretic Approach

Alfredo Marcos

Abstract Evandro Agazzi singles out one of the fundamental knots of Modernity: the demand for autonomy. Science was one of the first human activities in demanding its own autonomy. In a complementary sense, the autonomy of science has been one of the major factors contributing to the development of Modernity. The search for autonomy, as Agazzi suggests, has put in the hands of humanity positive and important results, but it has also led to some excesses. As a consequence of such excesses, at the decline of Modernity, a certain cultural uneasiness and a strong request for new balances and connections became manifest. Agazzi so proposes to activate the systemic approach to obtain such new balances. Science will thus fulfil at best its own constitutive goals, beginning with autonomy, certainly, but also in respect and consideration of other fields equally autonomous, like ethics. Finally, we must also clarify that the systemic approach proposed by Agazzi does not eliminate human freedom, nor practical rationality, but, to the contrary, makes them possible and powerful.

1 Introduction

Well known internationally as one of the most prominent philosophers of science of our times, Evandro Agazzi has also elaborated speculative researches in the most important areas of philosophy. Actually, the interest he developed in logic and philosophy of science did not stop him from entering the present debate on problems inherent in ethics, political philosophy, epistemology and ontology with great lucidity, depth and timeliness. Indeed, we may affirm that precisely his deep investigation in philosophy of science naturally brought Agazzi to extend his research also to other fields. Such dynamics emerges in the unique reflection he

A. Marcos (✉)
University of Valladolid, Valladolid, Spain
e-mail: amarcos@fyl.uva.es

M. Alai et al. (eds.), *Science Between Truth and Ethical Responsibility*,
DOI 10.1007/978-3-319-16369-7_20

has left us on Modernity, a reflection that borders on philosophy of history and on that of society and culture. In this respect, Agazzi has given a very fecund contribution. He has argued that the search for autonomy is one of the essential traits of Modernity. From this finding, he has detected one of the most decisive causes of our contemporary cultural uneasiness, i.e. the very excesses of autonomy. Besides that, he has also identified and developed an adequate cure for this pathology: the systemic approach.

I will try to present such original contributions of Evandro Agazzi to today's philosophy. To begin with (Sect. 2), I will discuss on Modernity as a search for autonomy. This claim for autonomy appears in largely heterogeneous fields. We will actually confront ourselves with the autonomy of the subject, and with that of nations structured in modern times, with the autonomy or separation of powers, as proposed by Locke and Montesquieu, but also with the reciprocal autonomy of the different spiritual and practical fields of human life, especially art, morality and science.

The central problem, at the dawning of Modernity, was how to obtain and increase the autonomy starting from the hierarchic relations typical of ancient and medieval times. In the present days instead, the problem has changed radically. It consists, actually, in finding the way to balance the excesses derived from the reciprocal isolation between different fields of human life, i.e. we try to overcome—in the words of José Ortega y Gasset—"the barbarism of specialization", we beg for integration procedures, but trying at the same time to avoid a return to hierarchical relations.

The matter, right now, is not so much the subjects' autonomy, as the possibility to avoid their isolation and pathological disaggregation, so that the success obtained in autonomy will not become its own nemesis; it is not so much a question of autonomy of nations, but, rather, of inserting the latter in a super-national or global order; it is not, again, a question of the autonomy of science, but, rather, of its integration in the whole building of knowledge and, on the other hand, in the whole of human life, avoiding to fall back into a hierarchical relation, which this time would inevitably conduct us to the supremacy of science. We are not confronted then with an eminently modern problem, but with a problem *generated by modernity*, and not by its failure, but by its very success, maybe even excessive, in pursuing autonomy.

It seems obvious that a comprehensive philosophy of science should be involved seriously in the correct integration between science and all the other aspects of human life, primarily because the most popular positions in this respect are not very satisfactory. On one hand, in fact, scientism asserts a sort of supremacy of science on life and expresses a superiority of the scientific vision of the world. On the other hand, its most radical critics foster an anti-scientific mentality which results like-wise inappropriate. It is necessary to consider if there is a midway between the extremes—both rebuttable, in Agazzi's judgment—of scientism and anti-science.

To attempt a reply to this question, in a second passage (Sect. 3), I will refer to the systemic approach proposed by Evandro Agazzi. This systemic perspective intends society just as a system, inside which there are various subsystems reciprocally related to each other. Therefore science and technology are both interpreted as social subsystems. With such premises, it is clear that the philosophy of science may develop itself as a social philosophy and a theory of society. The "toolbox" of the philosopher of science becomes much heavier: without setting aside logic, semantics and epistemology, it seems necessary to be equipped with an adequate knowledge of social theory and practical philosophy. Today, this convergence of domains appears unavoidable. On one side, techno-science seems a factor that cannot be set aside in the configuration of society. It is not possible anymore to understand contemporary society while ignoring the techno-scientific factor. On the other hand, instead, the very social aspects of techno-science demand attention. Science and technology would appear incomprehensible without an accurate interpretation of their social aspects and their links with other social subsystems. Agazzi's thesis may, therefore, help us today to overcome the challenge of integrating techno-science in the whole of human life.

2 Modernity as Autonomy

Autonomy is certainly a desirable value and a key concept in modern thought, particularly in Immanuel Kant. Nevertheless it could turn into a "slippery" concept if autonomy is constituted as a prime and absolute value and not compensated by the necessary connections with other desirable values.

In *Il bene, il male e la scienza* (1992) Evandro Agazzi underlines that the passage from the Middle Ages to Modernity was characterized by a series of demands of autonomy on behalf of different intellectual and practical domains, starting with demands set to theology, which tended to occupy, so to speak, the top of the pyramid of knowledge. It is historically true that the autonomous exercise of reason in philosophy had already been claimed in the 13th century by the theologian and philosopher Thomas Aquinas. In his *Summa Theologiae* (1, q.1, a.2), he stressed that it was necessary to remain in line with the "natural light of the intellect" for solving prevailingly philosophical matters. The autonomy of natural science was further claimed by Galileo, whose intellectual and personal adventure may be interpreted as a search for the autonomy of science, rather than a clash between science and religion. Machiavelli, in turn, sustained the autonomy of political science and inspired British liberal scholars, who widened this principle to the economic domain, likewise did Kant and the romantics as referred to art.

As in many other areas of knowledge, Kant brought this journey into autonomy to a critical point. Kant himself, moreover, argued clearly for the autonomy between the three parts of the sphere of knowledge: science, morality and arts. Actually, he dedicated each one of his three great *Critiques* to each one of these parts. According to Kant's intention, each of these three domains counts with independent objectives and values, and each with a specific argumentative style. The way chosen by Kant will be followed again both by Weber and by Habermas, who will also give an interpretation of Modernity as the mutual autonomy of these three large domains.

In a first sense, certainly the most basic, we could interpret autonomy as independence of judgement. In this sense, each field may judge on the basis of its own criteria and values. So, for example, a work of art may be valued positively in the light of merely aesthetic criteria and, at the same time, negatively concerning its political repercussions. A scientific research project may result excellent on the basis of scientific criteria and values, but at the same time, economically or ethically unsustainable.

In a more ambitious sense, instead, we may intend autonomy not only with reference to evaluation, but also referred to action. In this way, to remain with the previous example, both the artist and the scientist will be able to act independently inside their own operative area without considering the exterior criteria of political, economic or moral character.

A third and more radical interpretation of autonomy implies the total absence of external controls limiting, for example, scientific activities. As Agazzi (1992: 13) points out, the acceptance of the first level of autonomy "does not imply that of the second, as also the second level does not include the third".

We note, first of all, a parallelism between the demand for autonomy coming from different areas of knowledge and the search for autonomy in the political and social fields. A good example could be the origin of nations, and the healthy tendency to the internal separation of powers. The aspiration to autonomy in the political sphere was not actually experimented as a simple fact, but as a just demand. The same may be said on the intellectual level, for which the freedom of philosophical, scientific and artistic productive thought was perceived as a desirable good and as a sure form of human progress. Following in this example, the autonomy of nations and that of the powers of the State depended on an extremely delicate balance of forces and on the will of reciprocal respect. As we know, often this balance resulted too fragile, this will too feeble, and the temptations of overwhelming the other nations too strong and dangerous. The newborn entities frequently wanted to become the head of a new hierarchic order. It so happened that some nations born with Modernity imposed themselves on others, limiting or eliminating their autonomy. In a similar way, the judiciary power tries sometimes to colonize the executive power and vice-versa.

In the intellectual sphere, the scientific vision of the world and the rationality of technological efficiency tried to rule and colonize the life world (*Lebenswelt*).[1] Such subjugation arose in the wake of the thought of Descartes and Bacon and of Newton's physics, and found an enabling atmosphere in the illuminist and positivist mentality. We note, therefore, that an autonomy, which is fair at the beginning, can degenerate first in autarchy and then in new hierarchical impositions, both in the scientific and in the political field. In fact Agazzi writes:

[1]The concept of life world—in German *Lebenswelt*—comes from the phenomenological tradition, and has been recently used by Habermas under the meaning of "background horizon of experience" and of "pre-reflective" life, from which, as a starting point, we may give meaning to whatever may be affirmed.

> Today the tendency to discuss over again these different points is clearly evident [...] operating a critical revision of the concept of autonomy, without, after all, letting us get involved in forms of obscurantism, retrograde involution or negation of the positive aspects certainly contained in the declarations of autonomy and liberty that we have considered (1992: 13).

We, as postmodern, maintain the advantage of a more profound historical perspective. We also have, thanks to this very advantage, the obligation to be fair in our balances. Neither the ideology of scientism and technologism, with its tendency to the hierarchical dominion of techno-science, nor the anti-scientific and anti-technological attitudes, which deny autonomy to techno-science, are satisfactory from our point of view. We know also that the enterprise to make life more scientific and technical has made at the same time dreams come true and generated new monsters. Today we are conscious of the fact that techno-science, which is substantially a positive reality, has fostered some epistemic and non-epistemic values that deserve recognition: cosmopolitism, objectivity, rigour, liberty of criticism, precision, efficacy, judgement impartiality and even others that have been extended to different areas of human action, and that are not alien to the progress of liberty and justice.

The spreading in all directions of the techno-scientific domains, on the other hand, has not always brought positive results to human life and to the lives of the other habitants of the world. If, on one side, we have assisted to a growth of knowledge and wellbeing, on the other, we have also favoured the end of certain values and different traditions, we have triggered pain and suffering. The 20th century is the proof of how the most brutal totalitarian intentions have fed themselves with the most advanced scientific means to produce suffering and destruction. So, the century that met undeniable progress, like the improvement of anaesthesia and antibiotics, also saw the development of the most efficient techniques to produce death and destruction. It would be naïve or dishonest to give the fault to techno-science, though we must recognize, in any case, that similar results would not have been possible without its contribution. Should we wish to draft a temporary conclusion, we could say that it would not be reasonable to put at the helm of human life only techno-science. Likewise, we cannot consider it totally self-sufficient, as it must be inserted in a net of limits, counterweights and pondered controls, without cancelling its legitimate margin of autonomy.

If we accept that decisions in science may be good or bad, as they may be rational or irrational, we affirm—or we suppose implicitly—that the criteria of goodness and rationality are independent from science, and we place ourselves outside of scientism. We recognize that the identification of human reason with the sole scientific method, and the plain correspondence between techno-scientific progress and human development are naïve and misleading. On the other hand, we cannot reject, with a puritan attitude, the entry of techno-science in our life, as some people persist to maintain.

To sum it up: the ancient and the medieval world have pursued a hierarchical type of order. The modern world instead has been fascinated by the idea of autonomy, mostly brought to the extremes in terms of autarchy, so imposing at times new hierarchies. It therefore appears that today's world must find a different and more harmonic balance, with the just weight given to relations and horizontal connections amongst science, morality, art and other various areas of human life.

Both closed hierarchy and autarchy, are terms full of constraints and dangers, and find reciprocal nourishment and support. Today we are trying to balance these two opposite tendencies, trying to find a more efficient mediation, a form of connection between different areas of human life, each one with its own values, interests and criteria. This is, without a doubt, one of the main challenges of post-modernity, if not actually the main one of our time.

A prove of what I am saying is the constant presence in today's debate of terms like dependence, conciliation (of family life and work), solidarity, dialogue, links, web, net, globalization, European community, globalization, etc., all of them tending to compensate excesses in matter of autonomy (Marcos 2012). In my opinion, one of the most promising suggestions for reaching the longed balance is certainly the new theorisation of the systemic approach elaborated by Agazzi.

3 The Systemic Perspective of Evandro Agazzi

In Agazzi, we find an acute diagnosis of the pathologies of Modernity, partly similar to that of other philosophers of the 20th century, as for example Habermas (1968). In general, we could speak of a sort of isolation illness, of a disconnection between the different fields of life and knowledge, of a loss of balance in the search for autonomy. Such disconnection has engendered a hypertrophy of techno-science, identified by Agazzi as scientism and technologism, and by Habermas as the scientific colonisation of the life world.

We could in any case point out a further, and deeper, affinity between the thought of Evandro Agazzi and that of Jürgen Habermas, i.e. the philosophical use of the systemic theory. However, the different ways they use the same theory discloses very significant differences for the development of our argument. While in Agazzi, in fact, the general systems' theory is useful to build new links between different areas of human life, links that avoid opposite poles as undesirable as hierarchy and autarchy, in Habermas the systemic logic is seen rather as a threat for the correct links between science, art, morality and the life world.

In a certain sense these differences are decisive. In fact, if Habermas' interpretation of systems theory were the only possible one, then the use made by Agazzi should be reconsidered. We think, therefore, that a further interpretation of the systemic theory is possible, able to solve our initial problems—i.e., disconnection and colonisation—without triggering new ones.

Agazzi's systemic perspective opens a demand for the dignity of ethics, without falling into easy moralism. It is the very logic of each subsystem, particularly of the techno-scientific subsystem, that indicates how one should pay attention to moral criteria:

> It is necessary to take up again here the discussion on human beings and try to discover the whole scale of values that inspire their actions, recognizing that their profound freedom consists in their possibility of self-realizing by honouring such values. This does

> not imply to plead for an imperialism of morality or practical philosophy on science and technology, but simply for an autonomy inscribed in a context of meaning, where science may recuperate its complete human dimension. For this reason the demand for a practical philosophy invites us also to consider and introduce the great themes of a genuine philosophical anthropology (Agazzi 2001: 51).

Even further, ethics itself is seen as an integral part of the system, like a subsystem amongst others, set at the same level, not so much as an alien area to be layered over the others in a dominating position. The type of rationality presupposed in all these spheres is the same, that is, the human rationality that supports action both in the scientific and the ethical system as well as in any other one. The key to the success of such an integration resides in the fact that in Agazzi there is no rigid link between systems and necessity. That means that human action, conditioned as it is by some kinds of systemic limits, remains, however, free and undetermined by any systemic automatism. It is so for what concerns techno-scientific production, as well as for moral, political, economic and aesthetic aspects.

Thus Agazzi develops decisively the general systems theory, elaborated for the first time by Bertalanffy (1968) and born in the field of cybernetics and biology. As is well known, such theory, because of its very general and abstract character, retains sufficient plasticity to make possible its application in many areas of reality. The possibility of its application also to the relations between techno-science and other fields of human life was actually suggested to Agazzi himself by a book which Jean Ladrière contributed to the UNESCO convention in 1974. In Ladrière's prologue we read:

> This book is for those – university students, professors and people interested in culture – who wish to clarify, on one hand, the complex relations between scientific knowledge and the technology generated by such knowledge, and, on the other hand, the impact of science and technology on culture, in particular on morality and aesthetics. Impact with a double effect, as it were to say: a dismantling of culture and an attempt to restructure it (Ladrière 1978: 9–10).

This means to affirm that such restructuring will move forward from techno-science. We are in front of the same landscape we have often described: there is an excessive breakup amongst different areas of human life, which may be potentially colonised by techno-science. The book also goes all the way back from the sphere of knowledge to the life world, with the goal of considering the impact of techno-science on the industrialised societies and on those on the way to industrialisation. Furthermore, the clear intention here was to suggest the possible use of the general systems theory as a theoretical frame to think the relationships between different fields, characterised as subsystems of the social system.

We accept, to start with, the autonomy of techno-science as a desirable value.[2] In Ladrière's words: "The growth of the scientific field's autonomy means that this field has each time the necessary resources to ensure its own support [...] and

[2]Science and technology maintain each one its own peculiarities, but it is a given fact that nowadays they assume a symbiotic behaviour. In a systemic perspective we may categorize these two realities—science and technology—as subsystems of the techno-scientific system, which, in turn, can be considered as a social subsystem.

growth" (1978: 46). It seems positive that techno-science frees itself by and by from rather uncontrollable external circumstances. Such a freedom could though give the idea of techno-science as a closed system, entirely free from external influence and therefore coinciding with *the* global system.

Some philosophers of technology (Ellul, Mumford and Winner, for instance) have insisted on the danger of a completely autonomous technological system, tending to grow independently from life styles, traditions or any other external value, including those expressed by a democratic will. Even from the sphere of the philosophy of science and technology there arose voices of criticism against an uncontrolled spreading of the scientific vision of the world, and, in general, of scientism. Already Kant himself recognized our need of worlds other than the scientific one, especially those of morality and art. In a period when the problem of the links between these fields existed already, Kant affirmed the theoretical legitimacy of each one and argued for their mutual autonomy. We could then hold that if techno-science should acquire the characteristic of a closed system or if it were to become a global system, it would obviously enter into a conflict with other spheres of knowledge and with the life world itself.

Agazzi indicates the price of such extreme solutions. In the first place, the moral field is reduced to the intimacy of individuals (becoming so a simple act of will or a fideistic act), with the consequent removal of a public rational debate on its problems. Secondly, morality is reduced to an object of scientific explanation (and, why not, also of technical manipulation), examined by psychology, sociology, neurophysiology and genetics. In this sense, "the 20th century—writes Agazzi (1992: 146)—has known a fundamental eclipse of this branch of philosophy, namely of the philosophical research that looks for the clarification of the meaning and for the proposal of guidelines for an *ethical commitment*".

Not only does the moral field result so blurred, but also the natural world is colonised by artefacts, that bring also, at the same time, to positive and negative results. On the negative side, we can argue that the sacred dimension of nature is eliminated, the fine arts are bound within the dark realm of the irrational, various traditions and values are debased, religion, wisdom and everyday experience lose value, and common sense is imprisoned in the world of insignificance. As already pointed out, what is outlined here has also produced adverse reactions that are to be correctly diagnosed as symptoms. In particular, in the field of morality, Agazzi (1992: 146) recalls with satisfaction how "in recent years a certain interest for ethical problems has arisen, and this is a very significant signal, in fact it is the sign that the scientification of the ethical field does not attain a positive result, and therefore the moral aspiration of man rises again strongly and underlines its difference with respect to the scientific dimension".

In the light of such considerations, the very systemic perspective may help us to overcome the contrasts amongst distinct and autonomous fields, without cancelling their differences nor the conditions for their specific autonomy. The main conclusion we may reach from such systemic approach is that, for pure systemic reasons, techno-science must respect in its development the values inherent to human life. This perspective, capable of rehabilitating ethics, is certainly alien to any type of moralism. As we will see, in fact, the duty of respect just mentioned does not emerge from moral reasons, but from systemic ones. We do not ask scientists as such to identify themselves, driven by altruism and good intentions, with the very values of the political, juridical, economic or ethical system. We are rather making clear that also the specific values of techno-science are realized thanks to the respect for the other areas of human life. This approach consists in highlighting that the colonisation of human life by techno-science could be pursued only at the cost of a loss of faith in techno-science itself. The excessive imposition of the scientific vision of the world ends up being an obstacle for scientific development itself, and the lack of social control on technological development ends up frustrating the very technological progress.

Furthermore, in line with the systemic perspective, techno-science is seen as a system of human actions. We may consider this system as a subsystem of the social one, connected with other subsystems (political, economic, educational, military, religious, ethical, mass media …). We could amplify our discussion to affirm that all these subsystems form the framework, the social environment, where techno-science is generated and operates. There are also natural subsystems—for example, the ecosystems and the planetary system—that constitute the other side of the setting where the techno-science subsystem lives. The exchanges of techno-science with all such subsystems are evident. Just think, for instance, that the techno-scientific system releases a good part of its research outcomes to the educational system, which, in turn, is able to form many people who increase or support the techno-scientific system itself. We could mention other examples, both in the domain of social and natural subsystems, but what we wish to point out here is that when we speak of techno-science we must think of an open social system, which interacts with many other systems. It may therefore be defined as an adaptive system, able to modify itself and its environment, within certain limits, to balance and develop itself. In sum, we are proposing to see techno-science as a human action system, social, open and adaptive.

The characteristics mentioned above are also in common with many other systems. The specificity of techno-science consists in its *constitutive functions*, that is, in the goals it pursues. According to Agazzi, science follows, in fact, two essential purposes: the development of rigorous and objective knowledge, and the diffusion of it. From his approach, technology will tend, instead, to an efficient knowledge and, at the same time, to its application directed towards innovation. Agazzi refers to these goals as "essential variables" of the system.[3] They are *essential* in the sense that they must remain within the limits of a certain critical range in order for the system to function and survive. It is intuitively clear that, if techno-science stopped producing rigorous, objective and efficient knowledge, if it stopped

[3]Agazzi distinguishes two variables essential to science, i.e. production of rigorous and objective knowledge (v1) and its diffusion (v2). Further proof of the fruitfulness of Agazzi's ideas is the possibility of building a systemic theory of the communication of science through the simple expedient of registering v2 in a specific subsystem called "science communication". For a development of this idea see Marcos (2010: Chap. 6).

diffusing and applying it, then it would simply have ceased to exist anything that we could rightly name "techno-scientific system".

The failure to realize the essential functions of the system may be caused by *internal tensions* or *external pressures*. In the first case it occurs inside the system itself—we recall, in this sense, the title of one of Kuhn's books: *The essential tension* (Kuhn 1977). In fact, it is true that certain tensions must be maintained and are essential to the survival and functioning of the system. Let's think, for example, of the tension between tradition and criticism, or between simplicity and precision. If scientists were not educated in a certain scientific tradition, science itself would be impossible, but if criticism to this tradition was prohibited, then science would certainly come to an end. If a theory is totally imprecise, even if very simple, it is useless, but if precision is acquired to the detriment of understanding and intelligibility we are not in a better position. Techno-science, therefore, is determined by tensions, delicate balances, without which it could not work. At the same time these tensions imply a potential danger, from the moment that they may be shattered or lose balance in any direction.

For this reason we speak of "dynamic balances", for which a non-catastrophic deviation may be internally compensated within the same system, as it occurs, for instance, with the homeostatic capacity manifested by living organisms. So, in historical periods when the biggest risk was the loss of a certain tradition, scientists have chosen to insist more on traditional values, instead of underlining those of criticism. On the contrary, in front of a risk of stagnation of a certain discipline, the critical aspect was encouraged. In this perspective, certain historical cases show a form of rationality that, without such interpretation, would appear incomprehensible.

It may also occur that the system suffers external pressures in addition to the internal tensions. In front of such possibility, the system may operate internal modifications or even modify its environment in order to recuperate its balance point or find a new one. Therefore, the history and philosophy of science should not ignore these external pressures if they really want to understand the techno-scientific system. But, in general, we may interpret the interaction with other subsystems in the light of the concepts of *input* and *output*. The techno-scientific system receives from the environment various types of input, like demands, supports and obstacles. On the other hand, the system issues to its environment some outputs, like rigorous, objective and efficient knowledge, as well as technological applications. In principle techno-science has to satisfy social demands, earn supports and reduce obstacles, with the goal of optimizing its essential variables.

We must not in any case forget that between the system and its environment there is a *feedback loop*, so that actions undertaken by a given subsystem cause indirect effects on the very same subsystem at the end of the cycle. In this way, for example, a loss of efficiency in knowledge production occurred in the techno-scientific system could influence negatively the economic system, and, at length, damage the financing of the techno-scientific enterprise itself. Or again: if science promotes researches contrary to socially recognized values, such as human dignity, health or safety, probably this would cause negative effects in other social subsystems, which would, in turn, react putting legal, economical or other kinds of obstacles to the scientific enterprise. These cycles are not necessarily of a vicious kind. Obviously, also virtuous ones may be created. Here too the examples, both historical and fictitious, could be multiplied at will. We prefer, anyway, to proceed in a different direction, outlining two important conclusive consequences.

In the first place, we note that, like any other system, techno-science demands a sound environment where it may be located. If, in the course of the maximization of its own essential variables, techno-science suffocated the other surrounding subsystems, it could suffer negative consequences. Therefore scientists and technologists should tend to optimize, rather than to maximize, these variables. This means that the variables can grow only inasmuch as the functioning of the techno-science is compatible with the correct functioning of the other surrounding subsystems. To give an example, it is clear that if we were able to experiment freely on the pain of animals and humans, we could have more rapid information on the physiology of pain. However, scientific research must, at times, accept limitations and controls, otherwise social lack of confidence could decree the end of science, making it impossible. Research, supported economically by the citizens' taxes, cannot oppose openly the main social values. This would bring, actually, to legitimize forms of fiscal objection with respect to certain lines of research. If a government went crazy cutting funds for elementary education to favour research, techno-science would probably have benefits in the short term, but we would end up, in the long term, with substantial losses in the techno-science, because of the lack of generational replacement. The same may be affirmed for the relations of techno-science with many other fields.

In other words, techno-science will work out best if interconnected with other sound subsystems: with a democratic political system, a fair juridical system, a flourishing economical system, a good educational system, a healthy ecosystem, and so on, as well as with a right ethical system. Respect for values of the other subsystems, in line with the perspective we are developing here, is also useful for the very objectives of techno-science itself. Particularly, respect for ethical values and acceptance of external controls are necessary for the development of techno-science, in virtue of reasons that are not of ethical, but of systemic nature (aside from the fact that scientists and technologists, as persons, should act according to moral reasons).

Secondly, it is clear that, in spite of systemic necessity, there is in techno-science a margin, limited but real, for free decisions. The members of the techno-scientific system can actually influence intentionally other subsystems of the environment, directly or indirectly, and influence the functioning of their own system in line with their own goals, always in respect of some given limits. This means that techno-science will never be perceived as a simple instrument, as a means at the service of any intention. It is necessary, instead, to recognize that techno-science has its own goals. In consequence, its legitimate degree of autonomy must be respected.

4 Conclusions

Taking as a starting point the philosophy of science, Evandro Agazzi questions the role of science in the whole of human life. Through this question he supersedes the limits of classical philosophy of science, triggering a reflection that ends up in the realm of practical reason, of history and culture. So he singles out one of the fundamental knots of Modernity: the demand for autonomy. Science was one of the first human activities, in the wake of Modernity, to demand autonomy for itself. In a complementary sense, we should say that the autonomy of science has been one of the major factors contributing to the development of Modernity. The search for autonomy, as Agazzi suggests, has put in the hands of humanity positive and important results, but it has also led to some excesses. As a consequence of such excesses, at the decline of Modernity, a certain cultural uneasiness and a strong request for new balances and connections became manifest. Agazzi so proposes to activate the systemic approach to obtain such new balances. This approach enables us to point out the reasons that science has for respecting other fields, traditions, practices and values, especially those of ethical nature. Such reasons are not specifically of moral nature, but more appropriately systemic. In this way science will fulfil at best its own constitutive goals, autonomy in the first place, but at the same time it will also respect and pay due consideration to other equally autonomous fields, like the ethical one. Finally, we must also clarify that the systemic approach proposed by Agazzi does not eliminate human freedom or, much less, practical rationality; on the contrary, it reinforces and makes them possible.

References

Agazzi, Evandro. 1992. *Il bene, il male e la scienza: Le dimensioni etiche dell'impresa scientifico-tecnologica*. Milano: Rusconi.
Agazzi, Evandro. 2001. Filosofía técnica y filosofía práctica. In *Racionalidad científica y racionalidad humana*, eds. Marga Vega, Carlos Maldonado and Alfredo Marcos, 35-52. Valladolid: Universidad de Valladolid.
Bertalanffy, Ludwig von. 1968. *General System theory: Foundations, Development, Applications*. New York: George Braziller. (Revised edition 1976).
Habermas, Jürgen. 1968. *Technik und Wissenschaft als Ideologie*. Frankfurt am Main: Suhrkamp.
Kuhn, Thomas. 1977. *The Essential Tension*. Chicago: The Chicago University Press.
Ladrière, Jean. 1978. *I rischi della razionalità. La sfida della scienza e della tecnologia alle culture*. Torino: SEI.
Marcos, Alfredo. 2010. *Filosofia dell'agire scientifico*. Milano: Academia Universa Press.
Marcos, Alfredo. 2012. Dependientes y racionales: la familia humana. *Cuadernos de bioética* 23: 83-95.

Ethical Reflections on Science and Technology

Boris Yudin

Abstract Taking as reference point Evandro Agazzi's publications on the ethics of science, this contribution discusses the development of ethical reflections on science and technology during the last decades. The article shows that former debates on the ethical liability of science turned to understanding ethical appraisal as a necessary part of scientific and technological projects. Ethics of science is treated not just as moral reflection on scientific activity, its ends, means and consequences, but as one of the fields of science studies, which is involved in the search for answers to the fundamental question "what is science?"

1 The Liability of Science to Ethical Appraisal

My discussions with professor Evandro Agazzi started more than a quarter of a century ago, in the second half of the eighties. Our first meeting took place at the VIII International Congress of Logic, Methodology and Philosophy of Science in Moscow in 1987. The program of the Congress included an Intersectional Symposium "Science and Ethics". At that symposium Agazzi made a presentation "Ethics and Science" (Agazzi 1988). The title of my presentation was "The Ethics of Science as a Form of the Cognition of Science" (Yudin 1988). It turned out that both of us discussed the same issue: is it possible, and if yes, then how, to make scientific activity an object of ethical appraisal?

As is well known, the area of the ethical issues generated by the progress of science and technology is among the most important in the philosophical reflections of Agazzi. I have been involved in discussions with him on these issues during all these years, we had a lot of meetings and debates in Moscow, Genoa, Lecce, other Italian cities as well as in different countries over the globe.

B. Yudin (✉)
Institute of Philosophy, Russian Academy of Sciences, Moscow, Russia
e-mail: byudin@yandex.ru

M. Alai et al. (eds.), *Science Between Truth and Ethical Responsibility*,
DOI 10.1007/978-3-319-16369-7_21

The main ideas elaborated by Agazzi after that Intersectional Symposium in Moscow were presented in his subsequent publications, such as “Responsibility. The Genuine Ground for the Regulation of a Free Science” (Agazzi 1989), *Moral Dimension of Science and Technology* (Agazzi 1998, that was in fact the Russian translation of his book later appeared in English as *Right, Wrong and Science*), “How can the Problems of an Ethical Judgment on Science and Technology Be Correctly Approached?” (2003), *Right, Wrong and Science* (2004). Therefore, we can maintain that he has built a thoroughly elaborated conception for the analysis of complex, dynamic, historically changing interrelations between science and ethics.

In particular, in Moscow’s presentation as well as in some of the subsequent publications, Agazzi was interested in these two main problems: “On which grounds can an ethical appraisal of science be built? In which directions and to which limits it can be carried out to remain meaningful?”.

He built his understanding of the interrelations between science and ethics on an elegantly elaborated concept of *autonomy* that has been applied to different spheres of human activity, including science. Agazzi stresses the historical nature and changeability of the phenomena that emerge in the modern age “in different sectors of the spiritual and practical life of man” (Agazzi 1988: 49). He mentions some of such sectors: politics (its autonomy was vindicated by Machiavelli), science (by Galilei), economics (by British liberals), art (by Kant and the Romantics). All these authors stressed the specificity of the respecting domains, in the sense of the need to develop purely internal criteria to appraise the fulfillment of their restricted and specific goals. The novelty in delineation of these sectors after Renaissance consisted in “the fact that the borderlines were now meant to express clear-cut ‘separations’, rather than simple ‘distinctions’, and that the consequent ‘autonomy’ of the different fields has quickly turned into a search for a kind of ‘freedom’ or ‘liberation’” (Agazzi 1988: 50).

This transition from distinctions to separations, or from autonomy to freedom—continues Agazzi—led to the rejection of any form of outer interference in the respective domains. This freedom had been expressed in three different forms (or stages): firstly, as independence in the criteria of judgment, i.e. the criteria can be legitimately borrowed only from the domain under consideration and not from somewhere outside it. Such independence, according to Agazzi, can be interpreted as freedom from values, or as compliance only with those values which had been adopted inside the domain.

Secondly, freedom entails an *independence in action*. This means, according to Agazzi, that

> the politician ‘as a politician’, the business man ‘as *homo oeconomicus*’, the artist ‘as an artist’ - and we can also add the scientist ‘as a scientist’ - are *legitimated* in acting according to the *pure* criteria of their profession, at least to the extent that they are performing *within* this profession (Agazzi 1988: 50).

Evidently, the second requirement is stronger than the first; it deepens possibilities of autonomy for an actor.

Thirdly, freedom is expressed as rejection of controls over the exercise of autonomy that would be exerted by external agencies in order to protect, for instance, more general social goals and values which could be endangered by the realization of the goals and values specific for the domain under consideration.

I want to stress the high analytical mastery, so characteristic of Evandro Agazzi, which is clearly demonstrated in his scrutiny of autonomy. The next steps of his reasoning on these issues are related with the search for grounds that would allow limiting in some situations the autonomy of a domain. Generally speaking, these are situations in which the realization of freedom in one of the domains causes serious problems for other domains. Taking into account science, we can say, according to Agazzi, that historically the question of restricting its autonomy became meaningful when its developments and its applications turned out to have not only positive, but also negative consequences for humans and society.

So, science as a specific domain of modern society becomes liable to ethical restraints only in so far as some actions which are permissible according to its inner criteria of judgment become unacceptable according to some outer goals and values. It is here that the reasons for limitations and controls imposed from outside can (and even must) be sought for.

"Nowadays …—writes Agazzi—we are confronted with the outcomes of such a process of "liberation", which has led to several intuitively unacceptable results: the autonomy of many single domains, if pushed to excess, brings them conflict with other domains" (Agazzi 1988: 50–51). So,

> the delicate problem we now confront is that of effecting a critical revision of the said points [i.e. of the three constituents of the "autonomy" concept – *B.Y.*], without becoming involved in obscurantism, regressive involution, or negation of the positive aspects which are certainly contained in the claims of autonomy and freedom (Agazzi 1988: 51).

I am now going to discuss in more detail some aspects of the interrelations between autonomy of science and limitations imposed on its realization. From the ethical point of view the concept of autonomy is important in two different aspects. Firstly, autonomy—understood as the capability to act freely on the basis of one's own considerations and decisions—is a prerequisite for the very possibility to speak meaningfully about the ethical responsibility of a person or a social entity. But, secondly, the autonomy of a person or a social entity poses limits on the relevance and significance of any ethical appraisal made from outside. Both of these points are essential for Agazzi's deliberations.

I think that the autonomy of a profession must not be understood as something just given by some authority: usually there exist external forces infringing on the profession's autonomy, so that its maintenance demands from a profession various decisions and actions, sometimes even rather hard and far-reaching. On the other side, as stressed by Agazzi,

> the autonomy of many single domains, if pushed to excess, brings them into serious conflict with other domains. In particular, this is being recognized in the field of science: the needs of protecting the environment, of avoiding technological catastrophes and regulating genetic manipulations (to remain within the most common examples) are producing a demand for the regulation of science and technology (Agazzi 1989: 2).

Now I want to stress the interdependence between the autonomy of an actor and the possibility to appraise his/her actions from an ethical point of view. Autonomy is a prerequisite for the moral amenability of an actor or agency. Indeed, any attempt of moral assessment of a non-autonomous, i.e. involuntary, forced act would be morally misdirected. That means, by the way, that the wider is the space of autonomous decisions and actions for acting agencies or individual actors (i.e., the space of freedom), the bigger is the burden of moral responsibility which is laid on their shoulders.

I want also to point out that the autonomy of a domain—I am going to discuss here mainly science—can represent not just an instrumental value, but a value in itself as well. This is why the scientific community is usually very much concerned in maintaining and strengthening its autonomy. At the same time, for different reasons, other domains rather often strive to limit the autonomy of science. I think, it would be possible and interesting to consider the whole history of modern science—starting at least from Galilei or the Royal Society of London—as a history of the struggle for science autonomy. This means that the real degree of science autonomy at every moment of time can be understood as a measure of a compromise between the cohesion of the scientific community and its ability to defend its autonomy, on one hand, and the effectiveness of different external forces striving to breach the autonomy on the other hand. So, external interventions into the domain of science, even those which are directed to minimize negative consequences of some scientific advances, lead to a lessening of the moral autonomy of science.

Along with such historical account of the emergence of the ethical agenda in contemporary science Agazzi, departing from his previously elaborated three-staged construction of autonomy, discusses also the logical grounds of these newly arising ethical concerns. It is in this context that he introduces for the first time the concept of aim as a defining constituent of any domain of action. At the stage of the independence in the criteria of judgment, the autonomy of a domain is expressed in assigning to it "a well-determined specific aim, and by indicating the criteria for evaluating how particular facts, assertions, actions and products ought to be in order for this aim to be pursued in the most satisfactory manner" (Agazzi 1988: 51). And this aim (or goal) represents, according to Agazzi, a "value" which inspires the action.

I have some reservations regarding the identification of an aim (or goal) with a value; it is unclear for me in which sense an aim "represents" a value. To my understanding, the discourse about "values" characterizes our preferences of some objects, acts, states, conditions etc. over others. This does not mean that our aims in every case are determined by these preferences. Our values can direct us in the choice of our goals, yet in general they belong to a more fundamental, more stable level in the structure of a personality or a domain, whereas the activity of setting and achieving goals takes place closer to the level of everyday practices. Sure, some aims or goals more immediately linked with the values of a domain under consideration; we can name them *institutionalized* aims (or goals). I think that when Agazzi equates aims with values he has in mind just these, institutionalized aims of a domain.[1]

2 Asking "What Is Science?" from an Ethical Point of View

In previous pages I tried to elucidate two essential points: firstly, that in his article of 1988, as well as in some of subsequent publications, Agazzi tried to substantiate the idea of liability of science—naturally, only in some respects—to ethical scrutiny. Secondly, according to Agazzi, science as a specific domain can be ethically appraised only from outside, from other domains, when they infringe the autonomy of science.

Now I shall try to develop my own conception of the interrelations between science and ethics. I am going to suggest that the ethical agenda can come into science not only from outside, but from inside as well. This means that the ethics of science will be presented as one of the fields of science studies. I suppose that my approach is complementary to that of Agazzi.

Sure, ethics of science operates within the range of modalities that belong to the sphere of moral judgments. This determines one dimension of a two-dimensional space, another dimension of which is formed by cognitive judgments.

So, the ethics of science studies the phenomena occurring in science. One can be even more definite: the *ethics of science is a form of science studies*. It studies precisely science (of course, seen from a specific point of view).

Since we recognize the ethics of science as a form of the science studies we admit at least the *possibility* of a corresponding subject-matter existing within science. In other words, there exist moral judgments and assessments inside science, and these judgments and assessments as such deserve scientific investigation.

There are many different types of activity connected with the production and circulation of scientific knowledge in society (the so called "knowledge society") which are the privilege of scientists. This privilege makes them responsible for the production of scientific knowledge, its translation to the general public as well as to the next generations, and of its various practical applications. These types of activity are part and parcel of the processes through which new knowledge is obtained. To use a metaphor, they create the field of forces in which scientific cognition unfolds. Sure, one may single it out and oppose it to all other types and forms of scientific activity. This operation, however, will to a considerable degree be analytical, but in real life we are dealing with integral sets of specific scientific activities; each of them can receive ethical appraisal.

[1]Somewhere else I discussed other aspects of Agazzi's understanding of goals which are pursued by science and difficulties arising in this connection (Yudin 2004: 258). For the response of Agazzi on my comments see (Agazzi 2004: 329).

Here is an example. The educational context considerably influences the cognitive activity in science (for more details see Petrov 2006; Yudin 1986: 186–189; Frolov and Yudin, 1989; Yudin 2010: 387–468). This means that any newly obtained fragment of knowledge should fit into a textbook. The corresponding norms of cognitive activity (which, besides technological, have also ethical sense) help realize this condition. This condition itself can be subjected to ethical assessments. It goes without saying that in usual everyday practice these norms require no special awareness or reflection; they are "internalized" and act, so to speak, automatically. Reflection becomes necessary when changes occur either in the way the cognitive activity is organized, or in the system of education, or in the way they interact. Such changes may put to question the normative determination and call for its correction or revision.

Nowadays interrelations between science and society become rather complex and multifaceted. They require, among other things, more and more legal and ethical regulations: hence the need for the ethics of science. It provides special knowledge about scientific activity, including knowledge about the variety and effectiveness of different means for its regulations. This type of knowledge distinguishes ethics of science from all other fields of the science studies.

Here a question arises: what defines the place of the ethics of science as contrasted to the methodology of science, sociology of science and other disciplines which study science?

I think that the question "What is science?" unites all these fields of the science studies. What is more, this question is also relevant for the philosophy of science in general. For a long time philosophy of science limited itself to a positive stand *vis-a-vis* science. There was no doubt about the desirability of scientific progress. The task was to promote it. The methodology of science concerned itself with the ways and means of obtaining reliable scientific knowledge. The sociology of science saw its task in defining social conditions most conducive to scientific progress according to the so called "inner logic" of science development. As far as the study of the ethical problems of science is concerned, it was dominated by the Mertonian reconstruction of the ethos of science as a system of norms designed to ensure the stability of scientific progress. This means that science was maximally free to develop according to its inner logic.

No matter how important these problems are, the methodology, sociology and ethics of science should go farther than that. The wave of discussions (which has been constantly on the rise within last decades) goes on identifying newly recognized critical points in scientific development as well as in the impact of science on humans and society. These discussions clearly demonstrate that the question "What is science?" has acquired new dimensions. The philosophy of science of the past ages considered it to be settled, or at least, admitting of a straightforward answer; the task was to formulate the adequate criteria of demarcation between science and non-science (metaphysics, in the first place). This desire to identify pure science and the emphasis on what distinguishes science from other spheres of mental activity and action are characteristic of the philosophy of science of the preceding periods. It contrasted science to other spheres of human thought and action rather than look for correlations, ties and cooperation.

Today, however, there is a need to elaborate intensively and qualitatively, in many fields, new approaches to the science studies. Thus, the critical points outlined by the discussions on the ethical issues of science do not belong to science alone. So, the question "What is science?" occupies the crossroad between science and the human world. The starting point here is the existence of science within this world, not in isolation from it. This leaves the question "What is science?" open. Despite its extremely general wording it presupposes a search for historically specific, rather than abstract-universal, definitions and characterizations of science.

The openness of this question means that there is no (and there cannot be any) predetermined answer to be identified, explained and clothed in suitable wording. The answer is the result of disputes, criticism and self-criticism of different conceptions and views. Each of them is, inevitably, partial and one-sided. It fails to reflect science in its entirety and contemporary science especially, as a complex and multifaceted phenomenon.

The most important point is that our searches for an answer presuppose an understanding of the radical changes that science has brought into the world. The very boundaries between science and non-science are not fixed, they change with the course of time. They are, besides, the zones of intensive interaction between science and the phenomena that determine it; this interaction strongly affects both science and the life of humans and society.

In these circumstances the answer to the question "What is science?" leaves the sphere of purely academic interest and assumes practical importance. Today's philosophy of science faces the task of comprehending what science is in general and what science is today. This comprehension outlines the scope of people's expectations for science and, consequently, humans' positions, decisions and actions concerning some of the most significant spheres of their existence. What I have in mind here is not the prediction of specific scientific results but foresights of the general trends and general structures of the cooperation between science, society and its culture.

Nowadays the concept of science has become one of the basic concepts of contemporary culture as understood by Erick Yudin (1978). This is the case because science has assumed an extremely important, many-sided and rapidly expanding role in social life. To enable the culture of today to define itself and to clarify its major problems, we need an analysis of the concept of science (i.e. an answer to the question "What is science?"), realized through the varied manifestations of science. At the same time the efforts of the philosophy of science to answer this question are precisely the efforts to formulate a rational and thoughtful attitude towards science and towards everything connected with it in one way or another.

It is very important that one-sided and narrow answers are not taken as the final answers, and thus block further search. The position represented by M. Foucault, M. Douglas and their rather numerous followers is prone to this shortcoming. They equate science and power or, to be more exact, they regard science as an instrument of domination. Research based on this premise shows many interesting, essential and previously ignored aspects of the real existence of science. Nevertheless, I think that those who are inclined to absolutize this view, to regard it as the only one rather than an addition to other opinions, commit a grave error. Ours is an age which has demonstrated the negative effects of a one-sided and once popular stand which resembles in many ways the position we are discussing. Here I refer to the interpretation of science as a means of domination over nature.

If we commit ourselves to purely analytical purposes we will probably be able to reduce research activity as a whole, human activity in general and interpersonal relations to the pattern of domination /subordination. This operation, however, will undoubtedly provide a bleak and monotonous picture of reality.

Let us discuss a simple example. I publish an article. This fact can be interpreted as an attempt to impose on my colleagues my own understanding of scientific findings or, even more, my will. This simplified approach fails to explain the multitude of norms (their nature and essence) which guide me in writing the article. Neither has it explained the fact why people resort to such sophisticated methods to impose their will rather than to use a stick.

The domination/subordination approach cannot explain the distinctions between the desire to impose one's will on others contrary to their own will, the desire to prove one's point with arguments open to critical assessment, or the desire to communicate.

When examining the question "What is science?" as the pivot of the philosophy of science it is advisable to bear in mind another point of view. Those adhering to it hold that the question is senseless since in reality we are dealing with a multitude of different sciences which have little in common, rather than with one science. It is not my task here to disprove such a position. I would like merely to point out that in practical terms the search for an answer is more important than the answer itself. The eternal task of the philosophy of science, its debt to culture, is to formulate continually the ever new definitions of science, to criticize and reassess them, to make them more profound and more in line with a rapidly changing reality.

I think that this task also determines the place of the ethics of science in the realm of science studies. Though concentrating on certain critical points, the ethics of science is still concerned with science as a whole and not just with its individual aspects or fields. To be sure, the ethics of science has its own view of science: it provides nothing more than a projection of the multi-dimensional phenomenon of science onto the ethical plane. The methodology and sociology of science, like any other fields of the science studies, produce their own projections of science. No one of them offers an integral image. To be integral this image requires a combination of all these projections.

This means that ethics, methodology and sociology of science should cooperate. Significant achievements of this cooperation can be expected not at the level where the ready research findings obtained in one of these fields are applied to other fields. The cooperation which urges constant revision of the initial premises in the study of science and recognition of their limited and one-sided nature is much more fruitful. The only road to the multifaceted and integral image of the whole without which we cannot improve our theoretical constructs or (what is more important) orient ourselves in the world in which science exists, lies in matching all these projections in our imagination.

3 Ethics of Science in Flux: From Appraisals to Regulations

Current developments in science, in its interrelations with other domains of social life as well as with society at large brought about essential changes in the agenda of ethics of science. Nowadays a lot of conferences, workshops, expert panels and many other meetings and many other discussions on these issues are carried out, a lot of decisions and documents are adopted. We have all reasons to say that now discussions on the *possibility* to make science an object of ethical scrutiny give way to the *necessity* to elaborate effective mechanisms for ethical regulations in different areas of scientific activity.

One of the most striking examples of these changes is the establishment of closer ties between different domains of social life. The rapidly growing intensity of these ties makes more difficult to draw distinctions between interacting domains. This means, by the way, that in many practical cases we cannot assert unequivocally in which domain—science, technology, politics … we are acting. So, the question "What is science?" becomes even more complicated. It is especially evident in the interrelations between domains of science and technology, analyzed by Agazzi.

After posing and treating the problem of the moral liability of science, as I described above, Agazzi came to the similar problem, that is, the problem of grounds, conditions and limits of moral liability with respect to technology. We can reformulate both of these problems in Kantian-like words, "How is an ethics of science possible?" and "How is an ethics of technology possible?" Agazzi draws a distinction between science and technology in this way: "Science is essentially a search for *truth*; technology consists essentially in doing something *useful*" (Agazzi 2004: 55). Then he comes to discussing the difference between science and technology in terms of aims. The distinction, according to Agazzi, "is based on the different *specific aims* of science and technology: the specific and primary aim of science is the acquisition of *knowledge*, while that of technology is the realization of certain *processes and/or products*". As we can see, here Agazzi also argues in terms of aims (or else of "specific and primary functions")—the function of science being the gain of *knowledge*, "while that of technology is the performance of certain *procedures*, or the making particular *products*" (Agazzi 2004: 55).

Stressing my general consent with this rather clear-cut and consistent approach, I nevertheless would resume here my own line of arguments. I would prefer to speak not so much about aims as about motives or attitudes of actors involved in these activities. There are two interconnected reasons behind such choice. The first is that, to my mind, we can speak about ends of such complex socio-cognitive entities as science and technology only in some indirect, figurative sense. The second reason: it seems to me that reference to motives rather than to ends allows us to take into account more definitely the fact that these characteristics of science and technology are *institutionally* prescribed to the actors.

To substantiate the distinction between science and technology, Agazzi turns to such concepts of ancient Greek philosophy, as *episteme* and *techne*. "*Episteme* focuses attention… on the *truth* of what is known … it concerns *pure knowledge*", whereas "with *techne*, the focus is on efficiency", it concerns another type of knowledge, "*knowledge of doing or making*" (Agazzi 2004: 57). So, the distinction takes the form of the contraposition between to know and to do.

In this context the author introduces a thoroughly elaborated and profound distinction between technique and technology. According to Agazzi's clarification, technique is "essentially the competent application of certain *know-how* attained through the accumulation and transmission of concrete *experience*… without necessarily being accompanied or supported by a knowing *why* such concrete procedures are especially efficacious" (Agazzi 2004: 56). This means that technique does not presuppose any kind of rationalization which can allow more effective performance of any "technical" activity: such type of positive change can take place only by chance analogous to a useful genetic mutation in biological organism.

As a great advancement of the ancient Greek philosophy and science Agazzi estimates the efforts they undertook in order to make this technical activity an object of special reflection aimed at questioning "why" one form of such an activity has turned out to be more effective than another one. So, he makes a difference between two types of knowing: know-why and know-how. The first one is exemplified by episteme, and the second by techne. Unlike technique, that is mainly empirical in its origin and developments, technology is based on the specific type of knowing, which is directed not toward objects of outer reality, but toward actors and their activities. It is such specific directedness that gave birth to technology, which is the dimension "wherein efficient operation is conscious of the reasons for its efficacy and is based upon them, that is, where operation is nourished by its grounding in theoretical knowledge" (Agazzi 2004: 57). Techne of ancient Greeks, however, was only a prefigure of technology in the more habitual modern sense. The latter appeared much later, under quite different historical circumstances:

> It is typical of Western civilization to have established a correlation between science and techniques: first by investigating *why* certain technical procedures were successful (that is, by looking for an *explanation* of this success capable of giving the *reasons* for it, as occurred as early as in ancient Greece) and secondly by *purposefully designing* instruments appropriate for the reaching of certain results as a consequence of the application of previously acquired scientific knowledge … It is this second step which has led to the establishment of *technology* as something distinct from simple techniques, and which accounts for the strict interdependence of science and technology, which could lead one to believe (mistakenly) that they are one and the same thing" (Agazzi 2009: 12.)

So, now we have a bilateral interrelation between science and technology: on the one side, developments in technologies open new possibilities for scientific progress; on the other side, scientific advances become starting points for technological innovations. According to Agazzi, such interaction generates something like a loop of positive feedback due to which the development of science pushes ahead the development of technology and vice versa.

On my part, I would propose a kind of mechanism which allows understanding why modern science is so effective in the domain of technology. It is generally accepted that one of the most important traits of modern science consists in its footing on experiment. Now, an experiment could be interpreted as a design and then construction of some device, equipment etc., which is created to test some propositions derived from a hypothesis under consideration. So, our experimental facility gives us the opportunity to perform some transformation—be it physical, chemical, etc.,—the result of which will coincide with a prediction made on the basis of our hypothesis. It is crucially important that this result of transformation will be *reproducible* once all parameters of the transformation are constant. In other words, our experimental facility allows us to control the processes of its functioning and hence to reproduce the transformation which draws our attention. Suppose then that the transformation reproducibly performed by our facility turns out—after its adaptive resetting—to be useable from some practical instead of theoretical point of view. This means that our experimental facility now can be understood as a prefiguration of a *technological* facility which is capable to reproduce some practically (and commercially!) meaningful and useful effect. So, as we can see, research, carried out in experimental science and aimed at getting empirically grounded knowledge, i.e. a *theoretical know-why*—in slightly changed context can take up the role of *technological know-how*.

Relaying on his thoroughly elaborated understanding of science, technology and interconnections between them, Agazzi turned to the issues of ethical appraisal. It is essential, however, that the demand for regulation of science and technology converts problem of reasonable possibility of ethical appraisal with relation to science and technology into a more acute problem, namely, the problem of searching for points in which these ethical appraisals can be applied in the most effective way. In this respect the scheme developed by Agazzi seems rather suitable. He proposes to single out in scientific and technological activities such constituents as ends, means, conditions and consequences in order to carry out ethical appraisal of various scientific and technological projects.

"... the *first* step in the evaluation of the *moral* quality of a human action,—writes Agazzi—is the consideration of *the end* of this action" (Agazzi 2009: 5). Then he draws a distinction between subjective ends (based on actor's intentions) and objective (or *intrinsic*) ends of an action or activity. To my regret, I was not able to find out an explanation of the concept of the objective end as an intrinsic one; I guess Agazzi had in mind something close to institutionally determined, as I mentioned earlier. In any case, judgement on moral acceptability of an action, according to Agazzi, must be based on appraisal of this objective end. From this point of view, taking into account that "the specific aim of pure science to be the search for truth, it is clearly immune from moral objections in itself" (Agazzi 1988: 52).

With regard to the ends of pure science the only meaningful area of ethical assessment consists of problems which in present-day science are related to research integrity. Therefore, with respect to the ends of activity, ethical appraisal can have essential meaning mainly in those cases when we are going to discuss phenomena of applied science or technology.

Turning to means, Agazzi stresses that under this respect pure science can be morally objectionable. He writes:

> In fact, at least in the case of the *experimental* sciences, truth cannot be discovered simply by thinking, or by watching, but requires the performance of operations, and this implies the *manipulation* of the object which is submitted to investigation. Since manipulation is *action* and not knowledge, even when the acquisition of knowledge is its explicit aim, it may well happen that a particular manipulating action not be morally admissible in itself" (Agazzi 2009: 5).

The *conditions* of the action are the next point of discussion. "They are similar to the means, but differ from them mainly in that the means are tools for directly reaching the end as a terminus of a certain action, while the conditions are something which makes the action itself possible, and thereby serve the end only indirectly" (Agazzi 2009: 9). And, according to Agazzi, "an action seeking the realization of a morally legitimate goal through the adoption of morally acceptable means still remains open to moral questioning until its conditions have been analysed" (Agazzi 2009: 9).

The last point of Agazzi's analytical scheme is the possible *consequences* of such activities as scientific research. He stresses that

> This is quite often the *only* point which is taken into consideration in many discussions concerning the ethics of science.... one could not underestimate the relevance of the consequences in the moral evaluation of actions, since it is an obvious moral principle that one is responsible for the consequences of one's actions, and therefore has the duty of trying to foresee them to the extent possible (Agazzi 2009: 10).

Agazzi's article "Why has science also moral dimensions?" started with such a question:

> Discourses on 'science and ethics' or on the 'ethics of science' have become rather frequent in the last few decades, so that they do no longer sound a little inappropriate, as they initially seemed to be. Is this simply the consequence of the fact that we have become 'accustomed' to such discourses and, therefore, more tolerant in the use of our language, or is there some deeper and serious reason? (Agazzi 2009: 1).

I gladly confirm this observation by Agazzi. Actually, discourses on "science and ethics" now became much more frequent than they were, say, 30 years ago. More than that, now these discourses are not just glance from outside, they turned to become integral part of the whole process of scientific and technological development.

Agazzi notes (Agazzi 2009: 7) that "the real situation of our time is the overwhelming presence of *technoscience*, that is, of that inextricable mixture of science and technology, that no longer allows one to overlook the concrete ends in the moral consideration of scientific activity". To my understanding, technoscience presents something more complex than a mixture of science and technology. Technoscience is a complex which includes also many diverse activities on the part of business, mass media, general public, NGOs, politicians… It generates various fields for clash and reconciliation of different interests through development of deliberative procedures and practices. More than that, it requests and creates social technologies for such a type of deliberative activities. It is through technoscience that ethical considerations enter the inner side of technological development.

I think that the excellent theoretical job which was and is done by Agazzi in the area of ethics of science and technology, which represents the "know-why" for this field of research, now becomes significant and called-for also at the level of practical "know-how", say, in present-day discussions on responsible research and innovations.

References

Agazzi, Evandro. 1988. Ethics and Science. In: *Logic, Methodology and Philosophy of Science VIII. Proceedings of the Eighth International Congress of Logic, Methodology and Philosophy of Science, Moscow, 1987*, ed. Jens Erik Fenstad, Ivan T. Frolov, Risto Hilpinen, 49-62. Amsterdam – New York – Oxford – Tokyo: Elsevier Science Publishers B.V. (North-Holland).

Agazzi, Evandro. 1989. Responsibility. The genuine ground for the regulation of a free science. In *Scientists and their Responsibility*, ed. W.R. Shea and B. Sitter, 203-219. Canton: Watson Publishing International.

Agazzi, Evandro. 1998. *Moral Dimension of Science and Technology* (Russian translation of *Il bene, il male e la scienza*. Milano: Rusconi 1992). Moscow: Moscow Philosophical Foundation.

Agazzi, Evandro. 2003. How can the Problems of an Ethical Judgment on Science and Technology Be Correctly Approached? In *Tecnoética. Actas del II Congreso Internacional de Tecnoética*, ed. J.M. Esquirol, 15-25. Barcelona: Publicacions de la Universidad de Barcelona.

Agazzi, Evandro. 2004. *Right, Wrong and Science. The Ethical Dimensions of the Techno-Scientific Enterprise*, ed. Craig Dilworth. Amsterdam – New York, NY: Rodopi.

Agazzi, Evandro. 2009. *Why has science also moral dimensions?* Report, presented in Moscow, April 22, 2009.

Frolov, Ivan, and Yudin, Boris. 1989. *The Ethics of Science: Issues and Controversies*. Moscow: Progress Publishers.

Petrov, Mikhail. 2006. *Philosophical Problems of "Science on Science". Subject-matter of Sociology of Science*. Moscow: ROSSPEN Publishers (in Russian).

Yudin, Boris. 1986. *The Methodological Analysis* asa *Trend in Science Studies*. Moscow: Nauka Publishers (in Russian).

Yudin, Boris.1988. The Ethics of Science as a Form of the Cognition of Science. In: *Logic, Methodology and Philosophy of Science VIII. Proceedings of the Eighth International Congress of Logic, Methodology and Philosophy of Science, Moscow, 1987* ed. Jens Erik Fenstad, Ivan T. Frolov, Risto Hilpinen, 79-90. Amsterdam – New York – Oxford – Tokyo: Elsevier Science Publishers B.V. (North-Holland).

Yudin, Boris. 2004. Knowledge, Activity and Ethical Judgment. In *Right, Wrong and Science. The Ethical Dimensions of the Techno-Scientific Enterprise*, ed. Craig Dilworth, 255-260. Amsterdam – New York, NY: Rodopi.

Yudin, Boris. 2010. Ethics of Science. In *Philosophy of Science: General Course: Manual*, 6th edition, 387-468. Moscow: Gaudeamus (in Russian).

Yudin, Erick. 1978. *The System Approach and the Principle of Activity*. Moscow: Nauka Publishers (in Russian).

Bioethics

Gonzalo Miranda

Abstract In his long and articulated path, Evandro Agazzi has often faced issues and problems that are now counted in the discipline that we call bioethics. It is probably a kind of practical "landing" motivated, on the one hand, by his constant search for objective truth in philosophy of science, and, on the other, by his deep desire to contribute to the existential questions of our time. We find in Evandro Agazzi some elements that make his studies on the issues of bioethics solid and enriching: his profound knowledge of science, his strong attention to the ethical aspects of human life, his epistemological realism, his anthropological vision open to the complexity and the transcendence of the human person. Among his contributions we must mention his solid reflections on the concept of the human person. Some other issues, like the status of the human embryo and the moral reasoning in bioethics are also analyzed in this paper.

1 Introduction

Evandro Agazzi is a philosopher of science, and, as such, he is best known and recognized. However, in his long and articulated path, he has often faced issues and problems that are now counted in the discipline that we call bioethics.

It is probably a kind of practical "landing" motivated, on the one hand, by its constant search for objective truth in philosophy of science, and, on the other, by his deep desire to give his contribution to the existential questions of our time.

It is also interesting to note that Agazzi has been an active member of the Italian National Committee for Bioethics, in which he has made important contributions, such as when he was head of the commission that worked on the issue of the status of the human embryo.

G. Miranda (✉)
Pontifical Athanaeum Regina Apostolorum, Rome, Italy
e-mail: g.miranda@arcol.org

M. Alai et al. (eds.), *Science Between Truth and Ethical Responsibility*,
DOI 10.1007/978-3-319-16369-7_22

We could, perhaps, divide his writings on bioethics into two major groups: foundational issues and hot-button issues in current debates.

In the first group we can mention, for example, the following themes: ethical anthropocentrism and the animal world; the human being as a person; nature and artificiality; pain and suffering; bioethics and dialogue; and cultural dialog.

Among the specific topics we can mention: the status of the human embryo, informed consent, the ascertainment of death, and ethics in biotechnology.

2 The Elements for an Important Contribution to Bioethics

We find in Evandro Agazzi, in his thinking and in his personality, some elements that make his studies on the issues of bioethics solid and enriching.

First of all, as a philosopher of science, he studied complex scientific issues for many years. He also studied the biological sciences, and was often involved in bioethical issues.

On the other hand, he has always demonstrated strong attention to the ethical aspects of human life, and therefore also of human behavior in science and in its applications. His well-known book entitled *The Good, the Bad and Science* (Agazzi 1992), is a clear example of his "ethical concerns".

In my opinion, the epistemological realism cultivated and promoted by Agazzi throughout his intellectual career regarding bioethical issues is also very interesting - most recently expressed in his recent book *Scientific Objectivity and Its Contexts* (Agazzi 2014).

His anthropological vision seems equally fruitful to me, far from every kind of reductionism, open with intellectual honesty to the complexity and the transcendence of the human person.

As in all other disciplines he addresses, you can see the intellectual strength and rigor that also distinguish Evandro Agazzi when he deals with the issues of bioethics.

3 What Is Bioethics?

It's known that bioethics is a discipline born in the early seventies and still insufficiently mature with regard to its identity and its epistemological status.[1] Agazzi studies this theme when he proposes his reflections about the promotion of a shared bioethics (cf. Agazzi 2011).

[1]Francesco D'Agostino, a recognized Italian expert in bioethics, writes: "[...] even as an academic discipline, or more generally, as a mere field of interdisciplinary reflection, bioethics [has not] acquired in recent decades that epistemological status, unified and reasonably consolidated, which was, and continues generally to be considered desirable by all" (D'Agostino 2002).

3.1 The Definition of Bioethics

With his usual intellectual stringency, the author offers a definition of bioethics before he analyzes the characteristics of the discipline: "Bioethics is the study of moral problems that arise in the context of the biomedical sciences and their applications" (Agazzi 2011).

It seems to me that it is a good definition, for the fact of being very concise and at the same time enough informative and explicit.

However, I consider it appropriate to clarify that in referring to "moral issues," I do not think he means that the bioethics has a "negative" nature, which pays attention only to the "problems." In fact, while bioethics addresses the ethical problems, it should also be used in a positive and proactive way to foster the respect for human life and human dignity, as well as the responsible stewardship of the natural environment that surrounds us.

We will see later how, according to Agazzi, bioethics, like all ethics, must be based on the promotion and protection of *values* or *purposes* that deserve to be pursued for themselves. I think then that the term "moral problems" should be understood in an open way, as they are referred to in the "moral dimension" in the field of life sciences. We speak of "moral problems", therefore, in the sense of "moral issues".

The expression "biomedical sciences" may seem to reduce the scope of the analysis strictly to the medical practices. Agazzi, however, also addresses issues relating to, for example, respect of animals and other living beings, which means that he doesn't intend to reduce bioethics to *medical* bioethics.

3.2 The Characteristics of Bioethics

Our author presents some of the characteristics of bioethics, starting with its *interdisciplinarity*.

It is a characteristic perhaps more "chatted about" than understood. In fact, I think it's still a task of bioethicists to investigate what it means and to understand how to apply it in a methodologically rigorous and effective way.

For Agazzi, the interdisciplinary nature of bioethics is not reduced to the confluence of different disciplines in the analysis of its various issues:

> [...] the interdisciplinary nature of the treatment of bioethical issues arises from the fact that to discuss and search for a correct solution of each particular problem in a very specific field, you have to take into account the views and insights of different disciplines [...] The contribution bioethics has made to the culture of our time is that these 'points of view' [medical, biological, but also psychological, economic, legal, religious, and moral at the end] are not something optional and almost annoying that 'accompanies' reflection ... On the contrary, these various considerations weigh (and have the right to weigh) considerably in decision making (Agazzi 2011).

Perhaps here we can recall the concept of "system," so dear to Agazzi (cf. for example Agazzi 2008): bioethics could be seen as a complex system composed of many interconnected parts, while seen at the same time as a subsystem of the global system of culture and society in which it participates along with other disciplines.

Another feature of bioethics is its *moral specificity*. It seems to me that it is important to note that Agazzi underlines the strictly moral dimension of bioethics. Often bioethics is conceived and presented as a mere "procedure," a methodology of analysis and discernment for a purely deontological or legal kind of decision-making. Agazzi strongly affirms that bioethics is also called to operate a discernment of certain behaviors in order to understand if they are allowed, prohibited or due.

In this regard, he reminds us that even if the "prohibitive" feature is often present in bioethics (as in any ethics), it is not its principal dimension (and here we see his understanding of bioethics as a discipline which is also propositive, as I mentioned a bit earlier).

4 The Human Person

Perhaps one of the greatest of Agazzi's contributions to bioethics comes from his reflections on the concept of the human *person* (cfr. Agazzi 1993).

The author comments on the attempts at separating the concept of "person" from that of "human being". He then presents a detailed and punctual examination of the concept of "person" and the various logical possibilities of such a separation.

Starting from the premise that "every kind of separation is a denial", Agazzi discards the logical possibility of applying a *simple* or *absolute negation* ("man is not per se a person"), since pure negation does not say anything about the reality to which it refers. Similarly, we must exclude the logical possibility of a *dyadic negation*, understood as the opposition of two entities, since it is not possible to conceive of 'man' and 'person' as opposites.

Then he shows that we can perfectly understand that all human beings are persons, even those lacking certain distinctive characteristics of the person, like consciousness, if we consider what he calls the *triadic denial*, which is the negation involved in *deprivation* as well as in *potentiality*.

Regarding *deprivation*, one must consider the three elements that make up the deprivation of any one property: the holder (ontological substrate) according to its nature; the property itself (for example, consciousness); the factual possession of that property. In this way we can understand that

> Deprivation removes only the possession, but never the ontological substratum of the holder, and not even the property itself, which (being nothing in itself, but simply being real in the ontologically real substratum) remains necessarily that property only in reference to the capacity of that substratum (ibid.).

Potentiality, Agazzi explains, also involves a form of triadic denial "since it explicitly regards a property that is still *not* possessed by a certain being, but that *should* be possessed by it at a later time under normal circumstances" (ibid.).

This leads to the logical conclusion that "man is a person *in act* that, in the different stages of his development, is continually *in potentia* with respect to the full realization of his faculties and properties, including the conscience" (ibid.).

And this way, Agazzi can say:

> The dissolution of the artificial separation of man and person forces us to relocate at the bases of bioethics the respect for the man, that is for all human individuals, simply because they are human beings, without the pretentious exception that some of them may not deserve this respect because they are not persons (ibid.).

5 On the Status of the Human Embryo

The above considerations and statements about the concept of the person would seem to bring us to the conclusion that the human embryo should be considered a human being in all respects, and therefore a person, from the moment of fertilization. But it is not so for Agazzi, since it's not clear to him whether it is really a *human individual* from the moment of conception.

Agazzi clearly presents the problem of the identification of the moment of *individualization* of the human embryo when he presents the works and the document of the Italian National Committee for Bioethics on the theme: "Identity and Status of the Human Embryo" (1996; Agazzi 2005).

He explains how they came to the unanimous conclusion that the human embryo is a person from the moment in which it is a human individual; but he also shows that there were many difficulties with regard to the status of individual in the first two weeks of development: the characteristic of totipotentiality of its cells, the possibility of twinning, the lack of the primitive streak.

In another text he elaborates on this position of "perplexity," arriving at the conclusion that the human embryo is not a human individual during the first two weeks of its initial development (Agazzi 2007).

It is above all the possibility of the phenomenon of twinning and the formation of "chimeras" (the fusion of the cells of two embryos into a single one) that bring Agazzi to his position:

> [...] It is difficult to understand how a human individual can develop, dividing his individuality into two or more individuals or, symmetrically, how two different human individuals can be compacted into a single individual. This difficulty would disappear if we admit that during the period under consideration, the individual entities gradually formed are only cell systems that can lead to different results [...] (ibid.).

The key point is that of the *individual identity* that we can better grasp through the criterion of *re-identification*: the recognition of the same individual despite being in different circumstances and conditions. Proposing the thought experiment of twins that return to the past progressively in their minds, each of them could say, "It was me," but only up to the point at which they were established as two separate individuals: "[…] because the individual entity existing before that time gave rise to his embryo as well as that of his twin brother and, for this reason, s/he can say, 'It was me'" (ibid.).

The Author proposes, therefore, the solution of accepting the term (and reality) of the so-called *pre-embryo*: a "biological structure" resulting from fertilization that will give rise to the *embryo* around the fifteenth day of development.

In this way, he concludes: "The human embryo is a full individual (or person), but this ontological state can not be attributed to human pre-embryo."

I have summarized Agazzi's explanation in a fairly detailed way, as, to me, this seems to be a key point in bioethics today. For the same reason I will now make some critical considerations.

First of all, it must be noted that the term "pre-embryo," actually used in certain bioethical and scientific documents, has been "disgraced" for several years now, especially in the scientific field. Among the various possible quotes or testimonials, I recall here only that of R.O'Rahilly. It's especially interesting if you keep in mind that O'Rahilly is the embryologist who established the "Carnegie Stages of Human Embryological Development", used for many decades now by the *International Nomina Embryological* (now the Terminologica Embryologica) *Committee* which determines the scientifically correct terms to be used in human embryology around the world.

In the third edition of his textbook *Human Embryology & Teratology* (O'Rahilly and Müller 2001), he gives five reasons why it is not appropriate to use the term "pre-embryo"; one of them is simply that "it may convey the erroneous idea that a new human organism is formed at only some considerable time after fertilization" (O'Rahilly and Müller 2001).

From the philosophical point of view, it seems to me that the concept of *system* can illuminate the issue of the individualization of the embryo. I am a person because I am a human individual; and I am an individual because my body is a bunch of cells which are structured into a whole, which is a living, *organic system*. Since when am I an organic system structured as a whole? Embryology shows me that it's from the time of fertilization of that oocyte by that spermatozoon. In that moment, a "biological structure" is formed, which is a real living organism, a living organism of the human species.

Let us do another thought experiment. Imagine we pull out the four morula cells contained within the zona pellucida at a certain time. We replace those cells with four amoebae. Suppose that the amoebae could multiply as embryo cells can. We know, however, that those amoebae will never form, as the cells of the embryo do, the various biological structures of what we call a blastocyst (about five days after fertilization), with the embryoblast and the trophoblast, with the cells already differentiated into "polar" and "non-polar", leaving a large empty space in the center, etc. The amoebae would not do what the embryonic cells do because they would be merely an unformed cluster of cells, while the embryo is indeed a cluster of cells, but it is organically structured in a organic living system.

In 2001, a team in Cambridge led by K. Piotrowska demonstrated experimentally that at the very moment of the penetration of the spermatozoon into the oocyte, a morphological axis is established, from which the cells of the embryo multiply and differentiate in a precise systematic order (Piotrowska et al. 2001; Piotrowska and Zernicka-Goetz 2001).

Certainly, we must take the phenomena that Agazzi and others suggest seriously, that is that we cannot speak of individual identity in the earliest stages of embryo development, especially regarding the phenomenon of monozygotic twinning (and the possible formation of a chimera).

In my opinion, though, none of the reasons given, nor all of them together, demonstrate that the embryo is not an individual from the moment of conception. I will speak here only on twinning.

It is true that we do not know well the causes and mechanisms of monozygotic twinning, but we can say that it is due to the separation of some cells, which, still being totipotent, can generate a second individual, genetically identical to the first. Certainly we know that this can be caused artificially, as is frequently done with animal embryos from the fifties, with the technique of splitting. It has being also done with human embryos, in the famous experiment by Hall et al., at the George Washington University School of Medicine in 1993 (cf. Gourdon and Byrne 2002).

Actually, reproduction by twinning or by gemmation is very frequent in many species. In the case of many plants and some animals, you can separate one part of the organism from which a second plant or animal derives. This happens with the rose plant, the willow, and with some annelids (in a natural way, by binary fission, or even after forced cutting, for example in the case of some species of earthworms). If a branch of a rose plant generates a second plant, should I conclude that the original plant was not a living individual? Is this earthworm, from which a part comes off generating a second earthworm, not a living individual?

Human twinning can be considered a type of reproduction by fission. Perhaps thinking that a human being can derive from the body of another has a slightly odd emotional effect on us, but I do not see why we cannot understand it rationally.

Let's take the abovementioned thought experiment proposed by Agazzi again. I think that if I were a monozygotic twin, going back in time in my mind I would say, "It was me." Since the beginning of my existence, I have been the same living organism that I am now, with this size and these characteristics; I do not know if it was originated by the fertilization of an oocyte by a spermatozoon or after the separation of a part of the body of my twin brother; but that human living organic system "was me".

6 Moral Discernment in Bioethics

We might perhaps think that if we could convince Evandro Agazzi the validity of the above reasoning on the status of the human embryo, he would conclude that we must therefore always respect the life and integrity of every human embryo.

But it is not the case. He states that the determination of the ontological status of the embryo "is important but not decisive for the solution of moral issues relating to the management of embryos", for he says that "when it comes to concrete situations [...] we must realize that always a *plurality of values* are at stake", and sometimes we find ourselves in situations of *conflict of values* with the corresponding *conflict of duties* (Agazzi 2007).

In the article just quoted, the author recalls the distinction introduced by D. Ross (cf. 1930) between *absolute* and *prima facie* duties. He explains that "A *prima facie* duty is what imposes on us a moral obligation, but may, in a given concrete situation, yield to other duties or rights that would be considered as morally more obliging" (Agazzi 2007).

In another text he proposes another similar distinction: one between the understanding of values as "sacred" and conceiving them as "excellent." Comparing the concept of excellence to that of sacredness (or absolute), he explains that

> [T]he character of excellence [...] can be defined by saying that a given value represents a good of exceptional importance and therefore must be protected and promoted in every possible way, but at the same time may be subject to certain restrictions for a good reason (consisting essentially in the necessity of making such a value compatible with other values, those being also "excellent") (Agazzi 2011).

Here we cannot do an in-depth analysis of this complex and central issue of moral philosophy. In a way, we are faced with the classic debate about the existence of *moral absolutes*, which developed also within the Catholic Moral Theology. In fact, Pope John Paul II felt the need to intervene on the foundations of morality, and specifically on this point, with his encyclical *Veritatis Splendor* (II 1993). Reaffirming the existence of some "intrinsically evil" acts due to the immoral object freely willed by the subject, he explains that

> In the case of the positive moral precepts, prudence always has the task of verifying that they apply in a specific situation, for example, in view of other duties which may be more important or urgent. But the negative moral precepts, those prohibiting certain concrete actions or kinds of behavior as intrinsically evil, do not allow for any legitimate exception (Johannes Paulus 1993).

As I said before, Agazzi clearly states that morality involves the perception of what is allowed, prohibited or dutiful. Indeed, the experience of morality, of moral value is the experience of having to do or not do, irrespective of personal desire or sentiment. The choice of the action to take or avoid should not consist merely in opting for a value subjectively considered superior to others. Such a choice would not present itself with the character of moral obligation, which is characteristic of many moral choices, especially when it comes to "negative duty" (not to do this).

When we speak about "moral absolutes" we do not refer to absolute values but rather to the experience of the "absolute duty" not to act in a way that contradicts a value considered important. It is the absoluteness of the moral experience, the experience of the obligation in freedom: if I am convinced that this action is morally wrong, I *can* choose it, but I *cannot* make it a good choice (cf. De Finance 1984; Miranda 2001).

Agazzi refers to bioethics as public ethics: “All this is true as long as we do not forget that bioethics is basically a form of public ethics and, therefore, it cannot claim to morally bind the individual conscience, which maintains its indestructible freedom” (Agazzi 2011).

In my opinion bioethics should not be considered solely, or even primarily, a public ethics. In addition to its social and normative application (which gave rise to the so called “bio-law” jurisprudence), bioethics should try to illuminate the individual consciousness (doctors, patients, researchers…). The clinical bioethical committees have precisely this function.

Anyway, I think that even in the public sphere, if a person is convinced that a certain behavior should not be allowed by society because it goes against some inviolable good, he or she can (and perhaps should) speak out loud and clear. That person may give her reasons and may also act against the mainstream, trying to persuade others and possibly even to provoke legal changes. These are the individuals who, opposing vigorously certain aspects of the predominant ethos, provoke profound changes for a better society. Take, for example, the cultural and legal struggle against slavery carried out in the United States, in the mid-nineteenth century, by people like Abraham Lincoln; or in the twentieth century by of Martin Luther King, fighting against the discrimination of black people in that country.

What would have happened if those people had simply accepted that behavior and those laws because they knew that they were considered good for the majority, according to their “excellent” values?

7 Conclusion

With these last considerations we went back to the question of the identity of bioethics, mentioned at the beginning this text. I have presented only some of the many bioethical issues addressed by Evandro Agazzi, although I think they are the most significant. Also, I have clearly presented my critical remarks about some of his positions.

But, as I said, Agazzi has given and continues to offer an important contribution to this field called bioethics. Still more, he is giving an important contribution to the cause of the pursuit of scientific and ethical truth, fostering the respect for human dignity.

References

Agazzi, Evandro. 1992. *Il bene, il male e la scienza*. Milano: Rusconi.

Agazzi, Evandro. 1993. L’essere umano come persona. In *Bioetica e persona*, ed. Evandro Agazzi, 137–157. Milano: Franco Angeli.

Agazzi, Evandro. 2005. Lo statuto dell’embrione umano. In *Il Comitato Nazionale per la Bioetica: 1990-2005. Quindici anni di impegno*, ed. Comitato Nazionale per la Bioetica, 7–16. Roma: Presidenza del Consiglio dei Ministri, Dipartimento per l’informazione e l’editoria.

Agazzi, Evandro. 2007. El estatus ontológico y ético del embrión humano. In *Dilemas de bioética*, ed. Juliana González Valenzuela, 109–133. México: Fondo de Cultura Económica.
Agazzi, Evandro. 2008. *Le rivoluzioni scientifiche e il mondo moderno*. Milano: Fondazione Achille e Giulia Boroli.
Agazzi, Evandro. 2011. En búsqueda de una bioética compartida. In *En busca de una bioética compartida*, ed. Gerardo Martines Cristerna, 17–38. México: Ediciones Hombre y Mundo.
Agazzi, Evandro. 2014. *Scientific Objectivity and Its Contexts*. Springer.
D'Agostino, Francesco. 2002. Prefazione. In *Introduzione alla biogiuridica*, ed. Laura Palazzani, VII–VIX. Torino: Giappichelli Editore.
De Finance, Joseph. 1984. *Etica generale*. Cassano Murge (Bari): Tipografica Meridionale.
Gourdon, John B., and Byrne, James. 2002. The history of cloning. In *Cloning*, ed. Ann McLaren, Council of Europe Publishing.
Joannes Paulus, P.P. II. 1993. Litterae Enciclicae cuncti Catholicae Ecclesiae episcopis de quibusdam quaestionibus fundamenta-libus doctrinae moralis Ecclesiae. *Acta Apostolicae Sedis*.
Miranda, Gonzalo. 2001. *Risposta d'amore*. Roma: Logos Press.
O'Rahilly, Ronan R., and Fabiola Müller. 2001. *Human Embryology & Teratology*. New York: Willey-Liss.
Piotrowska, K, and M. Zernicka-Goetz. 2001. Role for sperm in spatial patterning of the early mouse embryo. *Nature*: 517-521.
Piotrowska, Karolina, Wianny, Florence, Pedersen, Roger A., and Zernicka-Goetz, Magdalena. 2001. Blastomeres arising from the first cleavage division have distinguishable fates in normal mouse development. *Development*: 3739-3748.
Ross, William David. 1930. *The Right and the Good*. Oxford: Clarendon Press.

Religious Faith, Natural Science, and Metaphysics

Juan José Sanguineti

Abstract This chapter presents Agazzi's considerations on the relationship between science and religion. In the first part I discuss Agazzi's analysis of natural and empirical science, which can be seen as a source of the modern and systematic conflict with religious belief, if science is interpreted according to logical positivism. This interpretation mainly means taking the scientific perspective (i.e. physical objectivity) as closed, precluding any metaphysical view. In the second part I focus on the positive relationship between metaphysical insight and religious faith, which is a necessary condition for the harmony between science and religion. I will especially highlight Agazzi's notion of rational faith as a helpful element in a correct understanding of the problem tackled in this chapter. Faith and reason are necessary interactive elements both in natural science and in philosophy or metaphysics.

1 Introduction

A considerable part of Evandro Agazzi's contribution to philosophy is devoted to the problem of the relationship between science and religion. This topic is very important and there is a long-standing and insightful tradition in many philosophical and scientific circles in the United States, Europe and other countries, dedicated to the challenge posed to religious belief (Christian, Hebrew, Muslim or other creeds) by modern science and technology.

The problem can be viewed from many different perspectives. In English-speaking cultures, where science is often taken more seriously than in other cultural domains, an academic concern for the theological implications dictated by certain scientific views is widespread in many authors and initiatives. One might recall, for instance, the efforts of the Templeton Foundation to foster research surrounding the big questions posed by science, placing them within the background of religious

J.J. Sanguineti (✉)
Pontifical University of the Holy Cross, Rome, Italy
e-mail: sanguineti@pusc.it

M. Alai et al. (eds.), *Science Between Truth and Ethical Responsibility*,
DOI 10.1007/978-3-319-16369-7_23

faith and theology. This special attention to the positive relationship between science and religion flourished in many other international academies and scientific groups in recent decades, in spite of the misunderstanding and tension which are at times translated into ideological positions, especially in the field of education.

With regard to an analysis of this issue, Agazzi can be said to hold the perspective of philosophy of science. If traditionally science was seen as an intellectual enterprise connected to wisdom and contemplation of God, why in modern times is science often associated with atheism or viewed as a rational enterprise that must inevitably conflict with religious faith?

This chapter presents Agazzi's considerations on this topic according to two main points. First, I discuss what I consider to be Agazzi's main contribution to the problem, namely his analysis of modern empirical and natural science which often became a source of systematic conflict with religious beliefs. Second, I focus on the relationship between metaphysical insight and religious faith, which in my view is a necessary condition for the harmony between natural science and religion. I will especially highlight Agazzi's notion of rational faith as a helpful element in a correct understanding of the whole of these points.

2 Modern Natural Science and Empirical Rationality

In an attempt to gather a general overview of the history of the conflict between science and religion, we could say that the central point, and Agazzi would agree, lies in the birth of modern natural science and its method. The new methodology, initiated by Galileo in a more explicit way than by other authors of his time, consists of a dismissal of the Aristotelian account of natural science as a search of the essence or nature of things, which, once grasped through a definition, would enable one to derive properties of things as a set of logical or analytical consequences. According to this framework, there is no difference between science and philosophy, contrary to our current distinction between physics and philosophy of nature, biology and philosophy of biology, and so on.

There is no place for this distinction in a classical account of science (Sanguineti 2002). Galileo's approach, instead, altogether renounces the possibility of uncovering the essence of physical things, and remains satisfied with the knowledge of some empirical properties adequately understood in terms of mathematical functions.

> For the first time in an explicit way, Galileo states that – at least in the case of objects of nature ('natural substances') – the pretence to answer the Socratic question is always in vain and illusory; 'testing the essence' is a task which he renounces, limiting himself to a more humble yet approachable goal – that is, the knowledge of 'some properties' (*affezioni*) of the natural things or, as we would say today, an accurate ascertainment of how some natural phenomena carry on (Agazzi 1974: 10).[1]

[1]This and subsequent translations from Agazzi are my own.

Some authors, like William Wallace, tried to show that the Galilean and Newtonian physical-mathematical discoveries might be translated into an Aristotelian syllogistic account of science as displayed in the *Posterior Analytics* (Wallace 1983, 115–143). This can certainly be accomplished, but it is purely a formal reconstruction. Nevertheless, Aristotle as a real scientist, especially in biology, does not follow an overly rigorous method such as the one delineated in the *Posterior Analytics*, which is elaborated more in accordance with geometry and some aspects of his physics. The systematic syllogistic method seeking to analytically derive all the properties found *per se* in a substance (which is the object of any scientific endeavour) from the essence as inductively grasped, is more typical of a scholastic method of working in science, against which Galileo battled throughout his scientific career.

Agazzi rightly remarks that Galileo follows an empirical method without concerning himself with pursuing essences, inaugurating the modern detachment of physics from philosophy. In doing so, Galileo was in a certain sense anticipating the times. His method is scientific in a modern sense, but two centuries were to pass before a clear distinction between experimental physics and philosophy would be reached. The ideal of a deductive science grounded in the intuition of essential properties still lingered in the classical account of rational mechanics. Agazzi is fully aware of the epistemological problems (phenomenalism) that emerge from the new methodology in physics (Agazzi 1974: 3–32). The problem of essence and substance in the new physical configuration in some sense justifies Kant's efforts to interpret the new science in a transcendental (idealistic) way.

The conflict with religion, at least at the epistemological level, appears whenever modern natural science is seen as a kind of scientific research that must exclude any illusion of intellectually attaining the realm of something that extends beyond matter, matter here being understood as a reality characterized by empirical properties whose intrinsic lawlikeness could be made an object of science. Aristotle looked continuously in his ontological endeavours for *separate substances* (separated from matter), which in other terms are the purely immaterial substances—such as separate intelligences, and above all God as transcendent Intelligence moving the cyclic universe. This is precisely what is precluded by modern empirical science according to the tenets of the conflict in question. Explanations, regardless of any mathematical and theoretical elements they may contain, can never make the leap to separate substances. Human reason must be confined to this physical world. Hence, the way to atheism is opened and the conflict with religion is elaborated in a systematic rather than a casual way.

Thus far, I have highlighted Agazzi's claim that a major occasion for atheism was modern empirical natural science, which is understood according to a rigorous positivistic closure when the moment arrived to establish the new scientific method. Historically speaking this was largely successful, a new normative standard for scientific operations, a standard purified—such was the pretence, though never fully satisfied—from all immaterial additions (called metaphysical or supernatural). This was mainly the task of logical neopositivism (the Vienna Circle).

The claim was that only empirical terms (i.e. referring to observable properties) are fitting elements that might integrate existential statements and keep in check unobservable hypotheses likewise referring to existent objects in order to explain their spatio-temporal properties and relations. Within these constraints, one can build a universe of discourse, or a special ontology constituted by objects, properties and relations, which is in turn the object of natural science, and more precisely the object that defines it as such.[2] The whole of this universe of discourse is by definition the field of objectivity of a given science (Agazzi 1974, 1981, 1983, 2014).

Agazzi and other non materialistic philosophers (Maritain 1932) maintained that this particular physical objectivity originates from the selective or abstracting operation of the mind that encapsulates an intellectual vision of the world within certain parameters, more or less like our senses perceive sensible objects from a certain point of view (light, sound, etc.). As a consequence, physics must be seen as a partial science, a partial way of viewing things and therefore of explaining phenomena and events of the natural world.

Positivists and materialists, instead, take this abstraction to be the only fruitful way to study natural objects and, consequently, to create technical objects. Indeed, only the physical Galilean approach enables us to physically intervene in nature, to produce changes according to its own laws. Empirical and experimental science, then, is the necessary condition for developing technology. It is no wonder that technical progress in Western civilization is due to this approach, unattainable by philosophical speculations about nature *qua* nature. It is likewise no wonder that nothing can be established by experimental views and techniques, as Agazzi often claims, when trying to discover what values might guide humans in their use of technology.

Modern science as we know it was born from a kind of restriction of our rationality. The decision to remain confined within the sensible realm of the universe transforms what is partial into something assumed to be the whole, yet this should be understood with an awareness of its partiality. People working in this area—scientists, professors, students—acquire an empirical scientific habit (habit in the Aristotelian sense), forgetting that their vision is partial, unconsciously treating science as philosophy, and in this sense accepting it as a kind of complete worldview.[3] This attitude can be called *scientism*, a form of reductionism.[4]

> Reductionism, lurking at all times, has been refuted with many proofs in the field of science, yet it survives in a residual way, though perhaps less consciously, whenever one claims that the *totality* of problems subject to human reason should be *reduced* to those which the sciences can properly address. Thus, scientism is the most radical form of

---

[2]The definition of such an ontology must contain the properties that are considered observable and the basic operational procedures that render observations public or intersubjective, normally through instrumental measurements.

[3]Not materially total, of course, but formally, in the sense that there is no other possible perspective to obtain a real knowledge of things.

[4]Scientism is tricky because it is a hidden philosophy. Scientistic authors, of course, deny this label. They take advantage of the prestige of science and make philosophical assertions beyond the possibility of the scientific method, such as "nothing can be true outside physical science", and many other conclusions concerning man, freedom, reason, truth, and the mind.

> reductionism inasmuch as it is not capable of eliminating the need of problematizing the whole, but it thinks that the whole is identical with an horizon which is actually partial, despite how vast it might seem (Agazzi 2007: 94).

This restriction is not supported by any evidence. It is a choice, perhaps emerging from a materialistic faith, or just from habit and *routine*, but not from rational knowledge. It renders an impression that anything alleged to exist out of the empirical worldview is incompatible with science (for example, human freedom, self-consciousness, values, the human person, and most certainly God). The reasons invoked according to this empirical closure beg the question. If one believes that nothing exists outside the realm of the empirical, the concrete reason for this is only a decision to remain closed within self-imposed empirical limitations. Thus, the epistemological empirical choice is consonant with ontological materialism or naturalism (which is today's name for materialism), the former being the basis for the latter.

The problem I address in this section is that of the root of the modern conflicts between science and religion according to Agazzi. I tried to assess up to what point the new empirical, or experimental, method typical of modern science may be considered a systematic source of this conflict, and something that ultimately is an occasion for atheism. The conclusion is that this is true only if one takes up this method as necessarily closed or detached from the metaphysical perspective.

Within this framework, it is possible to single out particular areas of conflict, as Agazzi does with respect to the Galileo affair and evolutionism. The specific conflict surrounding Galileo is historically important, but from a theoretical point of view, today it appears more as a misunderstanding—theological and exegetical—than a systematic problem, as Agazzi rightly suggests. Galileo's judges were surprisingly obtuse in linking a particular cosmology to Christian faith, namely to the Scriptures.

The problem of evolution is much more serious in that it touches upon origins—namely, the origin of the universe, of life and of man. When pursued, this topic naturally suggests a global view of nature and a certain doom, opening up to the question of sense, which is typically philosophical. Science concerning the origins of everything, if this is possible, is not too far removed from philosophical views and questions. The tricky feature of scientism as mentioned above dramatically emerges in the topics generally considered as not easily tractable in the purely scientific sphere, according to the strict empirical method as such. The need for a metaphysical interpretation seems crucial to these highly specialized scientific areas, such as cosmology,[5] in order to obtain an overall insight of the problem at hand (the sense of the universe, the sense of time, the place of man in cosmos). Thus, possible conflicts between evolutionary theories, scientific approaches to animal life, neuroscience, and religious faith may naturally arise, perhaps in a

[5]Agazzi touches upon the problem of the scientific status of cosmology, not corresponding to the Galilean constraints, in his paper on cosmology (Agazzi 1991). This status introduces philosophical assumptions in modern cosmology and suggests an enlargement of the notion of scientific rationality, far from the positivistic account.

confusing way when a clear distinction or a correct relationship between science and philosophy is not recognized.

Rather than being an isolated conflict, the general overview of nature emerging today from the whole of the natural sciences, and particularly biology (including evolutionary theories, neurobiology, bio-computational approaches, ethology, etc.), offers a framework which already belongs to the popular understanding of nature in our scientific culture, and one that in some way defies the theological account of divine creation. This overview is philosophical in the broad sense of the word. This synoptic panorama must be carefully studied, and it is not impossible to insert it into the Christian picture of a universe created by God in which human persons hold a very special place due to their reason and freedom.

Perhaps many people today do not care too much about epistemological restrictions, but in doing so they fail to capture the possible metaphysical implications in natural sciences. So the crux of the conflict between science and religion is simply reduced to the question about the need, rather than the possibility, of acknowledging a realm of existential being beyond matter, namely the spirit (the human spiritual dimension, or human soul) and God.

This need has collapsed for some persons for practical reasons. As Agazzi observes, human beings traditionally turned to God in order to obtain material security and spiritual relief, but many of these needs seem to be satisfied by modern technological and scientific achievements (medicine, economy, neurobiology, psychology) (Agazzi 1983: 122–124). So the sense of our dependence on God and of our impotence to solve our great problems has decreased on account of this new era of technical progress.

> While people of a time long past had the impression that recourse to God was necessary to solve their problems, science seems today to exonerate them from this need (Agazzi 1983: 123).

But a more careful analysis of the human situation in our contemporary world does not confirm the claim that science affords a solution to all human problems. The impression of not needing God is ambiguous. In the absence of God as a transcendent and personal being, people tend to replace him with some absolute dimension of life which in turn becomes the object of faith (science, nature, man himself). They thereby run the risk of being disappointed, which is often the gateway for depression and nihilism (Agazzi 1969: 178).

> Neither the religious dimension nor the problems to which it responds can be suppressed. If people are no longer able to find the answers to their problems in a positive historically determined religion, they will look for it elsewhere, for instance, in various ideologies which, in that moment, accomplish for them the role of the 'faith that saves', the religious faith (Agazzi 1983: 162–163).

Summarizing the points made in the previous pages:

1. The historical root of the conflict between science and religion lies, according to Agazzi, in the new experimental methodology of modern natural science when this method is taken as closed and precludes by definition (and decision) all possibility of knowing existential relevant truths. This decision is formally present in logical positivism, wherein God seems to be scientifically excluded.

2. It can be shown that this is a route which leads to self-refutation. The empirical way of perceiving things is partial. Taking this partial and abstract view as complete is the very definition of scientism, which is a kind of reductionism. Scientism is logically inconsistent. Philosophy, beyond the empirical closure, is not eliminable.[6]
3. However, the overall worldview emerging from natural sciences today is nevertheless impressive. Due to an overwhelming practical (technological) dimension linked to that worldview, which efficaciously deals with many human needs, many people today think that there is no need for God, nor religion, or that God is a pious invention subject to a scientific explanation (perhaps with recourse to psychology or neuroscience).
4. The last point (n. 3) easily overlooks the logical limitations of science and techno-science (n. 2). Of course, there are many other limits—anthropological, ethical, ecological, even physical—which I have not considered in these pages. In any event, when one presents a philosophical proposal that attempts to go beyond natural science and technology and reach a more metaphysical level—hence, open to the human spirit and to God—he immediately finds himself in conflict, according to the difficulties mentioned in n. 1. In more practical terms, people (scientists, professors, etc.) who on a daily basis deal with pure scientific conceptual instruments lack the conceptual instruments to deal with the immaterial, even if in some occasion they do feel the need of going beyond matter in order to speak of something immaterial. They only know scientific ontology, and all other ontology appears to them as inappropriate and awkward.

I would like to add a further historical point that contributed to the modern conflict between science and religion. Toward the late 18th and early 19th centuries, Enlightenment and Positivism promoted a strong confidence in the power of pure reason—in first place in philosophy, later in science—and spread a hostile attitude towards religious faith. Modern atheism based on science owes much of its existence to this historical trend. Many contemporary "scientific" atheists—such as D. Dennett, R. Dawkins or J.P. Changeux—today propose to return to the spirit of Enlightenment. But even in post-modern ideologies—ecologism and other anti-science movements, for example—one is prohibited from revisiting belief in God as Absolute even though, paradoxically, this belief for the above mentioned Dennett, Dawkins and Changeux seems akin to rationalism.[7]

Not everyone shares this view. Many scientists, movements and cultural initiatives are convinced of the real compatibility between modern science—for instance evolutionary biology or neuroscience—and the belief in God as Creator of the universe, though some matters that raise theological questions, for example the problem of evil, cannot easily be passed over and require rational explanations,

[6]See in Agazzi (1981, 326–327) an indirect demonstration of why metaphysical knowledge cannot be cancelled out, using the Aristotelian method of *elenchos*.

[7]This historical trajectory can be exemplified in Nietzsche's atheism or Heidegger's agnosticism confronted with Hegel's God or against the Absolute in idealism.

as is traditionally done in the philosophy of God and in theology. I do not here have in mind American creationism or the theory of the intelligent design, which in many aspects are problematic both in their religious claims as well as in their accounts of natural science. I am instead thinking of many authors who are simultaneously believers and scientists, for example Francis Collins,[8] creator of the BioLogos Foundation that deals with various issues concerning the harmony between science and religion. Many prominent modern scientists, from Galileo's time up to the present century, were believers in God (and sometimes were also religious persons), such as Newton, Kepler, Maxwell, Planck, Heisenberg, Mendel, Pasteur, Lemaître, Sherrington, Eccles, Ayala, and others.

There is no historical link between atheism or philosophical positivism and the great scientific discoveries in the modern history of science. The authors mentioned above did not perceive a particular difficulty in being believers and scientists, and being both coherently, for they did not share the "empirical closure" elaborated by some philosophers of science. Many of them felt the need for God as a transcendent being in order to ultimately explain the existence of our universe and its amazing and profound order and complexity. This feeling was due to a natural and implicit "metaphysical" inference from the physical order extended to some superior Intelligence, reason or personal spirit who could be thought of as the ultimate source of all that exists—namely, of mankind and the universe, when rightly seen as not self-sufficient in their contingent existence. Even if they were educated in a religious creed, which facilitates the perception of the invisible presence of God in creation, they could not fail to intellectually understand this as a matter of rational faith.

3 Metaphysical Insight and Religious Faith

Unlike in other periods, though not too long ago, people involved in natural sciences today seem more inclined to be materialists. Placing aside sociological analyses and statistical surveys of different cultural areas, one can rightly surmise that this phenomenon is due to the prestige and amazing development of physics, chemistry, and biology in our scientific culture. A person working closely with physics tends to see every problem in physical terms. Scientific information, classically reserved to a limited group of scholars and experts, is today available to everyone by means of education, news media, popular books and magazines, films, and the Internet. Natural sciences and technology constitute an essential part of daily life, even if simply for practical reasons.

[8]Collins guided the team of the Human Genome Project, successfully carried out in 2003, together with Craig Vender. A former atheist, Collins converted to Christianity when he was 27 years old during his practice of medicine. It was not a purely intellectual conversion, but fully religious, with an awareness that science and religious faith are compatible and that the latter gives a sense to the former (Collins 2007).

We could therefore think that this might be the reason—a cultural reason—for the particular difficulty that ordinary learned people experience in acknowledging anything immaterial, or spiritual, that transcends the material world. This often is, if not the unique reason, at least one of the relevant reasons for the current crisis, among many persons, of the faith in God, in the afterlife, in the human spiritual soul, i.e. in all the spiritual dimensions that touch upon religion. Western countries are living in a materialistic culture, although the great pioneers of modern science, as we have seen, usually were not themselves materialists and non-believers.

The problem is related to a massive eruption of atheism in contemporary culture and has been analysed from many points of view (historical, cultural, philosophical, pastoral, etc.). Science is just one of these aspects, and I will focus on this while highlighting several of Agazzi's suggestions on the topic.

There is little doubt that grasping in natural sciences aspects that might be helpful for the acknowledgment of God's creation and providence requires a metaphysical insight that extends beyond scientific methodology. This can be done spontaneously, though at times runs the risk of seeming naïve, or of inadequately mixing different dimensions (easily jumping from the Big Bang cosmology to God, from quantum's indeterminacy to God's interventions, etc.). Yet on the other hand, atheists and materialists are often too hasty in drawing conclusions upon the mere suggestion of science and hence run an analogous risk. We must ask ourselves: why do some so easily conclude to the existence of God (think of Francis Collins and his colleagues today) through contemplation of nature, whereas others, independently of philosophical positions, are not compelled to admit a connexion between the universe as disclosed by science and God as a Creator?

I think this question is essential, though very difficult to answer. It is nothing less than facing the problem of why there are believers and non-believers. This is a very personal matter, irreducible to syllogistic constraints or to rationalistic evidence. It is personal because the solution involves the whole of our existential attitude with respect to life, death, persons, and moral duties. To rationally acknowledge God's existence is not a matter of information, as when a scientist simply accepts or denies some explanatory hypothesis on logical or empirical grounds. The existence of God—especially as a personal God, or as a personal being endowed with the capacity of making voluntary choices concerning the universe and human existence—is not obvious. But, paradoxically, this non-obvious existence (or non-existence) is crucial for an absolute and overall sense of our lives. We may be attracted by the strong and at the same time soft thought of God's existence, or perhaps intimidated by it, or even afraid of being deceived.[9]

Agazzi usually comments on this point by remarking that what is "at stake" in these existential questions is nothing less than the whole sense of our existence (Agazzi 1981: 335–336). But there are no compelling answers, and certainly none

[9]Frequently these thoughts and feelings do regard God not in a purely private way, but in the context of institutional and historical religions in which God is worshipped. In Christian religion, the claim is that God himself has the initiative of addressing man with his personal commitment in the Incarnation.

of a compelling logical or empirical nature. The only existential pressure here is that any answer, positive or negative ("God exists", "there is nothing beyond matter", etc.) is for us a matter of "life and death". Many ethically and anthropologically deep problems share this quality. Epistemologically, human knowledge in these matters is not purely rational (and philosophers are not exempt from this fact), rather it engages and unites in a profound and mysterious way both freedom and reason, to which must be added the role of various virtues such as honesty, humility and sincerity. To perceive what is true and good in these matters, classically associated with wisdom and to science, implies a very personal commitment.[10] Even compelling evidence can be rejected if it clashes with an already rooted commitment, albeit irrational.[11]

I agree with Agazzi's claim that this personal conviction might be qualified as faith and that it is natural to every human person.

> Every human being manages to give a sense to his/her life by adhering to a faith, not necessarily religious, yet nevertheless capable of showing him/her something for which it is worth living and dying. Insofar as they are endowed with reason, humans tend to 'give themselves a reason' for their being in the world, and therefore also to check the content of their faith using reason (Agazzi 2008, 109).

The relationship between religious faith, metaphysical elaborations and human sciences must be seen within this perspective. According to Agazzi, this rational faith, though frequently embedded in religion, cultural traditions and education (sometimes also in ideologies), corresponds to the ultimate sense of human life.[12] Its referent is sometimes mentioned as the world of Life (in the sense of the Husserlian *Lebenswelt*) (Agazzi 2010). It is not a faith in a subjectivist sense, for it encompasses all the cognitive intellectual resources (intelligence, perception of the world, acknowledgment of one's self and of other persons), granting an ultimate meaning to what is known as a whole —we see this, for example, in relation to the value of human life and to other fundamental human tasks in science, politics, family, religion, etc.

Regarding the scope and the epistemological features of this "anthropological" faith, which cannot be confounded with a mere *Weltanschauung* or a prescientific general worldview, there is much to consider (Agazzi 1983: 155–156). It is a rational knowledge, though not in a rationalistic sense. It is metaphysical

[10]It is in this sense that I understand the usual and true assertion that the philosophical proofs of God's existence convince only those who already believe in God. Not because these proofs are necessarily invalid, but because the affirmation of God's existence requires a personal commitment. Some rational evidence can be helpful, but is not sufficient to convert an atheist.

[11]This point is touched upon in Christ's statement concerning the unmoved (stubborn or tepid) attitude of some persons in their moral lives: "If they do not hear Moses and the prophets, neither will they be convinced if some one should rise from the dead" (*Luke* 16, 31, in Gavigan 1988).

[12]Skepticism and relativism apparently refrain from having faith in something for which it is worth living. But this attitude, if not cynical, is often taken with a sense of pessimism and sadness, and not rarely it becomes the object of a kind of intellectual justification which in some way or other conceals an unconscious pursuit of meaning, and one that furthermore ends in frustration or disappointment.

inasmuch as it reaches beyond sensible perception and its object is not a material thing. Is it uncontroversial? It seems not, judging from the infinite debates and opinions surrounding the meaning of life. If it were indisputably obvious, it would not be faith. As *faith* (and not merely belief), however, it entails certainty and confidence, grounded in intellectual insight that invites one to believe, without the pressure of sensible perceptions or of analytic truths.

I would like to introduce a new element to this confrontation between science, metaphysics and faith, presupposing that the last two items touch upon ultimate questions "worth living and dying for" because they regard the meaning of human existence on earth. I refer to the first principles, which for Aristotle were objects of *intellectus* (*noûs)*, and not of *ratio* (*logos*). In my view, the first principles are not just a series of formal axioms, but profound and basic ontological and intellectual habitual convictions that ground all other knowledge, rational practise, language, and volition. The lively (not academic) certainty that one is a human person living in a world populated by other persons, and that our knowledge is capable of attaining the truth or of distinguishing between what is real, possible, potential, unreal, false, etc., are unshakeable convictions for any human person, even if the majority of people would be incapable of explaining the meaning of "real", "potential", "person", etc. Indeed, this is a task for metaphysics.

I do not think that the first principles, in this brief account, include the ultimate meaning of human life, which is the object of faith as discussed above following Agazzi's indications. If the first principles correspond to what Aristotle assigns to *noûs*, Agazzi's notion of faith, instead, seems to concern classical wisdom (*sofía*). Now, science and wisdom are rational developments of human knowledge on the basis of the first principles (here "on the basis" does not mean "automatically deduced"). The world according to primitive knowledge is not self-sufficient. It requires explanations in different areas. Many of these explanations are implications which human reason derives under the requirement of coherence (non-contradiction) and of the principle of causality, and this is why people raise many questions on the basis of what they observe.

But the use of reason is twofold. First, explanatory reason can systematically address the knowledge of particular things and properties of the world. This aspect can be called *scientific rationality*, which is not an absolute need for every human person. Secondly, people cope with the absolute need of articulating (or discovering) the ultimate meaning of their life in the world. What classically was known as *wisdom* is a response to precisely this aspect. This is not a privilege of philosophers, nor a way of knowing reserved only to those who are concerned with the ultimate questions. Not every person is a philosopher, but every person tends toward wisdom, or perhaps thinks that s/he has already sufficiently solved the problem regarding the meaning of life—this includes many convictions about what justice might be, and likewise injustice, values, the sense of doing science, or at least doubts and questions about problems such as what or who God is, or the meaning of human suffering and death. This is exactly the realm of rational faith as discussed by Agazzi in many of his writings concerning the problem of the relationship between science, religion and metaphysics.

Metaphysics as a scientific inquiry, or philosophy understood as a professional endeavour, deals with these universal, ultimate questions which every person necessarily copes at least with his rational faith. This is the platform upon which Agazzi situates the interplay between faith and rational knowledge.

> An important part (not all) of the metaphysical discourse can be seen as a relation established between faith and knowledge (*sapere*[13]). The metaphysical discourse, in this aspect, is revealed as an attempt, within faith, to use reason in order to clarify what is believed and render it a form of a knowledge (*sapere*) (Agazzi 1981: 337).

From this perspective, reason as an extension of our intelligence beyond what is directly perceived by human senses and by the immediate intellectual perception of certain fundamental truths—namely, the first principles—frequently begins by establishing beliefs and opinions (in natural sciences, these are hypotheses or conjectures), and only afterwards attemps to operate a transition from faith to rational knowledge (metaphysical knowledge in the philosophical field; corroborated knowledge in natural sciences; well grounded opinions in practical knowledge).

A rationalistic account of knowledge considers faith as a popular view, and one that is completely submissive to reason, whether philosophical or scientific. Agazzi, instead, conceives both metaphysics and science as interplays between rational faith and reason as demonstrative knowledge, interplays that cannot ever be abandoned and replaced by pure faith or a pure reason. This point is equally valid in philosophy and science, albeit in a different way.

> The metaphysical discourse is always a discourse which unfolds inside a 'metaphysical faith', which is not necessarily a religious faith, for it could very well be an atheistic faith, yet always a faith. Hence, metaphysics appears as a use of reason that tries to transform into rational knowledge (*sapere*), if possible, what is attained through faith (Agazzi 1981, 337–338).

It may happen that rational inquiry corrects some aspects of faith, or even forces one to abandon false beliefs.[14] Dogmatism, in a negative sense, is the attitude of always rejecting the possibility of a critical examination of one's faith, even if only to clarify it.[15]

Contrary to a rationalistic view, metaphysics and the positive sciences normally operate within the framework of a previous faith. This does not mean that metaphysics would be a mere hypothetical science. Philosophers often claim to be able to attain true rational conclusions, with the exception of relativists and the followers of the so-called "weak thought" (but it can be argued that even the latter hold

---

[13]In Italian, *sapere* means rational knowledge, science, whether philosophical or in the sense of particular sciences such as physics or mathematics.

[14]But it can never remove the unshakeable certainty of the first principles.

[15]Even the principles are open to rational clarification, otherwise they would be believed in an irrational way. In an analogous way, Christian faith, in its essential points, is never considered by a genuine believer as something that perhaps in the future might be abandoned, as it happens in scientific hypothesizing. Even so, it remains open to rational clarification, which is the task of theology.

convictions, for example the strong conviction that there are no absolute convictions). Agazzi, too, acknowledges the merit of various important metaphysical conclusions—to be found in many different philosophical schools—that claim to be "non hypothetical, unconditionally valid and irrefutable" (Agazzi 1983: 150), something than can be fully attained if the philosopher succeeds in proving that empirical reality would be contradictory unless there is a metaphysical reality (Agazzi 1983: 150).[16]

The general conclusion of these considerations is that the opposition between faith and reason, religion and science, claimed by some authors since the Enlightenment and renewed by the so-called new Enlightenment—sometimes taking advantage of the achievements of recent science—is not valid. Perhaps one might object to this that atheists normally do not place reason in opposition with rational faith, but only with religious faith, claiming that the latter (which believes in supernatural entities) is irrational. This difficulty mixes various aspects. If atheists think that believing in God is irrational, they should argue their points using philosophical arguments, but not in the name of science. Natural sciences do not have recourse to God in their explanations because this recourse falls beyond their competence, just as it would be irrational to think that God could intervene in a football match as one secondary cause among others. Agazzi's argument is that there is an interplay between faith and reason at the level of natural sciences, therefore within their competence, and another at the higher level of metaphysics. In the latter case faith might also be religious and not purely speculative, especially in religions like Christianity, which are by nature open to reason in all its aspects.

How, then, can these two levels be related? Each of them—namely the scientific level and the philosophical one—shows a dynamic interplay between faith and reason. The answer it that the latter provides an ultimate interpretation, which is both ontological and ethical, to the discoveries and technical achievements of the former.[17] This can be done more effectively, in my view, provided philosophy explores such fields as philosophy of nature, philosophy of science, and philosophical anthropology. Otherwise, the exchange between religion and natural science runs the risk of being rather extrinsic.

From the perspective of the sciences, in turn, when one considers scientific results not in a piecemeal way, but globally—as it happens in the framework of great theories, for example, in cosmology, or in evolutionary biology taken as a whole—a philosophical interpretation is more inviting. Natural sciences today, considering their complexity and interdisciplinary relations, present a sufficiently consistent scenario—from cosmology to biological sciences—that more than ever

[16]This idea could be applied to the affirmation of God's existence, as it is suggested in Agazzi (1983: 152).

[17]This interpretation is a human need. Even scientists consider their research and discoveries within a universe of sense and meaning, using their intelligence with the help of the first principles and a reasonable faith. They can also try to respond to ultimate problems, but in doing so they think as human persons, not as scientists (Agazzi 1983, 118). It would be illegitimate, however, to propose these reflections as if they were scientific conclusions.

seems to invite a philosophical overview (Vitoria 1994). This interpretation is a philosophy of nature or a philosophy of science that can be afterwards related to theological topics, such as God's creation and providence. Science is very useful, moreover, for furnishing the material upon which the philosophical reflection might be undertaken. In this sense, science and philosophy are not extrinsic, but complementary and necessary to each other, and this despite the usual changeable state of scientific theories.

In conclusion, philosophy is a necessary cognitive mediator between science and religion (or between science and theology). According to Agazzi, it can be shown that the whole of the empirical experience—the epistemological space given to natural sciences—does not equate to the whole of reality as such. This experience is open to metaphysical investigation as well. Hence, the latter opens the conceptual space to transcendence (in the sense of God and the human spirit), which is in turn the cognitive space given to religious faith (Agazzi 1983: 134, 153–154).

Summing up,

1. Natural sciences are neither purely empirical nor purely rational. They presuppose metaphysical aspects—the first principles—and they possess a particular dynamism according to the intellectual interplay between faith and reason with respect to their object, which is the world as captured by the totality of empirical experience.
2. The answer to the problem of the ultimate meaning of human existence in the material world does not come from natural sciences, but from a higher metaphysical perspective. This perspective in ordinary people is given through faith, normally embedded in religion, or at least in some existential attitudes. This faith corresponds to what classically was called wisdom.
3. A systematic rational inquiry on the ultimate meaning of human existence pertains to metaphysics—considered as a philosophical discipline—or to theology, which is the science of religious faith.
4. From numbers 2 and 3, we can conclude that the relationship between faith and reason, or between science and metaphysics—including religious knowledge—is positive and natural, and cannot be prohibited by any artificial caveats. Philosophy of nature and philosophy of science are cognitive mediators that facilitate a positive and fruitful relationship between those items.

References

Agazzi, Evandro. 1969. L'ateismo e la scienza. In *Ateismo sfida ai cristiani*, ed. Giuseppe Lazzati, 164-180. Milan: Vita e Pensiero.

Agazzi, Evandro. 1974. Temi e problemi di filosofia della fisica. Rome: Abete.

Agazzi, Evandro. 1981. Considerazioni epistemologiche su scienza e filosofia. In *Teoria e metodo delle scienze*, ed. Carlo Huber, 311-340. Rome: Università Gregoriana Editrice.

Agazzi, Evandro. 1983. Science et foi, Scienza e fede. Milan: Massimo.

Agazzi, Evandro. 1991. The universe as a scientific and philosophical problem. In *Philosophy and the origin and evolution of the universe*, eds. Evandro Agazzi and Alberto Cordero, 1-51. Dordrecht: Kluwer Academic Publishers.

Agazzi, Evandro. 2007. Fede, ragione e scienza. In *Scienza e fede: le nuove frontiere*, ed. Paolo Dell'Aquila, 83-94. Cesena: Società Editrice Il Ponte Vecchio.
Agazzi, Evandro. 2008. Scienza. Intervista di Giuseppe Bertagna, Brescia: La Scuola, Brescia.
Agazzi, Evandro. 2010. Science and Religion. In *Encyclopedia of Life Support Systems* (EOLSS), vol. 1, 149-184. http://www.eolss.net/sample-chapters/c05/e6-89-21-00.pdf. Accessed 30 June 2014.
Agazzi, Evandro. 2014. Scientific objectivity and its contexts. New York: Springer 2014.
Collins, Francis. 2007. The language of God. A scientist presents evidence for belief. New York: Free Press.
Gavigan, James, ed. 1988. The Navarre Bible, St. Luke, Dublin: Four Courts Press.
Maritain, Jacques. 1932. Les degrés du savoir. Paris: Desclée de Brouwer.
Sanguineti, Juan José. 2002. Science, Metaphysics, Philosophy: In Search of a Distinction. *Acta Philosophica* 11: 69-92.
Vitoria, María Angeles. 1994. A dynamic link between science and theology. In *Origins, time & complexity*, eds. George Coyne and Karl Schmitz-Moormann, part 2, 302-308. Genève: Labor et Fides.
Wallace, William. 1983. From a realist point of view. Boston: University Press of America.

Author Index

M. Alai et al. (eds.), *Science Between Truth and Ethical Responsibility*,
DOI 10.1007/978-3-319-16369-7

MIX
Papier aus verantwortungsvollen Quellen
Paper from responsible sources
FSC® C105338

If you have any concerns about our products,
you can contact us on
ProductSafety@springernature.com

In case Publisher is established outside the EU,
the EU authorized representative is:
Springer Nature Customer Service Center GmbH
Europaplatz 3, 69115 Heidelberg, Germany

Printed by Libri Plureos GmbH
in Hamburg, Germany